T0336939

GENOMICS AND SOCIETY

GENOMICS AND SOCIETY
Ethical, Legal, Cultural and Socioeconomic Implications

Edited by

DHAVENDRA KUMAR

Institute of Cancer & Genetics,
University Hospital of Wales,
Cardiff University School of Medicine,
Cardiff, UK

Genomic Policy Unit,
Faculty of Life Sciences and Education,
University of South Wales,
Cardiff, UK

RUTH CHADWICK

Healthcare Ethics and Law,
School of Law,
The University of Manchester,
Manchester, UK

Centre for Economic Social and Applications of
Genomics and Epigenetics (Cesagene),
School of Social Sciences,
Cardiff University, Cardiff, UK

AMSTERDAM • BOSTON • HEIDELBERG • LONDON
NEW YORK • OXFORD • PARIS • SAN DIEGO
SAN FRANCISCO • SINGAPORE • SYDNEY • TOKYO
Academic Press is an imprint of Elsevier

Academic Press is an imprint of Elsevier
125, London Wall, EC2Y 5AS.
525 B Street, Suite 1800, San Diego, CA 92101-4495, USA
225 Wyman Street, Waltham, MA 02451, USA
The Boulevard, Langford Lane, Kidlington, Oxford OX5 1GB, UK

Notices
Knowledge and best practice in this field are constantly changing. As new research and experience
broaden our understanding, changes in research methods, professional practices, or medical treatment
may become necessary.

Practitioners and researchers must always rely on their own experience and knowledge in evaluating
and using any information, methods, compounds, or experiments described herein. In using such
information or methods they should be mindful of their own safety and the safety of others, including
parties for whom they have a professional responsibility.

To the fullest extent of the law, neither the Publisher nor the authors, contributors, or editors,
assume any liability for any injury and/or damage to persons or property as a matter of products
liability, negligence or otherwise, or from any use or operation of any methods, products, instructions,
or ideas contained in the material herein.

ISBN: 978-0-12-420195-8

British Library Cataloguing-in-Publication Data
A catalogue record for this book is available from the British Library.

Library of Congress Cataloging-in-Publication Data
A catalog record for this book is available from the Library of Congress.

For Information on all Academic Press publications
visit our website at http://store.elsevier.com/

Typeset by MPS Limited, Chennai, India
www.adi-mps.com

Printed and bound in the United States of America

Working together
to grow libraries in
developing countries

www.elsevier.com • www.bookaid.org

DEDICATION

'Harmonizing peoples with different genetic and genomic identities'

The problem is not the science, the science is fantastic. The challenge rather is making sure that the benefits of science go to those in greatest need.
Tikki Pang, WHO and National University of Singapore

CONTENTS

LIST OF CONTRIBUTORS

Sura Alwan
Department of Medical Genetics, University of British Columbia, Vancouver, BC, Canada

Michelle Bishop
NHS National Genetics and Genomics Education Centre, Birmingham, UK

James Buchanan
Health Economics Research Centre, Nuffield Department of Population Health, University of Oxford, Headington, Oxford

Ruth Chadwick
Healthcare Ethics and Law, School of Law, The University of Manchester, Manchester, UK

Angus Clarke
Institute of Cancer and Genetics (Formerly Medical Genetics), Cardiff University School of Medicine, University Hospital of Wales, Cardiff, Wales, UK

Ellen Wright Clayton
Center for Biomedical Ethics and Society, Vanderbilt University, Nashville, TN, USA

Sandie Gay
NHS National Genetics and Genomics Education Centre, Birmingham, UK

Gill Haddow
Science, Technology and Innovation Studies, The University of Edinburgh, Edinburgh, Scotland, UK

Hanan Hamamy
Department of Genetic Medicine and Development, Geneva University, Geneva, Switzerland

Rachel Iredale
Genomics Policy Unit, Faculty of Life Sciences and Education, University of South Wales, Pontypridd, South Wales, UK

Gerardo Jiménez-Sánchez
Harvard School of Public Health, Department of Epidemiology, Harvard University, Boston, MA, USA; Global Biotech Consulting Group (GBC Group), Mexico City, Mexico

Alastair Kent
Genetic Alliance UK, London, UK

Maggie Kirk
Genomics Policy Unit, Faculty of Life Sciences and Education, University of South Wales, Pontypridd, South Wales, UK

Michiel Korthals
Wageningen University, Wageningen, The Netherlands

Atina Krajewska
School of Law, Cardiff University, Cardiff, Wales, UK

Jennifer G. R Kromberg
Division of Human Genetics, School of Pathology, Faculty of Health Sciences, University of the Witwatersrand and National Health Laboratory Service, Johannesburg, South Africa

Dhavendra Kumar
Genomic Policy Unit, Faculty of Life Sciences and Education, The University of South Wales, Pontypridd, Wales, UK

Marion McAllister
Institute of Cancer & Genetics, School of Medicine, Cardiff University, Cardiff, Wales, UK

Koichi Mikami
Science, Technology and Innovation Studies, The University of Edinburgh, Edinburgh, Scotland, UK

Rhian Morgan
Genomics Policy Unit, Faculty of Life Sciences and Education, University of South Wales, Pontypridd, South Wales, UK

Mitali Mukerji
CSIR Ayurgenomics Unit-TRISUTRA, CSIR-Institute of Genomics and Integrative Biology, Sukhdev Vihar, New Delhi, India

Denis J Murphy
Genomics and Computational Biology Research Group, University of South Wales, Pontypridd, Wales, UK

Jim Philp
Organization for Economic Co-operation and Development (OECD), Directorate for Science, Technology and Innovation, Paris, France

Bhavana Prasher
CSIR Ayurgenomics Unit-TRISUTRA, CSIR-Institute of Genomics and Integrative Biology, Sukhdev Vihar, New Delhi, India

Michael Ruse
Florida State University, Tallahassee, FL, USA

Himla Soodyall
Division of Human Genetics, School of Pathology, Faculty of Health Sciences, University of the Witwatersrand and National Health Laboratory Service, Johannesburg, South Africa

Stuart Sutherland
NHS National Genetics and Genomics Education Centre, Birmingham, UK

Emma Tonkin
Genomics Policy Unit, Faculty of Life Sciences and Education, University of South Wales, Pontypridd, South Wales, UK

Adrian Towse
Office of Health Economics, London, UK

Richard Tutton
Senior Lecturer, Department of Sociology and Centre for Science Studies, University of Lancaster, UK

Wei Wang
School of Medical Sciences, Edith Cowan University, Western Australia, Australia; Beijing Municipal Key Laboratory of Clinical Epidemiology, Beijing, China

Sarah Wordsworth
Health Economics Research Centre, Nuffield Department of Population Health, University of Oxford, Headington, Oxford

Ma'n Zawati
Centre of Genomics and Policy, Department of Human Genetics, Faculty of Medicine, McGill University, Montreal, Canada

FOREWORD

The important ethical and societal issues raised by genetics and genomics research and by its applications to families with inherited disorders have been recognized for some years, and those practicing in this field, especially clinical geneticists and genetic counselors, have been in the forefront of efforts to draw these issues and potential problems to a wider range of clinicians and scientists. By the formation of links with a wide range of workers in the social sciences and broader humanities, these practical issues have also been placed in a more general theoretical framework, giving many powerful examples that illustrate how these general principles may affect individual lives and wider society.

While several previous books have been written exploring this field, they have mostly been concerned with the social aspects of classical medical genetics practice in families with genetic disorders. But now, the entire field is rapidly changing, with the availability of the entire sequence of the human genome (and of other genomes) and our increasing understanding of the complex processes involved in its function and dysfunction producing the new field of *Genomics*, beginning to have widespread applications outside the area of individual families and impinging on numerous medical, scientific, and other disciplines.

This book, whose editors and authors bring a wealth of experience from both practical medical genetics, genomic medicine and from the humanities, shows how workers in almost every field will have to be aware of the new genomics and how it will undoubtedly affect both their specific work and their more general approaches. Some of the issues raised are rooted in those already explored, and in some cases at least partly resolved, for classical medical genetics; others are entirely new. Raising and thinking about them now, as this valuable and comprehensive book does, while applications are still in their infancy, should help the wise use of this new and powerful field.

Peter Harper
Emeritus Professor of Human Genetics,
Cardiff University School of Medicine,
Cardiff, UK

PREFACE: INTRODUCTION TO SOCIAL AND ECONOMIC GENOMICS

Towards the end of the twentieth century, several major discoveries and applications in the science of genetics and genomics were achieved. This is evident from the impact of genetics and genomics in many areas notably in biotechnology, bioengineering, and healthcare. The beginning of the twenty-first century witnessed landmark achievements including the sequencing of the human genome and many other biological genomes. This truly led to the commencement of the genome era. Not surprisingly, several researchers and organizations committed themselves with huge resources to advance further basic, applied, and translational research in all areas of genetics and genomics. While the majority of the funded genetics and genomics research and other resources were assigned to molecular biology, biotechnology, and health, a limited number of projects were allocated to study the societal, economic, ethical, moral, and legal implications. Several leading experts and philosophers on economics, bioethics, and law continue to work and advise the genetics and genomics community, politicians, religious, and community workers. This book, conceptualized and led by a medical geneticist and a social and bioethical genetics researcher, reviews and consolidates the outcomes of genetics and genomics in the context of social and economic applications. Here, we briefly indicate the scope of topics covered in this volume.

The introductory chapter sets out how genetics and genomics might impact on the society and life in general with evidence and experience from the applications of electronics and computations sciences citing examples of automobiles, air travel, home appliances, computers, mobile phone, etc. on which all of us have become heavily dependent overcoming many fears and reservations put forward practically at every stage of many new discoveries and developments. In this context, the science of genetics and genomics would be no exception, and we are beginning to harness many benefits in medicine and health, food and agriculture, bioengineering, bioenergy, and ecology. Societal fears and obstacles are more likely to be overcome with rising awareness and acceptance of lay people and community leaders.

A number of ethical and moral concerns, for example, consent, confidentiality, rights to withhold information, etc., are encountered in medical genetic practice. A dedicated chapter examines these issues in the broader societal and health context. Several legal issues might be faced both by the family and health/social practitioner when dealing with genetic/genomic information. A leading expert discusses existing laws that could be applicable in good genetic/genomic practice with reference to specific jurisdictions.

Forensic applications of genetics and genomics have dramatically changed the legal practice worldwide. The genetic and genomic information for personal identification (DNA finger printing and genomic signature) is routinely used in criminal and immigration matters. Other applications in this context include the use of new genetic tools in dealing with sudden unexplained death including sudden infant death and sudden cardiac death cases. In the wider context, major population level disasters (earthquakes, Tsunami, air crash, terrorist acts, etc.) have also seen benefits of genomic tools (genome sequencing and population-specific genomic databases) in rapid and accurate identification of the victims.

Other issues' reviews in this volume include those related to human rights violation, patents, individual or community ownership, and intellectual property rights of the genetic and genomic information at community and population levels. Like any other new technical and scientific field, adequate and appropriate teaching and training of genetics and genomics is vital for successful health and social applications. Information on several facets used in genetic/genomic teaching and training at various levels is included in this book. Lay people are major consumers of any service or product. A dedicated chapter reviews the various measures undertaken in information sharing, informing, and educating lay people about genetics and genomics.

Globally, several hundred rare diseases are encountered by medical practitioners. Most of these are either inherited or have a known genetic cause. There is a huge burden of these diseases in the context of scarce resources. The current state and efforts made in the multidisciplinary care of patients and family members is critically reviewed with evidence gathered from existing genetic healthcare.

Genomics is rapidly moving into biotechnology and bioeconomy applications with huge sociopolitical impact. The key areas of genomic bioeconomic applications include biosynthetic diesel, new drugs, efficient crop production, and livestock improvement are reviewed with evidence from healthcare, food and nutrition, bioenergy, and ecological balance and improvement.

Developing nations constitute around two-third of the growing global population. Among many hardships and hurdles faced by people in the developing world, health burden is huge. Diseases related to infection, malnutrition, and poverty are common, however, a small but collectively large proportion is due to genetic or inherited diseases. Progress and efforts made by few emerging economies (Brazil, Russia, India, China, and South Africa—BRICS nations) in adopting genomic applications are highlighted. Like any new science, the healthcare applications of genetics and genomics are no exception facing challenges of traditional and customary health practices. There is data available on genetics and genomics in the context of traditional medical practices in China (traditional Chinese medicine, TCM), India (Ayurveda), and the Arab world (Arabic and Persian system of medicine). Projects like TCM Genomics and Ayurgenomics are funded by China and India, respectively. The evidence from specific areas where genetics/genomics could be applied in indigenous medical practices is presented, particularly on Ayurgenomics. These developments also have a socioethical dimension, in relation to the impact on traditional medical ethics in those areas: for example, implications for how new scientific understandings affect the healer—patient relationship, especially in the context of the expectations of evidence-based medicine.

Each society and cultural group is expected to have firm beliefs and taboos related to health matters, for example, children born with a physical abnormality like orofacial clefts. Examples and information derived from field genetic surveys carried out in tribal groups of China, India, South Africa are presented here. With the rapid advances and new developments in genetics and genomics, a number of leading international agencies have started looking at major issues of regulation, finance, and distribution of genomic applications. The dedicated chapter carries out in depth analysis on the current state of affairs at international level. There is increased awareness and interest on human genomics and global health. The global genomic alliance of major international agencies (Human Genome Organization, HUGO; UNESCO/WHO; Organisation for Economic Cooperation and Development, OECD; Wellcome Trust) and many others are convened aimed at genomic applications for global health, for example, "Global Alliance for Genomics and Health"—*GA4GH*. The charter for GA4GH includes ethical, legal, and social issues apart from setting out ground rules for genomic databases on specific disease conditions and many other genomic applications.

Finally, it is anticipated that the majority of genetic and genomic scientists, healthcare professionals, food and nutrition experts, and biotechnology workforce will remain committed to positive and ethically acceptable gains for the benefit of society and all population groups. However, regrettably there might be a minority who might exploit this science for developing dangerous mutated organisms, harmful plants and crops, and toxic biosynthetic chemicals for act of war, bioterrorism, and socioeconomic chaos. Mahatma Gandhi once said, "That atomic energy though harnessed by American scientists and army men for destructive purposes may be utilized by other scientists for humanitarian purposes is undoubtedly within the realm of possibility. ... An incendiary uses fire for his destructive and nefarious purpose, a housewife makes daily use of it in preparing nourishing food for mankind." This seminal quote not only refers to the applications of the nuclear science for peaceful purposes, but also has clear wide ranging implications on many other scientific discoveries and developments notably in genetics and genomics for global social and economic benefits irrespective of any society or nation.

Dhavendra Kumar

Institute of Cancer & Genetics, University Hospital of Wales, Cardiff University School of Medicine, Cardiff, UK

Genomic Policy Unit, Faculty of Life Sciences and Education, University of South Wales, Cardiff, UK

Ruth Chadwick

Healthcare Ethics and Law, School of Law, The University of Manchester, Manchester, UK

Centre for Economic Social and Applications of Genomics and Epigenetics (Cesagene), School of Social Sciences, Cardiff University, Cardiff, UK

ACKNOWLEDGMENTS

This book is the product of passion and dedication of all contributors and their colleagues. We as editors had rather vague conceptualization of genomics in the societal context. When presented the opportunity, many of our colleagues came forward and shared their experience and grasp of medical, social, and other aspects of genomics. This is evident from the superb high-quality chapters that cover the whole emerging new field. We are truly grateful to all our contributors, in particular, Prof. Stylianos Antonarakis (Geneva, Switzerland), Prof. Angus Clarke (Cardiff University, UK), Prof. Maggie Kirk (University of South Wales, UK), and Prof. Gerardo Jimenez-Sanchen (Harvard University, USA).

Ms. Lisa Eppich, the lead member of the publishers' team deserves special thanks and appreciation for her hard work and dedication to ensure smooth and timely publication of this new book. Lisa was helped by the editorial and publishing teams for which we offer warm welcome and offer sincere thanks.

Last but not the least, this book would have been difficult to write, edit, and publish without personal support and encouragement from our immediate and the extended family.

Dhavendra Kumar
Ruth Chadwick
Cardiff and Manchester

CHAPTER 1

Personal Genomics and its Sociotechnical Transformations

Richard Tutton
Senior Lecturer, Department of Sociology and Centre for Science Studies, University of Lancaster, UK

Contents

INTRODUCTION

In 2011, plant geneticist Jonathan Latham published a short and stinging commentary on the state of genomics under the eye-catching headline "The failure of the genome." Noting how the billions of dollars spent on GWAS (genome-wide association studies) had yielded "only a handful of [gene variants] of genuine significance for human health," Latham concludes that this situation represents the "most profound crisis that the science has faced" [1]. This downbeat evaluation was not an isolated example: as the 10-year anniversary of completion of the human genome sequence arrived, a number of other authors came to similar conclusions about what had been achieved measured against what had been expected of genomics in the 1990s. Back then, scientists had made the case that sequencing the human genome would serve to accelerate biomedical research and provide tools to elucidate the mechanisms and causes of human disease. Supporters of the Human Genome Project claimed it would bring about a profound transformation in drug development and the practice of medicine. Geneticist Bodmer and journalist McKie [2] saw that the Human Genome Project "will become the mainstay of the pharmaceutical industry in the next century," and its public funding also helped to bring into being a new start-up sector of firms which aimed to leverage proprietary gene sequence databases, bioinformatics software, and other informational-based tools to do business with biotechnology or pharmaceutical companies by selling access to these resources with the promise that they would aid disease gene discovery and the development of new drugs. Walter Gilbert—founder of

Genomics and Society
DOI: http://dx.doi.org/10.1016/B978-0-12-420195-8.00001-X

Myriad Genetics Inc., one of the first firms created to commercially exploit human genome research—envisioned that "the possession of a genetic map and the DNA sequence will transform medicine. One of the benefits of genetic mapping will be the ability to develop a medicine tailored to the individual: drugs without side effects" [3].

Almost two decades later, Evans et al. [4] however note that: "some wonder what became of all the genomic medicine we were promised" and call into question the unrealistic expectations which fueled a "genomics bubble" in the 2000s. Collins [5] who served as the Director of the Human Genome Project, also asked in an opinion piece published in *Nature* "Has the revolution arrived?" He responds by acknowledging that the impact of genomics on clinical medicine has been "modest" but concludes that: "the promise of a revolution in human health remains quite real. Those who somehow expected dramatic results overnight may be disappointed, but should remember that genomics obeys the First Law of Technology: we invariably overestimate the short-term impacts of new technologies and underestimate their longer-term effects."

These debates among scientists and clinicians about the revolutionary and transformative effect of genomics has run in parallel with discussions within the social sciences about genomics and its anticipated social impacts. In the 1990s, social scientists entertained tenebrous imaginings of the future [6,7]: Lippman [6] warned about the effect of genetics on changing social relationships, values, and reinforcing inequalities in power whereby elites would use genetic knowledge and technologies on others in society. Other social scientists, however, such as Giddens [8], were more positive in envisaging that new genetic interventions would empower individuals to make their own reproductive, medical, and social choices. Geneticization as an anticipation of the significant social transformational effects that genomics would have in the near future gave impetus to a growing field of social science research, some of which was funded by the ELSI (Ethical, Legal, and Social Implications) component of the Human Genome Project itself. In Great Britain, as the completion of the human genome neared, and concerns about the social impact of genomics rose up the policy agenda, so significant resources were given to social science to address the wider social transformations which genomics would bring about. For funders, their particular concern was with how publics would respond to genomics given the difficulties experienced in relation to genetically modified (GM) food and other scientific controversies of the 1990s, set against the promise of genomics to deliver huge economic benefits [9]. For 10 years, the ESRC (Economic and Social Research Council) funded four different academic centers to conduct and disseminate social science and bioethical research on genomics.

For Science and Technology Studies scholars Hedgecoe and Martin [10], the "transformative power of genomics" in both its promising and tenebrous registers, assumes that the "scientific and technological developments [in genomics are] somehow revolutionary, encapsulating a break from what has gone before at both a technical and social level." However, they note that not all social scientists writing about various aspects of genomics would subscribe to a "transformative" account of genomics and indeed have challenged precisely this view by pointing to continuities with the past. Against their own skepticism about the scale of the transformations brought about genomics, Hedgecoe and Martin set out to explore two questions in their own wide-ranging discussion. They ask (i) what are the sociotechnical expectations and transformations associated with genomics and (ii) what is seen as new or specific to genomics and what is the extent of the sociotechnical change? In this chapter, I draw on these two questions to provide a framework for my own engagement with the sociotechnical transformations brought about by or through genomics, specifically, with reference to one particular case study, that of personal genomics.

The science journalist Stix first coined the term "personal genomics" in 2002 to describe efforts to develop greater knowledge of individual variation in the human genome. "Personal genomics" signified a move away from the "one genome fits all" of the single sequence produced by the Human Genome Project using the DNA of a very few individuals to the sequencing or genotyping many thousands of individual genomes to identify ways in which they differ [11]. Following his own involvement in the Human Genome Project, the scientist and entrepreneur George Church launched the Personal Genome Project in 2005. Beginning with the genomes of 10 academics and entrepreneurs, the initiators of the Personal Genome Project spoke of how "we foresee a day when many individuals will want to get their own genome sequenced so that they may use this information to understand such things as their individual risk profiles for disease, their physical and biological characteristics, and their personal ancestries" [12]. While the Personal Genome Project is an academic venture, the idea of people having access to and analyzing their own individual genomes became the basis of a consumer service in 2007 with the launch of three companies that offered users information about disease risk, drug response, ancestry, and other traits for a price. These services were marketed under the rubric of "personal genomics." My chapter therefore explores two questions in relation to personal genomics: (i) what were and continue to be the sociotechnical expectations and transformations associated with personal genomics? and (ii) what was seen as new or specific to personal genomics and what is the extent of the sociotechnical change brought out by developments in personal genomics?

PERSONAL GENOMICS AND ITS SOCIOTECHNICAL EXPECTATIONS

In 2007, three firms—23andMe, deCODEMe, and Navigenics—launched their "personal genome" services onto the US market. 23andMe LLC and Navigenics Inc. were both new players—start-ups funded by venture capital investment. 23andMe was cofounded by Linda Avey, whose background was in sales and marketing and by Anne Wojcicki who had a previous career as an investment analyst, while Navigenics was set up by an oncologist David Agus and Dietrich Stephan, a geneticist. Unlike the two start-ups, deCODEMe was an offshoot of deCODE Genetics Inc. which had been established in 1996 to develop a population-based genetic resource in Iceland and had actually designed and conducted a number of GWAS and developed its own line of diagnostics.[1] In 2007, it launched its own dedicated consumer service— deCODEMe—as part of a strategy to generate near-term revenues for the company which was incurring significant losses. These three services were followed by a number of other companies which also offered similar services, including Pathway Genomics that launched in 2009.[2]

The appearance of these services has attracted considerable comment and analysis from inter alia social scientists, lawyers, bioethicists, policy advisors, and clinicians about how these services are currently and should be regulated in the future, their scientific reliability and the lack of laboratory standards, concern about fraud, scams and incredulous offerings, the quality of consumer information produced by companies and how results of tests will be interpreted and acted upon by users, and the role of medical professionals as gatekeepers and interpretative authorities of disease risk. In what follows below I outline the principal sociotechnical expectations of personal genomics that emerged around the time of the launch of the first services on the market. These are that companies would "democratize" the genome by offering a direct-to-consumer (DTC) service, cutting out the clinician as gatekeeper, and that their users would be empowered by receiving genetic risk information in this way to make decisions about their future health. Conversely, regulators, clinicians, and policy advisors challenged these expectations as in part unrealistic and that they imperiled the future of genomic medicine; they also articulated their own darker expectations that, far from being empowered, users would suffer anxiety and make inappropriate use of finite healthcare resources.

---

[1] In 2001, deCODE Genetics had signed a major alliance with Roche to develop molecular diagnostics. In 2007, the firm launched deCODE T2, a reference laboratory DNA-based test for a gene variant which its scientists had identified to be associated with increased risk of type 2 diabetes. To market it as a consumer test, deCODE entered into an agreement with DNA Direct, Inc. The firm went on to market three other diagnostics for gene variants linked to risk of early onset cardiac arrests, stroke, and prostate cancer.

[2] Soon after their launch, Navigenics, deCODE Genetics, and 23andMe expanded their potential customer base beyond North America. 23andMe shipped to almost all EU member states, Turkey, and Russia; Navigenics stated that its services were available through physicians in a smaller number of countries: Australia, Brazil, Canada, Greece, India, Japan, South Korea, Mexico, New Zealand, Puerto Rico, Singapore, Turkey, and Great Britain. deCODEme Genetics, on the other hand, offered its scans to be bought from anywhere in the world. Pathway Genomics also accepted custom from outside the United States.

In 2007, *New York Times* journalist Hamon commented that: "the exploration of the human genome has long been relegated to elite scientists in research laboratories. But that is about to change. An infant industry is capitalizing on the plunging cost of genetic testing technology to offer any individual unprecedented—and unmediated—entree to their own DNA" [13]. In marketing their services to prospective consumers, these companies adopted the language of empowerment, claiming that genetic risk information renders the future health more calculable and therefore more open to individual control. The companies stressed that "these DNA tests don't reveal your destiny. But they do provide a map that can help guide your future" [14]; as the Navigenics web site declared: "your future is more in your control than you might think." deCODE Genetics explained that:

> Now, for the first time in history, you can embark on a novel journey of discovery, guided by deCODE Genetics' team of pioneers in human gene discovery. deCODEme allows you to study how state-of-the-art scientific knowledge about human genetics applies to a scan of your own genome and to compare your information with that of others. With this information you'll be empowered to discover more about your past, present and future. [14]

These companies were not diagnosing in someone the presence or absence of a disease but providing probabilistic information about their likelihood of developing a disease or diseases in some form in their lifetime based, in part, on the firm's analysis of their genome. They are producing "future truths" which users may or may not come to inhabit. For the firms, the prospect of the future turning out otherwise was central to how they were marketing their service: it was founded on the belief that affirmative choices of users today to reshape their own life practices could prevent, delay, or ameliorate whatever comes to afflict them in the future [15]. As Navigenics stated at its launch in 2007: "the company will help people understand their genetic predisposition to disease and arm them with information about what action to take to help them stay healthy" [16]. The CEO of deCODE Genetics Inc. was quoted in a corporate press release as saying: "in an era when we are encouraged to take greater personal control of our lifestyle and health, we believe we should all have the opportunity to learn what our own genome can tell us about ourselves" [14]. It chimed with the proposition that personalized medicine not only involved the actions of scientists and healthcare professionals but also the active participation of patients and consumers [17,18]. Access to personal genomic risk information would be empowering and enabling of change, permitting users to take action in light of the information they received to improve their future health. The companies asserted that they had the expertise and authority to provide this information to individuals without clinicians acting as gatekeepers. In so doing, they positioned themselves as legitimate actors in constructing the future of personalized medicine "beyond the clinic" [19]. deCODE Genetics Inc. for instance attempted to trade on its record as a reputable research organization which could be relied upon to provide an accurate and reliable interpretation of genetic risks: "Your genome is unique and it's yours. deCODEMe gives you the confidence of getting to know your genome guided by a world leader in human genetics" [14]. Navigenics, by contrast, stated that it had established alliances with well-known and respected US healthcare providers and other medical institutions to help its customers understand and act on their risk information. Navigenics therefore imagined that they would work in partnership with clinicians, especially on any follow-up care or diagnoses.

However, many regulators and clinicians were openly hostile to this fledging industry and the role that it imagined it would play in bringing about a potential reconfiguration of the sociotechnical relationships between healthcare organizations, professionals, and patients in which both the production and use of healthcare information is fundamentally changed. In keeping with their DTC approach and empowerment discourse, firms had not sought premarket approval by regulatory authorities and had positioned their products in such a way so that they evaded existing regulation. They did not define them as diagnostic or medical devices and their tests were developed as "in-house" (or laboratory developed) tests, which have not traditionally been subject to regulation in either the United States or European Union (see Ref. [20] for further discussion). However, the definition of personal genomics services— whether they should be regulated as medical devices or not—was soon a contentious point of disagreement between companies and regulatory authorities on both sides of the Atlantic. I will return to regulatory interventions in this industry later in the chapter.

For now, I would note that critics articulated their own negative expectations about this industry and the implications of the DTC marketing of genetic tests. In doing so, critics did not tend to dispute the idea that genomics would bring into being individualized and preventive medicine but contended that attempt to bring that future into being through the selling of personal genome testing services was the problem. In effect, the firms were accused of making excessive promises that went against the current science and what is positively known about genetic risk. One implication of this happening too soon, in the words of the HGC (Human Genetics Commission, now disbanded), was "a potential for providers of such services to undermine the credibility of genomic medicine, by making inflated or misleading claims in marketing their products" [21]. Some genetic scientists share this view: Jakobsdottir et al. [22] for example caution against "overhyping association findings in terms of 'personalized medicine' value before their time, lest we lose the goodwill and support of the general public." Left unchecked and unchallenged, these firms could undermine the credibility of what others said about the future promise of genomics and undermine confidence in the entire enterprise. In June 2008, the HGC expressed that:

> *The HGC is concerned with the premature commercialisation of genetic tests, which give pre-dictive health information, and it stresses the need for an independent system of pre-market review for these tests, which considers the claims made in relation to the analytic, scientific and clinical validity of such genetic tests before they are offered direct to the public [. . .] The HGC believes that without such a system, these tests may cause unnecessary anxiety, false reassurance, further costs to the individual and the NHS and loss of public trust and confidence in genetic testing services. [23]*

As the above quotation illustrates vividly, policy advisors and other critics held negative expectations about what personal genomics would bring. They feared that those who would use these tests would experience anxiety, would draw on healthcare resources more, and even lose their confidence in the wider genomics project to improve human health.

Having now set out some of the principal sociotechnical expectations—both positive and negative—that surrounded personal genomics at the end of the last decade, I turn to consider what is new and specific about personal genomics that distinguishes it from past practices that warrants such promising and tenebrous expectations about the future.

PERSONAL GENOMICS IN PERSPECTIVE

23andMe, deCODEMe, and Navigenics were not the beginning of the DTC genetic testing market. This can be traced back as far as 1994 when University Diagnostics Ltd (UDL) became the first company to offer carrier testing for cystic fibrosis (CF) on a commercial basis in the United Kingdom. The firm initially marketed its test through the Cystic Fibrosis Trust and chose not advertise it more widely. Its Managing Director, Paul Debenham, in giving evidence to a House of Commons Science and Technology Committee in 1995, suggested that CF diagnosis "will be commercially viable because [there is] a desire by the public to know their CF status" but also noted that uptake had been low [24].[3] However, within a few years, the company ceased to offer this service due to low demand. In 2001, another company called Sciona began selling genetic tests via high street stores most notably the Body Shop, providing consumers with advice about lifestyle and nutrition. In response to criticisms of its activities, Sciona switched to marketing its services through healthcare

---

[3] Paul Debenham is now a Director at LGC, the company which bought out UDL, and served as a member of the Human Genetics Commission before its disbanding. He noted to the House of Commons Science and Technology Committee that UDL operated on the basis of "proactive self-regulation" and observed that critical media coverage would ensure that should any company not conduct itself according to the highest standards, it would soon go out of business.

professionals only and withdrew from the DTC market and soon relocated to the United States.[4]

However, at the time of UDL's move into the DTC market, British parliamentarians had expectations about a rapid growth in commercial genetic testing services sold to consumers outside of the context of the NHS. This prompted them to call for the establishment of an entirely new regulatory system for the expected deluge of DTC tests. However, as the HGC noted in its 2003 report *Genes Direct*, there had not been a rapid growth in DTC testing in the late 1990s [25]. While the DTC market in health- and lifestyle-related genetic testing—at least in the British context—faltered, genetic genealogy or ancestry services through companies such as Oxford Ancestors, Family Tree DNA, and DNAPrint Genomics (now ceased trading) flourished. According to Wolinksy [26], by 2006, these three firms had "attracted more than 300,000 customers."[5] In contrast to UDL and Sciona, these firms marketed and sold their services on the web.

In the United States, other firms also entered the DTC market in the early 2000s, providing lifestyle and nutrigenetic tests as well as tests of medical significance on the web. This included newly created firms such as DNA Direct and Mygenome and more established operators such as Kimball Genetics and Cygene. In an opinion piece in *Nature* published in 2002, the founder of the DTC firm Mygenome Fred D Ledley called for a "consumer charter for genomics services," declaring that "every individual has a fundamental right to information to obtain genetic information about himself or herself [...] The consumer-focused model empowers consumers with direct and private access to genetic tests and services" [27]. By 2007, evidence indicated that there were at least 25 companies located in the United Kingdom, continental Europe, and the United States offering a variety of health-related testing services, ranging from nutrigenetics, carrier testing for single gene disorders, and susceptibility testing for various conditions [28]. A further survey of the industry in 2010 indicated that there were more than 60 firms marketing genetic testing services on the Internet.[6]

---

[4] According to letters published on the GeneWatch web site, one reason given by several retailers for either stopping or even starting to sell Sciona's product in the first place was the lack of credible scientific evidence to support the company's claims (see http://www.genewatch.org/sub-425647, accessed 27 April 2011).

[5] To this figure of 300,000 can be added that of the 160,000 individuals who signed up to the 5-year Genographic Project paying US$99 per test to both help fund the research and receive information about their ancestry in return.

[6] In summer 2010, an unpublished internship project in ESRC Cesagen surveyed the consumer genetic testing industry. Of the 69 services then available, it was found that about a third of services on offer were predictive, carrier, or diagnostic tests for disease conditions, tests for "genetic relatedness" (e.g., genetic paternity) and ancestry accounted for another third, while pharmacogenomic and nutrigenetic testing along with various behavioral or lifestyle tests constituted another third of all the services offered. Given the heterogeneity and volatility of this market, assessments tend to be out-of-date quite quickly.

While the vast majority of companies were located in North America and a number of European countries, firms also emerged in South Korea, India, Japan, Singapore, and Australia. The 23andMe, Navigenics, and deCODEMe services were launched in November 2007 into this heterogeneous and volatile market. While each adopted varying business models, they offered something different to existing DTC firms that sold only individual tests. These new firms utilized the latest genomic sequencing technologies and data from ongoing GWAS to test several thousand different polymorphisms, providing risk estimates for the onset of over 100 disease conditions, information on drug metabolism, as well as on (ostensibly) nonmedically significant questions such as ancestry.

The genome-wide association approach had only just been consolidated as the primary way to investigate the role of genes and their variants in the development of common, complex diseases in 2005. Utilizing commercial genotyping technologies, this approach marked a departure from previous ways of finding disease genes because it interrogated "the entire human genome at levels of resolution previously unattainable, in thousands of unrelated individuals" [29]. Unlike candidate gene studies, it was also said to be "hypothesis-free," although this is not entirely the case as the GWA approach was predicated on the understanding that common diseases can be explained with reference to "common variants" [single nucleotide polymorphisms—or SNPs], which have a frequency of 1−5% in the population [30]. Using a case−control design, most GWAS have aimed to identify variants shared by groups of people diagnosed with a particular disease condition and which distinguish them from others who are disease-free.[7] One of the most notable groups to use this approach is the Wellcome Trust Case−Control Consortium which received UK£9 million in funding in 2005. Its first study published in June 2007 assessed 14,000 cases of seven common diseases (bipolar disorder, coronary artery disease, Crohn's disease, hypertension, rheumatoid arthritis, type 1 and type 2 diabetes) and 3000 controls, which were drawn from the 1958 British Birth Cohort and from donors in the UK Blood Service [31]. By 2013, the US NHGRI (National Human Genome Research Institute) catalog of GWAS (which assay at least 100,000 SNPs) contained 1653 studies and shows that more than a 100 loci for 40 common diseases have been identified and replicated [32]. In 2008, a review of the field summarized that: "GWAS have progressed from visionary proposals made when neither the sequence of the human genome nor many variations in this sequence were known, to routine practice of screening 500,000−1,000,000 SNPs in thousands of individuals" [33].

---

[7] Case−control GWA studies are based on the assumption that individuals can be sorted with a high degree of confidence into two groups: those affected by disease and those who are disease-free. This relies on the process of clinical diagnosis at a time when disease categories are becoming unstable and multiplying and the absence of recognized symptoms in individuals does not mean that they are necessarily disease-free.

Despite the fact that few large-scale GWAS had been published prior to their launch, the appearance of 23andMe, deCODEMe, and Navigenics in 2007 therefore represented a significant commercial and technical innovation. These firms transformed basic research into the role of gene variants in common, complex disease conducted on different population cohorts into a consumer service, offering users direct access to findings from the latest science that promised to elucidate their individual susceptibility to disease. To do so, firms genotyped customers' DNA and compared the variants found in their genomes to the variants in the published literature and their statistical association with certain disease states (to give the customer their "relative risk" of a future disease occurring) and statistical data on the incidence of diseases in the population (to determine the "average population risk") to calculate the average person's lifetime risk of developing a disease based on certain sociodemographic characteristics. Customers then receive their absolute risk based on these two calculations.[8]

Therefore, personal genomics firms, while they were not the first to offer DTC genetic testing, were the first to commercialize information from GWAS. However, they were not the first to see that genetic risk information from GWAS could empower individuals to take health decisions about their health. Since the late 1990s, a number of scientists had discussed the idea of using associations found between genetic variants and disease risk to create individual "genomic profiles" that could in turn predict individuals' chances of developing certain diseases or response to medications [34,35]. In 2003, Collins and colleagues at the US NHGRI published their vision of the future of genomic research. For Collins and his colleagues, developing GWA approaches to the study of genetic variants would indicate to individuals their probable risk of future disease before clinically measured factors such as high blood pressure were even recorded. They saw that this knowledge had the potential to form the basis of an "individualized preventive medicine." They speculated that:

The steps by which genetic risk information would lead to improved health are: [1] an individual obtains genome-based information about his/her own health risks; [2] the individual uses this information to develop an individualized prevention or treatment plan; [3] the individual implements that plan; [4] this leads to improved health; and [5] healthcare costs are reduced. Scrutiny of these assumptions is needed, both to test them and to determine how each step could best be accomplished in different clinical settings. [36]

Therefore, the NHGRI vision was based on two propositions: the first was that GWA approaches would produce new knowledge about disease risk which could be

---

[8] Alongside innovations in research on genetic associations, the commercialization and standardization of gene chips, and the use of the web to deliver personal genomics services, we might also add the often overlooked sociotechnical innovation in saliva collection and DNA analysis. This is the subject of a forthcoming paper by the author.

applied in a clinical context at the level of the individual patient; and that people would wish to gain this information and act on it accordingly. This vision conjured up the image of a rational person who acts in a thoroughly rationalistic way, seeking out information about their future health and amending their behavior in the present to minimize or avoid the risk of potential disease. As they admit, however, whether people would actually behave in this imagined way required further scrutiny. Yet, some biotechnology entrepreneurs did imagine that some people might behave in this way and would avail themselves of genetic risk information. Fred Ledley, for example, in founding his company Mygenome Inc., envisioned that:

> *Within this decade, the identification of discrete genetic factors involved in healthy development and disease will become routine, laying the foundation for truly personalized medicine in which individuals are empowered not only with self-knowledge of their genetic risk, but also with the ability to take informed actions to prevent disease and preserve health. [27]*

For Ledley, "truly personalized medicine" was not just a matter of pharmaceutical research; it was about individuals being empowered with self-knowledge of their genome to act in certain rational ways to reduce their risk. While Collins and colleagues imagined that "individualized preventive medicine" would happen within a clinical context, Ledley feared that traditional healthcare systems would only slow the uptake of genomic technologies and looked instead to a consumerist model to ensure their rapid application. By receiving information about their genetic risks directly without the need to go through a doctor, consumers would gain personal control over this form of self-knowledge. The leaders of 23andMe, deCODEMe, and Navigenics came to echo and reinforce these expectations when they launched their services 5 years after Ledley's call for a "consumer charter for genetics" was published. Whether the leaders of these firms were aware of or had been influenced by Ledley's intervention is an intriguing question.

However, in one further twist, it is of note that as 23andMe, deCODEMe, and Navigenics were preparing to launch, geneticists working on GWAS had already begun to explore the question of whether the associations generated from these studies could be used to predict for individuals their likely future experience of disease [34]. Prior to the first GWAS being published, genetic epidemiologists argued that individual risk is "essentially impossible to predict" [37] from population-based studies and that risk factors identified in populations have little "discriminatory power at the individual level" [36]. This was a problem manifest in relation to clinical risk factors such as cholesterol or blood pressure. However, would the same also apply to genetic risk factors found by GWAS? In a historical review of the field, the geneticist Leonid Kruglyak noted that given studies had only identified variants with small effects and which even when taken together could only explain a small part of what was understood to be the heritable component of disease, "we also have to ask how we can piece together individual risk from so many small genetic contributions."

The possibility of categorizing "individuals into groups with regard to risk of specific common diseases" was an open question [33]. With reference to cardiovascular disease, Humphries et al. [38] did not see that individual risk estimates using GWAS data offered any advantage over conventional risk factors such as blood pressure, age, family history, or blood lipid levels. Other geneticists, however, argued that the knowledge produced by GWAS could be harnessed so as to provide risk estimates to individuals even when the variants involved were of small effect [39]. In sum, then, despite the prospect of "individualized preventive medicine" having been one of the major justifications for this research and its funding, scientists were divided on whether it really would deliver on this prospect.

Away from the scientific and technical dimensions to personal genomics, what was new and specific to personal genomics to the way that it delivered genetic risk information to users?

Sociologists have recognized since the turn of the century that the Internet has had a transformative effect on how people access and engage with medical and health information [40,41]. The sociologists Sarah Nettleton and Roger Burrows suggest that, with the emergence and expansion of the World Wide Web from the late 1990s onward, a new form of medicine which they call "e-scaped medicine," has taken shape [40]. The notion of "e-scaped medicine" represents a purposeful play on words, to indicate not only that medical information is now ubiquitous online but that, as a result, it has escaped "and is thus no longer something that can be accessed and, more importantly perhaps, produced and regulated by medical experts" [40]. From this perspective, the emergence of online of DTC genetic testing services and personal genomics are very much a product of this wider transformation in how health and medical information is produced and circulated through online networks. However, the web has not remained the same and in the last 10 years what is commonly known as Web 2.0 has established infrastructures for the creation and sharing of content where users are then expected to actively create, upload, and share content, often at no immediate cost to them. Personal genomics companies—23andMe and deCODEMe in particular—went further than existing online DTC services by bringing together predictive genetic testing and the practices of interactive, "social" media of Web 2.0. Both 23andMe and deCODEMe included features on their web site to allow consumers to share information with others such as friends or family and provide opportunities for social networking. For Lee and Crowley [42] and Levina [43], it is the social networking aspect of how these web-based services are organized that mark them out from what has gone before. Levina [43] argues that personal genomics is emblematic of new forms of "network subjectivity" that encourages and even requires the constant sharing of information about oneself with others in the network so that the network grows.

Despite a number of significant differences from existing DTC services in terms of the information provided and how it was delivered, I would contend that regulatory and policy bodies did not see a fundamental break between personal genomics and what was on offer already to consumers. Both British and US policy advisors and regulators have tended to see that DTC genetic testing services should be subject to regulation and to some kind of premarket review. Back in 1995, when the British House of Commons Science and Technology Committee learned about the DTC selling of CF genetic tests, it warned that: "there is a very real danger that unscrupulous companies may prey upon the public's fear of disease and genetic disorders and offer inappropriate tests, without adequate counseling and even without adequate laboratory facilities" [24]. In the United States, too, a NIH-DOE (Department of Energy) Task Force on Genetic Testing in 1997 and the US Health Secretary's Advisory Committee on Genetic Testing (SACGT) in 2000 both came out against the direct advertising and selling of predictive genetic tests to the public [44,45]. Moreover, individual US states such as New York and California had also taken action against DTC companies even before the emergence of personal genomics (for more detailed discussion, see Ref. [46]).

However, a further way in which personal genomics is different to earlier forms of DTC genetic testing is not in terms of the consumer service offered but the way that some firms, 23andMe especially, has used Web 2.0 techniques to establish "research communities" dedicated to various health-related conditions such as Parkinson's disease. 23andMe stressed that the firm's services would not only enable individuals to "see what genetics research means for them" but also to share information with others with the idea that "ultimately, they will become part of a community that works together to advance the overall understanding of the human genome" [47]. The emphasis of 23andMe has been more on the potential of social networking and forms of online community-building that would take concrete form with the launch of its 23andMe' features in 2009 [48].[9] Unlike deCODE Genetics and Navigenics, its business model combined both a consumer service and the establishment of a database comprising genetic, phenotypic, and lifestyle information from customers to provide a platform for further research [49]. 23andMe therefore made use of similar techniques to Facebook with its launch of what it calls "Research Revolution," a consumer-led research model whereby individuals voted for studies to be undertaken into specific conditions and were offered a hugely discounted price on their tests if they agreed for their data to be used in these studies. The ultimate aim of 23andMe is to create a new

---

[9] In July 2009, the company launched the "Do-it-yourself Revolution in Disease Research" which called upon actual and potential consumers to "pledge allegiance" to a particular disease (out of a predetermined list of 10 conditions) on which they would help to support research, and to provide relevant personal data in connection to these diseases.

model of doing research. The company and its scientific collaborators went on to publish academic papers to demonstrate that what they called "web-based, participant-driven" research was a viable and valid model for undertaking research on gene association research linked to health and other phenotypic characteristics [50]. The firm has also worked with other research organizations on further projects on the role of genetic variants in conditions such as Parkinson's disease and allergies. In 2014, it was the recipient of an NIH grant valued at US$1.3 million to run a 2-year long project to further develop its web-based database and "research engine for genetic discovery" [51].

In sum, then, I argue that there are various aspects to personal genomics that are new and specific to personal genomics in relation to what we have seen in the context of DTC genetic testing more broadly: these include the use of GWAS data to predict individual risk of future disease, the use of Web 2.0 techniques to facilitate and encourage sharing of personal genomic data, and the creation of web-based genomic database for research. Given that, how then should we assess the scale of the sociotechnical transformations brought about personal genomics?

PERSONAL GENOMICS AND THE EXTENT OF SOCIOTECHNICAL CHANGE

This question can be responded to this question in a number of different ways. I will begin by considering the size of the industry and the uptake of these services. In the first few years in which personal genome firms operated, public information about the number of spit kits which they had actually sold as opposed to have given away as part of their corporate promotions was nonexistent. Wright and Gregory-Jones from the Public Health Genetics Foundation in Cambridge estimated using Internet traffic data that the three firms—23andMe, deCODE Genetics, and Navigenics—had received just over 660,000 unique visitors in the calendar year of 2009 and concluded that these firms had about 20—30,000 consumers in that period. On the basis of these figures, they suggest that: "in the absence of data to the contrary from the relevant companies, we therefore conclude that current demand for genomic profiling tests for susceptibility to common complex diseases is fairly small" [52]. Journalistic commentators came to similar conclusion to Wright and Gregory-Jones that there was an apparent lack of demand for these services [53], while others have pointed to a general lack of awareness of these services outside of certain elite groups as a potential explanation for why this might be the case [54—56]. At the time of writing, 23andMe reports it has in the region of 700,000 genomes in its database.

Regulators, most especially the US Food and Drug Administration (FDA), have also played a significant role in shaping the organization and size of the market for personal genomics services. When a firm called Pathway Genomics, which launched in 2009, sought to go further than 23andMe, deCODEMe, and Navigenics to market its services to a wider consumer market beyond the web by entering into an alliance with a pharmacy chain called Walgreen's to provide saliva collection kits in in-store, the FDA staged a significant intervention. The FDA contacted Pathway Genomics stating that, within the terms of the 1976 Federal Food, Drug and Cosmetic Act, it determined its product to be a "medical device" that required premarket review and approval by the FDA. A further five such letters followed in June to 23andMe, deCODE, and Navigenics among other firms setting out a similar case to each: they were making or selling products that were medical devices without premarket approval. In July 2010, the FDA issued a second series of letters to a wider range of DTC genetic testing companies. In that same month, in what appeared to be a coordinated action on both sides of the Atlantic, the FDA and European authorities announced wide-ranging regulatory reviews of DTC testing services and the regulation of medical devices. One result of the FDA's intervention was that many firms modified how they marketed and delivered their services [19]. Pathway Genomics switched to a system of requiring only registered physicians to order their testing kits on behalf of their patients, while 23andMe also began offering customers the option of consulting a genetic counselor.

In the 2 years following the FDA action, companies began to leave the personal genomics market as new firms also appeared. In August 2012, after its acquisition by Life Technologies Corporation, Navigenics reported that it would no longer accept new orders from potential customers and switched its business away from the consumer market to focus on the clinical sector only. After its 2011 bankruptcy, the biotech firm Amgen bought out deCODE Genetics and closed its deCODEMe service in December 2012. However, in that same period, 23andMe cut its price to US$99 and declared the ambitious aim of enrolling a million customers. It also became the first personal genome company to seek regulatory approval for its services. It submitted a premarket submission to the FDA for 7 out of its total of 240 health reports. In a press release to announce this submission, 23andMe [57] noted that: "our ongoing conversations with the FDA in the last year, in particular, resulted in a focused approach that resulted in our ability to compile a comprehensive analysis of 23andMe's DTC testing for FDA consideration." However, by November 2013, FDA issued a strongly worded letter to 23andMe, which complained that the company had not provided information when requested and failed to maintain communication with the FDA over the regulatory process. In the letter, the FDA asserted again its view that personal genome services constitute a medical device and therefore fall under its regulatory regime; it also stated plainly that the company must "immediately discontinue marketing the personal genome service until such time as it receives FDA marketing authorization for the [medical] device."[10] In response, 23andMe suspended its health report service for the US market.

However, a year later, the firm had successfully sought regulatory approval from authorities in Belgium and The Netherlands, registering its service as an *in vitro* diagnostic medical device. With this approval, the firm relaunched its health report service onto the UK market. In contrast to its US counterpart, the MHRA (Medicines and Healthcare Regulatory Agency) did not require premarket submission as it regarded its service as "low risk." However, the MHRA required 23andMe to include warnings to its customers not to change drug therapy or take medically significant action without medical advice. At the time of writing, 23andMe is doing what no other firm in the United States has been permitted to do—running television advertising. Time will tell whether this move will prove to be commercially successful for the firm.

In addition to thinking in terms of the size of the industry and the popular uptake of personal genomics services, there is also the question of whether the genetic risk information the firms produced for customers was transformative for them, empowering them to take decisions in relation to their future health. Against the backdrop of extensive debate about both the positive messages about the benefits of personal genomics for users and the fears expressed by others about its problems, there was only limited evidence gathered on actual users of personal genome services [54]. This prompted a number of organizations to fund research on how people actually engaged with and responded to genetic risk information of the kind provided by personal genomics companies. A number of studies have been completed investigating whether the genetic risk information of the kind provided by personal genomics companies had any effect on modifying behavior. By 2013, a review of this field of research listed that 21 different studies had been published using a range of methodologies, which claimed to shed light on the interest in, attitudes to, uptake and impact of genetic susceptibility testing for multiple common, complex diseases among a much wider group of people [58].[11] In the terms set by the investigators, this research provides very much a mixed picture of the impact of personal genomics and there is no evidence to support the view that gaining this particular form of self-knowledge is a transformative experience for most people who have been included in these studies.

However, given 23andMe's success in gaining research grants to explore further its database, it is certainly a possibility that the web-based research model it has pioneered may prove to be influential in the medium term. At the time of writing, 23andMe has announced two agreements with major pharmaceutical companies, Pfizer and

---

[10] The FDA warning letter was sent on 22 November 2013 in the name of lberto Gutierrez, Director Office of In vitro Diagnostics and Radiological Health at the FDA to the 23andMe CEO Anne Wojcicki. The letter can be read at the FDA's web site."

[11] It is also of note that other areas of investigation such as people's response to pharmacogenomic risk information also remain to be explored more fully [54].

Genentech, to share its customer database to inform research on irritable bowel disorder and Parkinson's disease [59,60]. In some respects, then, 23andMe is adopting a similar business model as the genomics industry that emerged in the early to mid-1990s following the funding of the Human Genome Project: these firms also leveraged their genomic databases in pursuit of profitable alliances with pharmaceutical companies to utilize them in drug discovery and development programs.

CONCLUSION

In this chapter I have offered up the emergence of personal genomics—a development that has caught the attention of social scientists, ethicists, and lawyers who have investigated and written about the consequences of genomics in its various forms for the last decade—as a case study for reflecting on genomics and its expected sociotechnical transformations. In conclusion I would argue that, partly in the wake of regulatory opposition in the United States and criticism from clinicians, personal genomics has not brought about the kinds of sociotechnical transformations that its investors and supporters hoped for at the end of the last decade. Personal genomics has not—as yet—achieved mass appeal—and indeed US regulators and policymakers have, on the whole, been concerned to keep firms from expanding their markets too much. Personal genomics is of course a changing area of commercial and scientific activity. Aside from 23andMe, other firms have come and gone from a volatile market. Investment in whole-genome sequencing could overcome the current limitations of GWAS and begin to provide information that is more actionable for individuals to have when making decisions about their health. Beyond genomics, self-tracking and "mobile health" apps have proliferated and the field of "digital health" has now become the focus of sociotechnical expectations and social science research [61—63]. It remains to be seen whether this will prove to be more transformative than personal genomics in terms of bringing about sociotechnical changes to the practices of health and medicine.

NOTE

This chapter reproduces material from my book *Genomics and the Reimagining of Personalized Medicine* (Ashgate, 2014).

REFERENCES

[1] Latham J. The failure of the genome, The Guardian, 17 April 2011. Available from: <http://www.theguardian.com/commentisfree/2011/apr/17/human-genome-genetics-twin-studies>; 2011 [accessed 01.02.15].
[2] Bodmer W, McKie R. The book of man: the quest to discover our genetic heritage. London: Abacus Books; 1994. p. 227.

[3] Gilbert W. A vision of the grail. In: Kelves D, Hood L, editors. The code of codes: scientific and social issues in the human genome project. Cambridge: Harvard University Press; 1992. p. 94.

[4] Evans JP, Meslin EM, Marteau TM, Caulfield T. Deflating the genomic bubble. Science 2011;331 (6019):861−2.

[5] Collins FS. Has the revolution arrived? Nature 2010;464(7289):674−5.

[6] Lippman A. Led (astray) by genetic maps: the cartography of the human genome and health care. Soc Sci Med 1992;35(12):1469−76.

[7] Flower M, Heath D. Micro-anatomo politics: mapping the Human Genome Project. Cult Med Psychiatry 1993;17:27−41.

[8] Giddens A. Modernity and self-identity: self and society in the late modern age. London: Polity Press; 1992.

[9] Diamond I, Woodgate D. Genomics research in the UK—the social science agenda. New Genet Soc 2005;24(2):239−52.

[10] Hedgecoe A, Martin P. Genomics, STS and the making of sociotechnical futures. In: Hackett E, Amsterdamska O, Lynch M, Wajcman J, editors. The handbook of science and technology studies, 817−840. Cambridge: MIT Press; 2007. p. 820.

[11] Stix G. Personal pills: genetic differences may dictate how drugs are prescribed. Sci Am 1998;279:10−11.

[12] Personal Genome Project. Personal Genome Project home page [online]. Available from: <http://www.personalgenomes.org/>; 2014 [accessed 07.01.14].

[13] Hamon A. My genome, myself: seeking clues in DNA, *New York Times*, 17 November 2007. Available from: <http://www.nytimes.com/2007/11/17/us/17dna.html?pagewanted=all&_r=0>; 2007 [accessed 01.02.15].

[14] deCODE Genetics Inc. Press release: deCODE Launches deCODEMe™: deCODE Genetics Inc. 2007.

[15] Sunder Rajan K. Biocapital: the constitution of post-genomic life. Durham and London: Duke University Press; 2006.

[16] Navigenics Inc. Press release: Navigenics launches with pre-eminent team of advisors, collaborators and investors. Available from: <http://www.navigenics.com/visitor/about_us/press/releases/company_re3, jlaunch_110607/>; 2007 [accessed 20.08.12].

[17] Hood L. Systems biology and systems medicine: from reactive to predictive, personalized, preventive and participatory (P4) medicine. Paper presented at the Engineering in Medicine and Biology Society, 2008. EMBS 2008. 30th Annual International Conference of the IEEE; 2008.

[18] National Institutes of Health. Personalized healthcare: opportunities, pathways, resources. Bethesda: National Institutes of Health; 2007.

[19] Prainsack B, Vayena E. Beyond the clinic: "direct-to-consumer" genomic profiling services and pharmacogenomics. Pharmacogenomics 2013;14(4):403.

[20] Hogarth S, Javitt G, Melzer D. The current landscape for direct-to-consumer genetic testing: legal, ethical and policy issues. Annu Rev Hum Genet 2008;9:161−82.

[21] House of Lords Science and Technology Committee. *Genomic Medicine, 2nd Report of Session 2008−09*, Vol. I. London: House of Lords Science and Technology Committee; 2009. p. 163.

[22] Jakobsdottir J, Gorin MB, Conley YP, Ferrell RE, Weeks DE. Interpretation of genetic association studies: markers with replicated highly significant odds ratios may be poor classifiers. PLoS Genet 2009;5(2):e1000337.

[23] HGC. A common framework of principles for direct-to-consumer genetic testing services: principles and consultation questions. London: Human Genetics Commission, Department of Health; 2010.

[24] House of Commons Science and Technology Committee. Human genetics: the science and its consequences. House of Commons Science and Technology Committee. 1994−1995. London; 1995.

[25] HGC. Genes Direct: ensuring the effective oversight of genetic tests sold directly to the public. London: Human Genetics Commission, Department of Health; 2003.

[26] Wolinsky H. Genetic genealogy goes global. EMBO Rep 2006;7(11):1072−4.

[27] Ledley FD. A consumer charter for genomic services. Nat Biotechnol 2002;20:767.

[28] HGC. More Genes Direct: Report on developments in the availability, marketing and regulation of genetic tests supplied directly to the public. London: Human Genetics Commission, Department of Health; 2007.

[29] Pearson TA, Manolio TA. How to interpret a genome-wide association study. JAMA 2008;299 (11):1335−44.

[30] Lander ES. The new genomics: global views of biology. Science 1996;274(5287):536−9.

[31] Wellcome Trust Case−Control Consortium. Genome-wide association study of 14,000 cases of seven common diseases and 3,000 shared controls. Nature 2007;447:661−78.

[32] Hindorff LA, MacArthur J. (European Bioinformatics Institute), Morales J (European Bioinformatics Institute), Junkins HA, Hall PN, Klemm AK, and Manolio TA. A Catalog of Published Genome-Wide Association Studies. Available at: < www.genome.gov/gwastudies > . [accessed 07.09.15].

[33] Kruglyak L. The road to genome-wide association studies. Nat Rev Genet 2008;9(4):314−18.

[34] Khoury MJ, Evans J, Burke W. Personal genomics and personalized medicine. Nature 2010;464 (7289): 680.

[35] Lindpaintner K. Genetics in drug discovery and development: challenge and promise of individualizing treatment in common complex diseases. Br Med Bull 1999;55(2):471−91.

[36] Collins FS, Green ED, Guttmacher AE, Guyer MS. A vision for the future of genomics research. Nature 2003;422(6934):835−47.

[37] Buchanan AV, Weiss KM, Fullerton SM. Dissecting complex disease: the quest for the Philosopher's Stone? Int J Epidemiol 2006;35(3):562−71.

[38] Humphries SE, Drenos F, Ken-Dror G, Talmud PJ. Coronary heart disease risk prediction in the era of genome-wide association studies. Circulation 2010;121(20):2235−48.

[39] Wray NR, Goddard ME, Visscher PM. Prediction of individual genetic risk to disease from genome-wide association studies. Genome Res 2007;17(10):1520−8. Available from: http://dx.doi.org/10.1101/gr.6665407.

[40] Nettleton S, Burrows R. E-scaped medicine? Information, reflexivity and health. Critical Social Policy 2003;23(2):165−85.

[41] Nettleton S. The emergence of e-scaped medicine? Sociology 2004;38(4):661−79.

[42] Lee SS-J, Crawley L. Research 2.0: social networking and direct-to-consumer (DTC) genomics. Am J Bioeth 2009;9(6−7):35−44.

[43] Levina M. Googling your genes: personal genomics and the discourse of citizen bioscience in the network age. J Sci Commun 2010;9(1):1−8.

[44] NIH-DOE Task Force on Genetic Testing. Promoting safe and effective genetic testing in the United States, final report of the task force on genetic testing. Available from: <http://www.genome.gov/10001733>; 1997 [accessed 01.02.15].

[45] SACGT. Enhancing the oversight of genetic tests: recommendations of the SACGT. Available from: <http://osp.od.nih.gov/sites/default/files/oversight_report.pdf>; 2000 [accessed 01.02.15].

[46] Groves C, Tutton R. Walking the tightrope: expectations and standards in personal genomics. BioSocieties 2013;8(2):181−204.

[47] 23andMe. 23andMe democratizes personal genetics. [press release] 8 September 2008. Available from: <https://www.23andme.com/about/press/20080909b/>; 2008 [accessed 22.08.12].

[48] Tutton R, Prainsack B. Enterprising or altruistic selves? Making up research subjects in genetics research. Sociol Health Illn 2011;33(7):1081−95.

[49] MacArthur D. Cheap personal genomics: the death-knell for the industry? Available from: <http://scienceblogs.com/geneticfuture/2008/09/cheap_personal_genomics_the_de.php>; 2008. [accessed 01.02.15].

[50] Eriksson N, Macpherson, JM, Tung, JY, Hon, LS, Naughton, B, Saxonov, S, et al. Web-based, participant-driven studies yield novel genetic associations for common traits. PLoS Genet, 2010; 6 (6):e1000993.

[51] Anon. 23andMe Scores NIH Grant to pump up genetic discovery database. Available from: <http://www.genengnews.com/gen-news-highlights/23andme-scores-nih-grant-to-pump-up-genetic-discovery-database/81250168/>; 2014 [accessed 01.02.15].

[52] Wright CF, Gregory-Jones S. Size of the direct-to-consumer genomic testing market. Genet Med 2010;12(9):594.

[53] Pollock A. Consumers slow to embrace the age of genomics, New York Times, 20 March 2010. Available from: <www.nytimes.com/2010/03/20/business/>; 2010 [accessed 07.01.14].

[54] Bloss CS, Schork NJ, Topol EJ. Effect of direct-to-consumer genomewide profiling to assess disease risk. N Engl J Med 2011;364(6):524—34.

[55] McGowan ML, Fishman JR, Lambrix MA. Personal genomics and individual identities: motivations and moral imperatives of early users. New Genet Soc 2010;29(3):261—90.

[56] McGuire A, Diaz C, Wang T, Hilsenbeck S. Social networkers' attitudes toward direct-to-consumer personal genome testing. Am J Bioeth 2009;9(6—7):3—10.

[57] 23andMe LLC. 23andMe Takes First Step Toward FDA Clearance. [press release] 30 July 2012. Available from: <https://www.23andme.com/about/press/fda_application/>; 2012 [accessed 22.08.12].

[58] Roberts JS, Ostergren J. Direct-to-consumer genetic testing and personal genomics services: a review of recent empirical studies. Curr Genet Med Rep 2013;1(3):182—200.

[59] 23andMe. Press release: 23andMe announces agreement with Pfizer Inc. to research genetics of ulcerative colitis and Crohn's disease. Available from: <http://mediacenter.23andme.com/blog/2014/08/12/pfizer_ibd/>; 2014 [accessed 01.02.15].

[60] 23andMe. Press release: 23andMe and Genentech to analyze genomic data for Parkinson's disease. Available from: <http://mediacenter.23andme.com/blog/2015/01/06/23andme-genentech-pd/>; 2015 [accessed 01.02.15].

[61] Lupton D. Quantifying the body: monitoring and measuring health in the age of mHealth technologies. Crit Public Health 2013;23(4):393—403.

[62] Lupton D. The digitally engaged patient: self-monitoring and self-care in the digital health era. Soc Theory Health 2013;11(3):256—70.

[63] Swan M. Emerging patient-driven health care models: an examination of health social networks, consumer personalized medicine and quantified self-tracking. Int J Environ Res Public Health 2009;6(2):492—525.

CHAPTER 2

Genetics, Genomics, and Society: Challenges and Choices

Angus Clarke

Institute of Cancer and Genetics (Formerly Medical Genetics), Cardiff University School of Medicine, University Hospital of Wales, Cardiff, Wales, UK

Contents

INTRODUCTION

Health care and rare disorders

Biology has become data-rich and has moved away from the focused testing of specific hypotheses toward a systematic approach to the recognition of patterns in massive arrays of data. This is fully compatible with the now classic scientific method of Popper, with its goal of progress through "the refutation of conjectures," but the procedures adopted by scientists are changing so that they are less likely to design and perform key, hypothesis-challenging experiments and more likely to invest in generating a large set of data and then working on different approaches to its analysis. As a part of biology, medicine will become *the* data-rich enterprise.

We can take for granted the generation of vast quantities of information about patients—we are all "patients"—and its application in health care. This information will include the sequencing of the individual patient's genome, or at least large sections of the genome, but also other "*omics*" data and indeed information about each patient's internal and external environment as well. The "genome" will come to include not only the DNA sequence but also its methylation status and its chromatin configuration in perhaps several tissues. The data concerning a patient's environment will include a systematic record of environmental exposures from before birth, of the person's diet, and of the composition of their microbiome, the set of microbial commensals (and potential pathogens) in the gastrointestinal tract, the respiratory tract, the skin, and genitalia. Furthermore, there is a need for a detailed phenotypic description that can be included in the analyses *in silico*. We are increasingly challenged by the bulk of data in need of interpretation; there will be no opportunity to relax as this problem will grow at least as fast as our capacity to handle the data over the next 15–20 years.

Attempts to analyze the genome sequence plus associated "*omic*," environmental and phenotypic data, and then apply it in health care will make progress but will inevitably begin with incomplete data sets. Only once there is some stability in the data required for a "full" analysis, so that complete data sets exist on many patients of different ages and in different states of health, will a clinically applicable interpretation of a full data set become feasible. While there is much agreement in practice about how raw sequence data from a person's constitutional genome can be stored and transmitted, this does not apply to the broader set of "*omic*" data, and does not even apply to sequence data obtained from a tumor, where complex decisions must be made about how many separate samples to take from a tumor. For tumors, single-cell analyses may be required with further decisions to be made about how many cells of which types are to be analyzed within a tumor and any metastases; to what extent can circulating DNA in plasma give an overview of the malignancy? In relation to environmental data, there is little consensus as to what data should be captured. As the information to be caught should be defined in advance, and one may not know in advance what disease will affect a person in the future, the environmental data should be as "complete" as possible. The very nature of genomic data—its linear and essentially digital structure—means that the problems of data collection and interpretation for purposes of health care will be much less than for these less predefined domains whose analytic structure remains wide open.

We will first turn to consider the rare, genetic disorders and later address the range of the more common, "complex" diseases. Rare disorders will be diagnosed much more rapidly, accurately, and cheaply once whole-genome sequencing (WGS) is available in routine health care. In addition to the substantial benefit this confers of some understanding and explanation for an individual's or a family's problems, and information about the potential risk of recurrence, this will also lead to a slow and piecemeal but steady increase in the range of conditions for which a rational therapeutic intervention may be expected to improve prognosis. Knowledge of the gene in which a mutation causes the disease can open up avenues to increase the understanding of the disease and thereby identify "rational therapeutic opportunities," even though gene therapy—in the sense of gene editing, gene substitution, or gene augmentation—may not often be feasible.

Two rare disorders may serve as examples of such progress: tuberous sclerosis (TS) and X-linked hypohidrotic ectodermal dysplasia (XHED). In TS, therapeutic opportunities have arisen following the recognition that the tumors and tissue disorganization that result from mutation in TSC1 or TSC2 occur because of the disinhibition of the mTOR signaling pathway. The use of mTOR inhibitors has been shown to cause the stabilization or regression of many of these tumors and it is hoped that the early use of such treatments will also greatly improve the neurocognitive outcomes for affected individuals and the quality of life for affected individuals and their families. In XHED, a modified version of the protein encoded by the EDA gene—absent or malfunctioning in affected males—can effectively cure the disorder if given in several doses early in life to affected mice and dogs. It remains unclear whether similar treatment will be of benefit in human patients if started after birth but it may be possible to start treatment *in utero*, when it is likely to be safe and may well be much more effective. However, in both disorders, there are serious challenges that prevent the rapid development of new therapies, not least the difficulty in recruiting sufficient numbers of patients to clinical trials, given that these are both rare diseases.

A third disorder, Rett syndrome, shows another situation. This condition is usually caused by inactivating mutations in the X chromosome gene, *MECP2*, but there are no ready "solutions" to remedy this. Gene augmentation has the potential problem of toxicity associated with duplication of this gene, especially if the extra copy of the gene were functional in cells that had inactivated the defective copy of the gene. A wide range of remedies is being developed; while some are working to "correct" the mutation, others are directed at the downstream physiological consequences of the mutation. None has yet become established as an effective treatment.

If these three diseases illustrate some of the possible futures for other disorders, there will be a gradual, case-by-case *creep* toward the enhanced understanding of the pathobiology of each rare genetic disease … leading to the slow and opportunistic development of rational therapies, depending upon the specifics of each disorder. This will indeed be a long, slow process, to be followed by clinical trials to establish safety and efficacy. The rate and quality of such progress will depend upon the effective coordination of diagnosis and management for those affected by rare diseases. Whether or not the care for those with rare diseases will be coordinated effectively, as envisaged by the rare disease frameworks within Europe, will depend upon the political will of those in power in different countries [1].

An opportunity to apply genomic knowledge rather sooner arises in pharmacogenetics, where testing patients for a rare predisposition to a potentially serious adverse drug reaction could be most helpful, if the genotypes at such loci were accessed by doctors or pharmacists at the point of prescription. This might apply to rare variants associated with Stevens—Johnson syndrome in response to flucloxacillin, myopathy in response to statin drugs, the risk of acute porphyria, or even the less infrequent drug reactions such as glucose-6-phosphate dehydrogenase deficiency. The IT connectivity required to enable such patient safety measures should be feasible: it does not require instant access to a person's full genome sequence at any health center, hospital, or pharmacy. There are other examples where the choice of drug treatment may be influenced by changes of the constitutional genome or, in the case of a malignancy, the tumor genome. We will refer to some such cases (following sections) but we will not attempt to review such applications of genetic knowledge as they are already entering regular clinical practice.

HEALTH CARE AND THE CHALLENGE OF COMMON, COMPLEX DISEASE

In contrast to the rare, Mendelian disorders, there will be a slow process of elucidation of the risk of the common, complex diseases conferred by the genome-wide scatter of low penetrance loci that modify disease risk but only very slightly.

Recognizing the small proportion of common disease cases that are associated with a high penetrance mutation at a specific locus will continue to be of great value in recommending disease prevention and surveillance programs, and will also present opportunities to develop rational therapies, as in the rare disorders. However, such benefits will apply to only this small proportion of the cases of the common disorders, such as those with a mutation in *BRCA1* or *BRCA2* or one of the Lynch syndrome genes. Decisions about treatment can already be influenced helpfully by the knowledge of whether a patient has a high penetrance risk allele for a *BRCA* gene mutation, as with the choice between lumpectomy, unilateral mastectomy, or bilateral mastectomy at the point of diagnosis of breast cancer and decisions about the type of chemotherapy to use. More generally, however, the knowledge of the precise pattern of disease susceptibility seems unlikely to be of much clinical value.

An individual's risk of the common, complex, and mostly degenerative disorders of western society is influenced by genetic variation at many loci—many loci of highly significant but usually very small effect—and by environmental factors of many types, accumulated over a lifetime. While it is helpful to understand these effects at the population level, the clinical utility of determining an individual's personal risk of developing these disorders—coronary artery disease, hypertension, stroke, diabetes mellitus types 1 and 2, Alzheimer's disease, the common cancers, rheumatoid disease—is much less clear, and the value of such risk estimations in arriving at a diagnosis of overt disease is also uncertain. The principal reasons for this lack of clinical utility are (i) that the genetic variants known to influence the risk of these complex disorders account for only a very modest fraction of the genetic contribution to these risks (see below), (ii) the recommended healthy behaviors do not depend upon the precise level of risk or the pattern of risk-conferring variants but is broadly the same for almost everyone, and (iii) the response to "personalized" health promotion advice is generally very poor and there is even scope for paradoxical effects mediated through a sense of fatalism or invulnerability. Only for those at very high risk of disease, as with the Mendelian forms of cancer or heart disease, would the prospect of improved compliance with health professional advice warrant genetic testing. There are very few patients for whom the usual advice of a conventionally varied but healthy diet and regular exercise would be inappropriate.

There are additional reasons for caution in promoting genetic testing to give everyone (or, perhaps, anyone sufficiently interested to pay) their pattern of risk-conferring genetic variants for a range of common diseases. First, if they understood the irrelevance of the genetic results to the behavioral and lifestyle recommendations appropriate for them, they would probably be unwilling to pay for the tests and the health services should certainly be unwilling to do so. More important than this, however, is the matter of people's—and society's—orientation to the very question. Given the rather poor response of many populations to health promotion advice, and the effectiveness of collective, government action (through both selective taxation and direct legislation) in curbing the consumption of tobacco and alcohol and enforcing road safety measures, it is not at all obvious that health promotion is likely to be [1,2] effective in achieving behavior change in the individual [2,3]. Indeed, population-level measures may achieve much more, although this would only be possible with a degree of public acceptance of, or acquiescence in the face of, government intervention. If public acceptance can be achieved then such an approach may be greatly preferable to the personalized promotion of healthy lifestyles to a vast sea of atomized individuals [4].

The sort of public health measures that could address these issues would include a coordinated government policy toward transport (encouraging the use of public transport, walking, and cycling to commute to work), the quality of air and water, and the promotion of healthy food (through a stick-and-carrot system of taxation and subsidy). One major obstacle to the adoption of such policies is the simple fact that they are collective in their approach and therefore run counter to the Thatcher-Reagan *Zeitgeist* that still reigns despite the massive harm it has inflicted, globally, on both "the west" and "the rest." The fact that all benefit from these public health measures, and that they work to reduce rather than aggravate still further the curse of health inequalities, is another reason why they should be adopted but are likely to be rejected by neoconservative politicians.

The excessive focus on individual rather than collective approaches to the public health is compounded by an excessive focus on genetic rather than environmental or behavioral contributors to the causation of the common, complex, degenerative diseases. Indeed, these two trends go hand-in-hand and mutually support each other. As a result, both errors need to be combated together, as two aspects of essentially the same confusion. A focus on genetic factors naturally and inevitably leads to a focus on the individual and his or her genetic constitution but for many disorders this individualization of health care is simply an expensive irrelevance. Rather worse, it can distract from those important contexts where a focus on the relevant genetic factors is genuinely very helpful [5]. In the context of heart disease, for example, there are a number of Mendelian disorders where a focus on the genetic basis of disease can be very helpful in determining the risk of disease in relatives, as in the cardiomyopathies, familial hypercholesterolemia, and the familial dysrhythmias. But at the population level, measures to discourage smoking and to improve diet and the general pattern of exercise will be of much greater value than the analysis of each citizen's pattern of alleles at the numerous but minor risk-conferring SNPs. Similarly, someone with a strong family history of breast or bowel cancer may be falsely reassured by being given a reduced risk when tested with a direct-to-consumer SNP-based health risk package. They may then decide **not** to come forward for formal risk assessment and the offer of mutation testing in the relevant Mendelian genes, which would be the appropriate step to take and could be life-saving.

If we fail to reject this inappropriate focus on the individual, we will be in danger of following the logic of health services that operate "Payment by Results." Such an approach, if pursued with firm consistency, will lead inexorably to doctors being held accountable for the behavior of their patients. At present we see some small signs of this in the United Kingdom, such as primary care being rewarded financially for achieving a high uptake of childhood immunization and other targets, but there are many difficulties with such schemes. These include the inevitable off-target and often predictably paradoxical effects—the unintended consequences of such attempts to manipulate a complex system in a simple-minded fashion. Under such a scheme, a medical career in more affluent and more compliant areas becomes more lucrative than work in more socially deprived communities; this effect is in turn likely to exacerbate existing health inequalities. Furthermore, medical practice risks becoming an exercise in motivational interviewing with the accompanying overtones of behavioral manipulation leading to two sets of ethical quandaries: (i) whether it is right for health services to provide incentives for patients for act in their own interests (such as losing weight or stopping smoking), with commitment devices [6] or even the making of direct payments for compliance, and (ii) whether health services may legitimately withdraw cover from "self-inflicted" disease, such as alcohol- and tobacco-related disease and injuries acquired through hazardous sports. It is not feasible to address these topics fully but only to indicate that such questions arise in the context of how genetics is applied within health care.

GENOME DATA: CHALLENGES OF INTERPRETATION, DISCLOSURE, AND DATA STORAGE

It is often assumed by those who discuss the clinical application of genomics in health care that the model set out by Biesecker is inevitable: once a patient's genome has been sequenced, then the sequence data will be kept in a secure database to be accessed by the relevant healthcare professionals whenever it may be useful [7]. A person's genome sequence will be a lifetime resource. However, there are several reasons why this model may not be a good guide to the future. First, the problems with interpretation of sequence data are likely to remain an active issue for many years. In the long term, these problems may be seen as belonging to a phase of transition but now (in 2015), at the start of that phase, the problems loom large and it seems likely that they will remain substantial for many years. Second, the notion of data storage over a patient's lifetime faces important challenges.

The clinical problems that arise from the interpretation of sequence data relate to three particular areas: genetic variants of uncertain significance (VUSs), incidental findings (IFs), and the aggregation of risk information for an individual. The first two have been discussed elsewhere (reviewed in Clarke, 2014 [8]), but an overview will be helpful here so that the implications for the organization of genetics services become clear.

The problem of VUSs has become familiar over the last decade as more DNA sequence results have been generated in the genetics diagnostic laboratory but the scale of the problem is immensely greater in genomic analyses (covering many or all genes, or the entire genome) as opposed to the targeted testing of one or two genes. It is not always clear whether a genetic variant identified on DNA sequencing will be benign (entirely innocent), or pathogenic (disease-causing), or perhaps disease-associated and contributing to disease. This uncertainty reflects our ignorance and is both growing and simultaneously shrinking: it is growing as the number of new variants identified increases, and it is shrinking as more evidence accumulates and, albeit slowly, variants that were VUSs are shown to be either harmless or pathogenic. Over years, even decades, as our knowledge catches up with the spate of variants identified, our ability to interpret the large majority of these will be much better than it is now. This problem can be seen as largely transient—a problem of the transition to genomic analyses—although it seems set to remain an important clinical issue for many years.

The problem of IFs refers to the finding of clinically important information that has not been actively sought or, at least, is not related to the primary reason for performing the investigation. A child may have a genomic investigation to see if the cause of their developmental difficulties can be identified. Whether or not a cause of these problems is identified, some completely different finding may emerge that could be relevant to the future health of the child or even to the current health of one of their parents or other family member. Which IFs should count as important findings that warrant disclosure to the patient or their family is being discussed very actively. While one would generally prefer not to generate information about a child's future risk of cancer or dementia in later adult life, especially when many adults prefer not to know such information, some such information may nevertheless be helpful if it allows the child's parents to remain in good health. Thus, if array comparative genomic hybridization (array CGH) reveals a small deletion within the BRCA1 gene in a 4-year-old boy with serious developmental problems, that is likely to be irrelevant to his developmental issues but may have been inherited from his mother, who may benefit from enhanced screening and perhaps chemo- or surgical prophylaxis. And that will be helpful to the child as well. However, of course, not everything is so straightforward: the devil will be in the detail of the circumstances of each case.

A helpful and influential document in this debate is a policy statement from the American College of Medical Genetics and Genomics (ACMG), which presents a list of 56 genes in which it recommends that pathogenic mutations should be actively sought in anyone having exome or genome sequencing because they are likely to lead to helpful health interventions. If such mutations are found, it recommends that they should generally be disclosed [9]. The ACMG now also recommends that this approach should be discussed when the test is being organized, so that disclosure should only happen after prior discussion and consent [10]. While the details of the list can be—are—contested, it has been really helpful to have the proposal: it has provided a focus for really useful discussions. Of course there is a potential for VUSs to be found in one of these clinically important genes, so that the two issues are not always distinct.

Genome sequence data storage

A major issue will be the question of who, if anyone, holds or can access a person's genome sequence once it has been generated? First, we will consider the situation in which health services generate the genome sequence and store the data. This presents serious challenges for data storage that has been broached in a report from the Human Genome Organisation (HUGO) [11], although this document has not yet triggered the broad discussion that is warranted. First, the information generated by a WGS today is likely to be regarded as outmoded and inadequate within 5 years, as a WGS will soon be expected to offer a read depth and coverage far superior to that currently achieved in diagnostic laboratories. Furthermore, the simple DNA sequence will by then need to be supplemented by the methylome and the chromatome, by which I mean that the pattern of CpG methylation and some information about the chromatin configuration (in the tissues sampled) will be expected as part of the analysis. Thus, the patient will need to have another sample taken, analyzed, and interpreted just as if it had never been done before.

Not only will the stored information most likely be seen as inadequate for clinical use if the analysis was performed a few years previously, but there will increasingly be costs associated with the data storage and data access. As the software and then (still more costly) the hardware systems need to change to ensure compatibility with future IT systems, the entire data set from all previous WGS analyses will need to be rehoused in new systems.

Along with the question of data storage comes that of acting on the stored information. Is there an obligation on the part of health services to reanalyze the stored data or even to perform an updated genome sequence and reanalyze the fresh sequence data? And then a further obligation to contact the patient again whose genome interpretation has changed in any significant fashion? [12,13] Such requirements will be difficult to cost into any system, public or private, as these obligations could become open-ended, perhaps with a need also to track down family members. How can we build a reasonable structure that can take on these responsibilities? One response is not to attempt to do so but to conduct the best analysis one can at the time, to report that and then to destroy the sequence information (and perhaps the sample too) to remove the "risk" of obligations extending into the future. If it is felt clinically that the analysis should be repeated then that can be done with a fresh sample and a fresh report can be issued.

Another, or additional, response to this challenge would be to pass the genome sequence to the patient. Indeed, this is an option being incorporated as an option into the 100,000 Genomes Project of National Health Service (NHS) England. This sharing of the raw data distributes the burden of storing the data and of responsibility for sequence analysis between the health services and the patient/family. While this could work out well it leaves open the possibility of additional unplanned, and very possibly overcautious or frankly unhelpful, demands being placed on the public healthcare system. The private interpreter of genome sequences may recognize a multiplicity of weak and/or unproven disease associations, which could trigger recommendations for additional disease surveillance or disease prevention measures of little or no established worth. Where helpful recommendations are made, they may distract health services from other activities of a higher priority; this would serve to increase health inequities through the distraction and opportunity costs of meeting the demands from analyses that the health services would not have performed.

On balance, my preference would be for health services to sequence, analyze, and issue the best report feasible at the time but then to destroy the raw sequence data. If and when updated analyses were sought by patient or clinician, then the WGS (plus add-on "omic" analyses) should be repeated and a fresh report issued. The sequence data could be passed in de-identified form and with appropriate consent, to a research database so that information about population frequencies of the patients' variants is not lost, but the diagnostic laboratory would not have taken on the impossible burden of the storage of data and the open-ended need to reinterpret the increasingly unsatisfactory original WGS data. However, which approach will be adopted in the longer term remains unclear.

Finally, we return to the question of how risk information for the complex diseases is to be aggregated and interpreted for an individual—perhaps simply from genetic data but *a fortiori* if environmental and phenotypic data are to be incorporated too. This question will be considered further in the section on future genetic research but can be introduced here. For most complex diseases, the contribution of genetic variation to the disorder in question is assessed through the estimation of its heritability. For many complex disorders and traits, under most circumstances, heritability is of the order of 50%, but only about 15—20% of the specific variation that accounts for this has been identified even through very large studies. Much of the unaccounted ("missing") heritability may then arise as a result of gene—gene (GxG) and gene—environment (GxE) interactions that are difficult to dissect out in human population studies, while some (e.g., in some psychiatric disorders) may be artifactual, as when new mutation events in monozygotic twins lead to an inflation of the heritability as estimated through twin studies [14]. Whereas GxG and GxE interactions can be assessed readily in plants and many animals, with specific breeding studies, they are much more difficult to assess in humans and their importance has been understudied and therefore often underrecognized.

REPRODUCTION

What effect will the new genomic technologies have on society, as they become embedded into routine midwifery and obstetrics practice? The technologies in question are (i) WGS as discussed earlier but applied to the identification of those at risk of having a child affected by a recessive disorder for which both parents, or perhaps just one parent in the case of sex-linked disorders, are carriers, and (ii) the application of next-generation sequencing to the free DNA in maternal plasma, which contains a fraction of DNA from the conception (the placenta); this is known as noninvasive prenatal testing (NIPT). Carrier screening for recessive disease can be applied at any time—before or during a pregnancy—to detect parents who carry an autosomal or a sex-linked recessive disorder [15—17]. NIPT is currently being used in Britain to determine fetal sex, Rhesus incompatibility and to identify fetuses affected by an autosomal trisomy, although the technology is intrinsically capable of identifying any genetic alterations to chromosome structure or number or to changes in DNA sequence: what it is used to test for is entirely open and there is no particular reason why it should be limited to the conventional targets of antenatal screening [18].

The consistent and whole-hearted application of these technologies could lead to major societal changes, which may be viewed very differently by different people. Virtually all serious congenital malformations, chromosome disorders, and autosomal recessive diseases, and a proportion of sex-linked diseases too, could be eliminated by carrier screening to identify carrier couples (and carrier mothers) in advance of a pregnancy and by NIPT and a thorough fetal anomaly ultrasound scan in the course of any pregnancy. This societal impact would depend upon the willingness of parents (especially mothers) to accept the offer of a termination of pregnancy in the event of a serious problem being identified. Such a development could be seen as very substantial progress, with the avoidance of suffering for patients and families and of costs for society. What is there not to like about this?

One reason for suspicion, or at least for questioning this model of progress, is the thought of how these technologies would have been used by the eugenicists of the German National Socialist party from 1933 to 1945. Of course this may be dismissed as unhelpfully emotive but it is worth more than a passing thought. Is it only the means that we object to in the Nazis' approach to Racial Hygiene and not their goals?

Further, what do we think of decisions made by some developing countries to introduce carrier screening programs with the clear goal of reducing the birth incidence of "expensive" disorders such as beta-thalassemia? In a poor country that has only recently passed through the demographic transition, so that children with thalassemia are surviving to be diagnosed, there is the agonizing question of what treatment to provide. Whereas wealthy countries can afford a high standard of health care for all those born, a poor country may be challenged if it attempts this. If they treated all affected children with blood transfusion and iron chelation, the costs would mount progressively, year on year, as the cohort of survivors continued to expand. The health budget could soon be completely swallowed in caring for this one group of patients. How can such a country afford to treat the patients without at the same time stemming the annual increase in their numbers by a program of carrier screening, carrier couple identification, prenatal diagnosis, and the selective termination of affected pregnancies? If introduced and actively promoted, how would such a program be merely an offer of screening without becoming coercive?

While we reflect on that question, it will be worth considering the situation in wealthy areas, such as Europe, North America, and Australia. With the development of new, rational therapeutics for more and more rare diseases, will our healthcare budgets be challenged in a very similar way by the great expense of such treatments? Consider some of the enzyme replacement treatments for metabolic storage disorders, which cost $>£100K$ p.a. per patient. If a large number of children were being treated for these and comparably rare disorders, the cost could become crippling even in a wealthy country. How will we balance the offer of carrier screening with the effort to diagnose affected infants and start them on costly therapeutics, at a time when an ideological commitment to budget austerity effectively prevents a public discussion of the (good) reasons for growth in government expenditure on health care and the (damaging effects of) health inequity that will result from continuing pressure on the national healthcare system, as those who can are driven to use private medicine?

This question applies to poor countries now but may well become a real issue in the west as well, in the course of the next few decades.

Returning now to the opportunities presented by the prevention of suffering through avoiding the birth of infants affected by severe genetic disorders, let us consider the difficulties that may arise as a program is introduced and becomes operational. For those who do not have a framework of religious beliefs that instruct them in what to think of this substantial societal change, there may be conflicting thoughts and feelings. While the avoidance of suffering is to be welcomed, the cost may be high for individuals caught up in screening and indeed for all of us. Those led to terminate a dearly wished-for pregnancy constitute only a small fraction of those presented with difficult and distressing decisions that may lead by small steps toward that final decision. The burden even of initial screening, when one wishes to approach a pregnancy positively and without reservation, may be weighty and oppressive. We commend parents who commit unconditionally to being a loving father or mother [19] but antenatal screening then constructs social settings in which such complete commitment to one's baby is undermined, as in the modern, genomic manifestations of Katz Rothman's "tentative pregnancy." Such contradictions and mixed messages sow the seed for a potentially bitter harvest (i.e., the consequences for parental commitment, parent—child relationships, and the willingness of society to support those affected and their families are unclear). Will care for those affected by these conditions improve in quality, because there will be fewer born and more resources will be available, or will it decline because of a bundle of connected problems, such as increased stigmatization and discrimination, reduced familiarity with these conditions within society, the reduced provision of services, and consequently the reduced expertise available within the health services?

If we could avoid so many genetic disorders, what will be the consequences for those who choose not to terminate affected pregnancies? Will children with Down syndrome and their parents be unsupported, as the frequency of the condition becomes less and less? [20] Will insurance systems refuse to cover the cost of medical, social, and educational support for such children who "could have been prevented"? Within the United Kingdom, will the NHS still provide care for such "optional," "lifestyle" conditions? Will similarly punitive policies come to be applied both to the decision not to accept antenatal genetic screening and to the decision to smoke?

At a time of public austerity, the confidence that parents might have in the willingness of society to support families in caring for a child affected by a serious malformation or genetic disorder may be undermined. That, in turn, may have consequences for the decisions parents make in the course of antenatal screening or during a pregnancy. It may also have consequences for the confidence parents have in facing the future with a child damaged by factors other than genetic conditions. If society becomes less accepting of disability because it is increasingly seen as "preventable," how confident will parents of a child damaged by preterm birth or birth asphyxia be in facing the future? Might this impact on decisions taken about (dis)continuing the care of a critically ill neonate?

Within the United Kingdom we have not so far had evidence of a decline in standards of care for infants with genetic disorders—indeed, the opposite may be true. We have not had much evidence of increased stigmatization or discrimination, although remarks to the parents of children with Down syndrome by passing strangers are sometimes crass and disrespectful, referring to the preventability of the condition. However, with the apparent adoption of a long-term strategy of austerity, so that state-funded health care will fall increasingly short in terms of quality, the public may become increasingly critical of those who bring "unnecessary" (and unnecessarily expensive) children into the world. This is difficult to predict but reassurance on the basis of past trends is unconvincing when the social ethos of inclusion/exclusion can change so rapidly.

Another reason to reflect upon societal pressures on genetics and reproduction emerges from some recent work in the United Kingdom on decisions being made about reproduction in the face of genetic disorders. The fact of stigma impacting on those affected by spinal muscular atrophy [21] and hypohidrotic ectodermal dysplasia (22, submitted) has been demonstrated as an important factor influencing the decisions being made within families, both by those affected and by their close relatives. Stigmatization is a reality that can be as heavy a burden for those affected as the more obvious physical effects, and this reality feeds into the decisions being made by those who know about life with such conditions.

The impact of the stigmatization to which an affected child may be subject is also likely to be an important factor in the decisions being made by those told they are at risk of having a child affected by a serious condition of which they have no prior awareness or experience. Indeed, the fantasies of such stigma may be very powerful but research into decision-making in the context of ultrasound or genetic information that has come to light in the course of a pregnancy has not attended very much to this aspect of the subsequent decision-making. While such factors will be difficult to access in the course of a family's responses to prenatal diagnosis, it should be a priority for such considerations to be explored, albeit gently and with the full recognition of the associated sensitivity.

FAMILY COMMUNICATION ABOUT GENETICS

An area that has been investigated within the United Kingdom and Australia in particular, but also elsewhere, is that of the communication of genetic information within the family. How will the difficulties of family communication be impacted by genomics? Two principal effects will be on the scale of the problems—arising much more often as much more sequence information is generated, both diagnostic information and the IFs that arise in the course of testing—and on the ease of interpretation of the information that is generated, as so many VUSs will be found. While much uncertainty will be generated, much of this will resolve over time; until that happens, families will be uncertain how to respond and whom in the family to tell. They may be guided actively by professionals where samples from other family members may aid in interpreting the VUSs, but approaching one's family members with a request to engage with genetics services for something carrying such uncertainty—that might or might not have serious implications for them—will not be easy.

An aspect of family communication around genetic questions that has not been explored much except in the anthropological literature is the pattern of marriage customs within different communities and whether marriages are likely to be arranged. From a detached and objective stance, it makes sense to be open about recessive disease in a society with customary consanguinity, whether or not marriages are often arranged, as marriage choice can take genetic risks and genetic tests into account if marriages are arranged, and preconception carrier screening can identify carrier couples and give them reproductive options. This may not always be how communication works out in such families but it could do so. In communities where consanguinity is uncommon or avoided, and especially if marriages are often arranged, then openness about genetic disease within a family may be difficult to achieve, especially in the context of an autosomal dominant or sex-linked disorder. As in the early nineteenth century England of Jane Austen's fiction, rumor of a family illness or weakness may ruin the marriage prospects of one or more generations of a family. The strength of feeling—the desperation—to keep a genetic condition as a secret may therefore be overwhelming, even though the secrecy may be to the long-term detriment of the entire family.

Another question concerning the family management of genetic information relates to information about children. The ACMG advice about the disclosure of actionable genetic information that emerges from genomic investigations of children as well as adults makes good sense as a family will often be unaware of their risk of a serious condition. The children will benefit through the benefits conferred on their parents and others from this knowledge when it emerges in the form of an IF (see earlier). However, when additional information about a child is released that is neither related to the reason for testing the child nor potentially actionable for the child or relatives, then unfortunate situations are likely to arise. Even more inappropriate, many would argue, would be newborn screening of a healthy infant by WGS with a full release of the findings [23]. This risks revealing very personal information about the child that she, in the future, might well have preferred others not to know. The targeted sequencing of loci of specific interest in a neonate would be an entirely different matter. The issue is one of maintaining an Open Future for the child as far as possible, although not, of course, if that would lead to avoidable medical harm.

THE FUTURE OF GENETICS RESEARCH IN MEDICINE

Where will genetic/genomic research be heading over the next few decades? The next phase of activity will be the further opening up of the "other omics," such as the methylome, the transcriptome, the proteome, and the metabolome. These extensions of the reach of genomics into the full spread of cellular processes in health and disease will comprise an enormous undertaking that reorients much of human biology and medicine. However, the applications to medicine will be slow to emerge and there will be other areas requiring active work, especially in the rare disorders. In the rare conditions, locus heterogeneity, pleiotropy, and novel mechanisms of disease will continue to surprise us and the opportunistic development of rational therapeutics will slowly but steadily bring real benefits to our patients, to the extent that society can afford to push forward the research.

In contrast to those areas, which may be seen as predictable and perhaps as "more of the same"—a consolidation of the ground opened by the Human Genome Project—the area that is going to be most exciting and which will utilize genomic approaches but require innovative thinking beyond the methods so far developed is the dissection of the GxG and GxE interactions discussed earlier. Without the specific breeding experiments feasible in plant genetics and the genetics of simpler animals, such as Drosophila, our ability to recognize the selective forces that maintain the enormous stock of genetic variation in human populations is greatly restricted. We will need to use unusual settings, such as arise in the intermixing of specific human populations, to generate data that can give us some of the equivalent insights. We need to search actively for circumstances in which antagonistic or disruptive selection would become apparent, with different alleles favored in different environments or in different genders, or by different genotypes at other loci. It is such mechanisms that will reveal the contrary selective processes often concealed because the net effect of selection at the locus in question may be near zero, and therefore invisible to most investigations, even when selection on specific combinations of alleles or of alleles and environment is very powerful. Such selective forces maintain our human genetic diversity, ensuring that there is no single optimal genotype at loci contributing to variation in many important and valued traits such as mental health and intelligence. These processes will slowly be revealed through these new approaches but we cannot even speculate about the range of applications likely to emerge from this work for the treatment of disease and the promotion of human self-understanding.

CONCLUSION

While human genetics has made rapid strides in the past three decades, the application of genomic approaches will further develop our understanding of human biology and disease and present opportunities for rational therapeutics, especially for some of the many rare disorders. The potential for the overemphasis on genetic factors in accounting for human disease is real, especially in the common, complex conditions, and needs to be resisted by those with a better grasp of disease mechanisms in their social and political context. The best approach to generating and interpreting genomic information for human health care is contentious but will be resolved over the next decade. It would be simple but probably wrong to accept at face value the model of storing a person's genome sequence as a lifetime resource for their health care as what is meant by a genome sequence is itself unstable and still under development, and the problems associated with storing and accessing WGS are likely to exceed the costs of repeat analyses. The opportunities for genomics to be applied in human reproduction are immense—holding out the prospect of almost eliminating many types of genetic disorder—but they raise complex social and ethical issues that are not going to be resolved easily. Another problematic area is the question of communicating "difficult" genetic information within the family and generating such information about children, which may not always be in their best interests. Finally, the question arises as to whither genetic research is likely to go in the next decade. In addition to the further development of rational therapeutics, another suggestion is that research into complex GxG and GxE interactions may become more tractable through the smart utilization of genomic approaches.

REFERENCES

[1] Rare Disease UK. London: Improving lives, optimising resources: a vision for the UK rare disease strategy; 2012.
[2] Marteau TM, French DP, Griffin SJ, Prevost AT, Sutton S, Watkinson C, et al. Effects of communicating DNA-based disease risk estimates on risk-reducing behaviours (review). The Cochrane Library 2010;10:1−74.
[3] McBride CM, Koehly LM, Sanderson SC, Kaphingst KA. The behavioural response to personalised genetic risk information: will genetic risk profiles motivate individuals and families to choose more healthful behaviours? Annu Rev Public Health 2010;31:89−103.
[4] Nuffield Council on Bioethics. London: Public health: ethical issues; 2007.
[5] Nuffield Council on Bioethics. London: Medical profiling and online medicine: the ethics of "personalised healthcare" in a consumer age; 2010.
[6] Rogers T, Milkman KL, Volpp KG. Commitment devices: using initiatives to change behavior. JAMA 2014;311(20):2065−6. Available from: http://dx.doi.org/10.1001/jama.2014.3485.
[7] Biesecker LG. Opportunities and challenges for the integration of massively parallel genomic sequencing into clinical practice: lessons from the ClinSeq project. Genet Med 2012;14:393−8.
[8] Clarke AJ. Managing the ethical challenges of next generation sequencing in genomic medicine. Br Med Bull 2014;111(1):17−30.
[9] Green RC, Berg JS, Grody WW, Kalia SS, Korf BR, Martin CL, et al. ACMG recommendations for reporting of incidental findings in clinical exome and genome sequencing. Genet Med 2013;15 (7):565−74.
[10] American College of Medical Genetics and Genomics (ACMGG). Incidental findings in clinical genomics: a clarification. Bethesda, MD: A Policy Statement of the American College of Medical Genetics and Genomics; 2013.

[11] Chadwick R, Capps B, Chalmers D, et al. Imagined futures: capturing the benefits of genome sequencing for society. Human Genome Organisation; 2013.

[12] Hunter A, et al. Ethical, legal and practical concerns about recontacting patients to inform them of new information: the case in medical genetics. Am J Med Genet 2001;103:265−76.

[13] Pyeritz RE. The coming explosion in genetic testing—is there a duty to recontact? N Engl J Med 2011;365(15):1367−9.

[14] Clarke A, Cooper DN. GWAS: heritability missing in action. Eur J Hum Genet 2010;18:859−61.

[15] Human Genetics Commission. Increasing options, informing choice: a report on preconception genetic testing and screening. London: Department of Health; 2011.

[16] Grody WW, Thompson BH, Gregg AR, Bean LH, Monaghan KG, Schneider A, et al. ACMG position statement on prenatal/preconception expanded carrier screening. Genet Med 2013;15 (6):482−3.

[17] Edwards JG, Feldman G, Goldberg J, Gregg AR, Norton ME, Rose NC, et al. Expanded carrier screening in reproductive medicine—points to consider. A joint statement of the American college of medical genetics and genomics, American college of obstetricians and gynecologists, national society of genetic counselors, perinatal quality foundation, and society for maternal−fetal medicine. Obstet Gynecol 2015;0:1−10. Available from: http://dx.doi.org/10.1097/AOG.0000000000000666.

[18] Wright C. Cell-free fetal nucleic acids for non-invasive prenatal diagnosis: report of the UK expert working group, PHG Foundation. Available from: <http://www.phgfoundation.org/download/ffdna/ffDNA_report.pdf>; 2009.

[19] McDougall R. Parental virtue: a new way of thinking about the morality of reproductive actions. Bioethics 2007;21:181−9.

[20] Skotko BG. With new prenatal testing, will babies with Down syndrome disappear? Arch Dis Child 2009;94:823−6.

[21] Boardman F. The expressivist objection to prenatal testing: the experiences of families living with genetic disease. Soc Sci Med 2014;107:18−25.

[22] Clarke A. Anticipated stigma and blameless guilt (submitted); 2015.

[23] Howard HC, Knoppers BM, Cornel MC, Clayton EW, Senecal K, Borry P. Whole-genome sequencing in newborn screening? A statement on the continued importance of targeted approaches in newborn screening programmes. Eur J Hum Genet 2015. Available from: http://dx.doi.org/10.1038/ejhg.2014.289.

CHAPTER 3

Genomics and Patient Empowerment

Marion McAllister

Institute of Cancer & Genetics, School of Medicine, Cardiff University, Cardiff, Wales, UK

Contents

INTRODUCTION

Genetic information has been provided to patients of healthcare services through the specialty of clinical genetics since the mid- to late twentieth century [1,2]. In the United Kingdom, clinical genetics services (CGS) comprise laboratory services as well as clinical services, with clinical services offering diagnosis of genetic conditions, genetic risk assessment, genetic counseling, and genetic testing, with a focus on conditions with a significant genetic component [3]. In the twenty-first century, these services have rapidly expanded and will continue to expand because technological advances, for example, "next generation" sequencing are enabling many more disease-causing mutations to be identified [4].

Genomic sequencing involving whole genome and whole exome sequencing and has the potential to reveal unanticipated (incidental) findings (IFs) that are not related to the clinical question for which the test was performed. When genomic testing is done to try and achieve a diagnosis in a child, this can identify IFs that may relate to adult onset conditions with risk implications for both the child and the child's parent(s) as well as genetic variants of unknown clinical significance (VUS). IFs can be for conditions that are treatable or untreatable. As the costs of these technologies come down, and their use in clinical practice becomes more widespread, there are challenges to be faced in ensuring that the information generated is provided to patients and their families in ways that empower, rather than disempower them.

Many of these technologies have been exploited commercially to provide genetic risk information (GRI) to the general public through "direct-to-consumer' (DTC) genetic testing. DTC tests tend to provide information about genetic susceptibility to common chronic diseases, rather than about conditions that "run in families" in a recognizably Mendelian fashion. This chapter will summarize research on the benefits or otherwise of high-risk genetic information provided by CGS and compare this with reports of people's responses to genetic susceptibility testing for common chronic diseases to provide some insight into the benefits of GRI more generally. This may enable some useful speculation about future developments in genomic testing.

WHAT ARE THE BENEFITS OF ATTENDING A CLINICAL GENETICS SERVICE (CGS)?

This section will summarize research evidence regarding known benefits of genetic information and genetic counseling to patients of CGS. Over the last 20 years, there has been a growing need for healthcare services to demonstrate benefits (positive changes) from the perspectives of the patients they serve. Patient-reported outcome measures (PROMs), short questionnaires that capture patients' (subjective) patient-reported outcomes (PROs) from using health care, are set to become a key part of how healthcare services are evaluated, managed, and funded across the world [5,6]. One challenge for CGS, particularly important in the current climate of cutbacks, is ensuring that these services can clearly evidence and articulate the benefits provided to patients.

Patients attending a CGS, seeking genetic information, differ from patients in mainstream health care in two important ways. First, many genetic conditions are still neither treatable nor curable. Although interventions to improve patient mortality and morbidity may be available for some patients, for example, cancer screening and risk-reducing surgery for some people at risk for hereditary cancer (HC) syndromes, and pharmaceutical interventions and implantable devices for those at risk for inherited cardiac conditions, these are still many genetic conditions that cannot yet be treated or cured. Second, a genetic diagnosis may have implications for other currently living or future unborn relatives, who may be at significant risk of developing the condition, or of transmitting it to subsequent generations. This makes diagnosis of an inherited genetic condition particularly devastating, and the responsibility to communicate GRI to at-risk relatives adds to the burden of a genetic diagnosis, given the lack of any treatment or cure in many cases.

The key interventions offered by CGS therefore center on provision of information and support, and referral to other services, for example, for cancer screening. Genetic information provided to patients can be generated by diagnosis of a genetic condition, by genetic testing, or by genetic risk assessment based on accurate family history information. Support can be provided to patients in the form of counseling to facilitate optimal decision-making about and adjustment to the genetic information provided.

NHS CGS are currently evaluated using process measures only, such as patient waiting times. These indicators have been used for the last 20 years as proxies for outcome measures because robust evidence of patient outcomes has not been available [7]. For a variety of reasons relating to the nature of the problems families bring to the genetics clinic, and the fact that many patients are themselves well, though they may be at risk of developing or transmitting a condition to their children, conventional measures of mortality and morbidity are thought to be neither relevant nor appropriate [8,9]. However, health policy pressure is mounting. Conventional health outcomes such as health status may be useful downstream long-term outcomes from some genetic information, particularly for inherited cancer and cardiac conditions. But are there short-term outcomes, directly attributable to the interventions offered by CGS that are valued by patients, even those to whom pharmaceutical, surgical, and screening interventions cannot be offered?

One approach to answer this question is to assess whether the goals of genetic counseling have been achieved, from the patients' perspective. Goals of genetic counseling include educating patients, and supporting them to make informed decisions and adapt to genetic (risk) information [10].

EDUCATIONAL IMPACT OF GENETIC COUNSELING

Evidence of accurate information recall after genetic counseling is contradictory, with some early studies reporting good information recall, but other studies showing that patients are not always able to recall specific recurrence risks after genetic counseling [11−13]. This may be partly explained by the use of heuristic thinking. It is well recognized that people use mental "shortcuts" called heuristics to help them solve problems and make decisions quickly, often referred to colloquially as rules of thumb, educated guesses, intuitive judgments, and common sense. Heuristics are prone to errors because they short-circuit the process of "thinking it through." Heuristics are *experience-based* methods that are influenced by factors such as past experience, recency, and familiarity [14,15]. A 2014 systematic review demonstrated that educational interventions that aim to improve subjective cancer risk perception are not effective in people with cancer or at high risk of cancer, with the strongest predictor of postintervention risk perception across the included studies identified to be baseline perceived risk [16].

Others, notably Gigerenzer, have argued that heuristics can be effective and adaptive, rather than being seen only as "errors" in cognition [17]. Gigerenzer emphasized ways in which heuristics can enable people to adapt and make decisions in a complex world without having to engage in complicated probability calculations, and constant reassessment in light of new information. Under natural conditions of limited information and cognitive resource, heuristics enable people to process information and make decisions quickly, and furthermore, people will often be happy to make a decision that is merely good enough: "satisficing" [18].

Furthermore, individuals may perceive risk as comprising not just the numerical probability of an event occurring, but also take account of factors such the severity of the event and their personal experience of it [19,20]. In the context of high-risk inherited conditions, GRI can be very personal information, because those receiving it in genetic counseling are very likely to have close personal experience of the condition in their family, having witnessed much loved relatives, indeed sometimes their own children, suffer and perhaps die from the condition. So they are acutely attuned to the severity and consequences of the risk, and perhaps less attuned to the numeric probability than their genetic counselor is.

This is further complicated if the Mendelian model of inheritance and associated disease risks provided in genetic counseling are not neatly reproduced in the patient's own family history, because, of course, their family is subject to the same sort of ascertainment bias as any small sample. Although the objective risk of inheriting a high-risk: cancer-predisposing mutation from a carrier parent may be 50% for males and females, a given family may only include, in living memory, affected women. There is some evidence that people from families with multiple affected members may develop their own predictions about who in their family is likely to develop the condition [20,21]. These predictions can be based upon the pattern of occurrence in their own family history (e.g., "in our family, only the women get it"), which they may perceive to contradict the predictions of a Mendelian model of inheritance, notwithstanding the problem of small samples. Indeed, these beliefs may form part of a process of coping and coming to terms with genetic risk [20]. Genetic counseling patients may therefore develop a composite "sense" of risk that is influenced by their highly salient personal experience of the condition in their family, as well as by the possibly less salient (to them) numeric risk figures provided by the genetic counselor. But only the numeric probability part of this is assessed in self-report measures that capture subjective risk perception. Indeed there is some evidence that the informational and educational elements of genetic counseling provide less benefit to patients than the supportive elements that address emotional issues, for example, grief and loss [22].

I have also argued elsewhere that measuring knowledge, or recall of GRI, is more a test of memory and understanding than an assessment of patient benefit [23]. So it would appear that recall of genetic information including GRI after genetic counseling is problematic and not straightforward as a way of measuring patient benefits from GRI.

PSYCHOLOGICAL IMPACT OF GENETIC COUNSELING AND TESTING

Does GRI promote adaptation and informed choice? Adaptation to chronic disease and disability more generally has been described as conquering challenges in the following spheres: stress, crisis, loss and grief, body image, self-concept, stigma, uncertainty and unpredictability, and quality of life[24]. In genetic conditions, measures of psychological distress (anxiety, depression) have been used to assess adaptation to genetic testing for Huntington disease (HD) and HC. The evidence has identified some consistent trends, in that studies have repeatedly found that people identified to be mutation carriers tend to experience an increase in psychological distress shortly after they obtain their genetic test results, but their distress tends to return to pretesting levels over time [25–29]. People who are identified to be noncarriers and those who obtain inconclusive test results tend to experience a decrease in psychological distress over time. A small proportion of people tested continue to experience distress after obtaining their genetic test result, particularly those who had a history of anxiety or depression before they took the test.

These studies grew out of efforts to determine whether predictive genetic testing causes significant psychological harm. The findings are very reassuring because they demonstrate that genetic testing does not cause significant long-term psychological harm to most people tested. It has been suggested that those who choose to go forward with predictive testing may be a self-selected group of people who believe themselves to be well equipped to cope with "bad news," and there is some evidence to support this [30]. However, these findings do not demonstrate a clear psychological benefit from genetic testing for those found to be mutation carriers. Although mutation carriers are unlikely to experience a long-term increase in distress resulting from their mutation status, this is not the same as achieving positive change.

Is there evidence that genetic counseling and/or testing promote informed choice? Although there have been calls for research in this area [31], the existing evidence is limited. In the context of prenatal screening for Down syndrome, there is some evidence that genetic counseling can reduce decisional conflict about specific testing decisions [32–36]. There is also some evidence that decision aids used alongside genetic counseling to support specific genetic testing decisions can increase decisional comfort [37]. However, it is unclear whether genetic counseling and genetic testing promote informed choice more generally.

It has been known for a long time that genetic counseling has little impact on reproductive outcomes [38]. Reproductive intentions prior to genetic counseling appear to be the strongest predictor of postcounseling reproductive outcomes, although interestingly, despite this, many patients report that they were influenced by genetic counseling. Women who undergo reproductive genetic counseling when carrying an at-risk pregnancy may have clinically elevated levels of anxiety before genetic counseling. However, anxiety tends to quickly return to normal levels with waiting period anxiety scores similar to scores in nonanxious samples, and posttest anxiety scores showing some evidence of a relief effect [39–41]. These studies also demonstrated that anxiety prior to genetic counseling is influenced by previous personal experience of prenatal testing, elevated perceived risk, and positive attitudes toward pregnancy termination. Women who would not consider a pregnancy termination may be considerably less anxious. Similar to findings with predictive genetic testing, women who are more anxious precounseling tend to be the ones who remain most anxious postcounseling, suggesting that women undergoing prenatal counseling and testing, particularly those who would consider terminating an affected pregnancy may benefit from interventions designed to address anxiety [41].

In summary, psychological studies investigating patient responses to genetic counseling and genetic testing, while reassuring health professionals that offering these interventions does not cause long-term psychological harm, do not demonstrate clear benefits to all patients. Psychological distress does not seem useful as a PRO that could be measured to capture positive change after genetic counseling or prenatal, predictive or presymptomatic genetic testing.

ARE THERE PATIENT BENEFITS TO BE DERIVED FROM USING A CGS?

Historically, CGS in the United Kingdom (UK) have been evaluated using process measures only, for example, patient waiting times and numbers of patients seen in the clinic. In an attempt to move toward evaluation of CGS using PROs, this question was tackled by a team based in Nowgen (A Centre for Genetics in Healthcare), Manchester. The "Valuation and Evaluation of Genetic Services team" conducted a program of research funded by the UK Department and later by the UK Medical Research Council to explore valuation and evaluation of CGS.

The first task tackled was to identify what outcome measures had been used to evaluate CGS and to establish the quality of those measures using systematic literature review methods [42]. Identifying outcome measures that had been used to evaluate CGS, and the key outcome domains captured by those measures, provided some insight into what evaluators considered to be important outcomes for patients.

PROMs used to evaluate CGS

The review identified 67 validated PROMs capturing in total 19 different PRO domains. Thirty-seven measures were generic, and the remaining 30 measures were developed specifically for use in clinical genetics. The range of PRO domains captured by the included measures was very broad, and included health status, psychological distress, coping, decision-making, family functioning, knowledge and risk perception, perceived personal control, quality of life, self-esteem, and spiritual well-being. No single measure identified in the review captured all of these possible PROs. Forty-six of the 67 measures identified had been used in just one study each. This review clearly identified the degree of discordance about what PROs are important to measure in clinical genetics.

The review assessed the extent of psychometric validation of the included measures and found this to be very limited. Published criteria for assessment of studies reporting psychometric properties of health measurement scales recommend assessment of internal consistency, reliability, measurement error, content validity, structural validity, hypothesis testing, cross-cultural validity, criterion validity, responsiveness (sensitivity to change over time), and interpretability [43,44] (see Table 3.1 for definitions of these). Although 63 of the 67 measures identified in the review were assessed for internal consistency, only 25 were assessed for test−retest reliability and only five for responsiveness.

Internal consistency assesses the extent to which questionnaire items relate to a particular dimension in a scale (e.g., cognitive control in the Perceived Personal Control questionnaire) [45,46] and is easy to assess. All that is required is a sample of the target population to complete the questionnaire once [47]. Statistically, the more items capturing that dimension, the higher the internal consistency will be. A measure with lots of different versions of the same question will have a high internal consistency score, whereas a parsimonious measure that is less burdensome for respondents is likely to have lower internal consistency.

Test−retest reliability is a more practical property than internal consistency, because it determines whether the questionnaire gives (on average) the same score if the underlying construct, for example, anxiety, has not changed. But it is harder to assess than internal consistency, because a sample of the target population must complete the questionnaire twice, separated by a time interval long enough that they don't remember how they responded the first time, but not so long that the construct being measured will have changed, usually about 2 weeks [47].

Responsiveness refers whether a questionnaire can detect change over time in the construct measured. It is also harder to assess than internal consistency, because, again, a sample of the target population must complete the questionnaire at two time points, during which interval the construct measured *would* be expected to change [47].

Table 3.1 Definitions of measurement properties for health measurement questionnaires

Internal consistency	The degree of interrelatedness among individual questions in the questionnaire.
Reliability	The proportion of the total variance in measurements due to "true" (error-free) differences between respondents; represents the degree to which the questionnaire gives the same score when the "quantity" of the construct to be measured is the same.
Measurement error	The systematic and random error of respondents' scores that are not attributed to true changes in the construct measured.
Content validity	The degree to which the content of the questionnaire is an adequate reflection of the construct to be measured.
Structural validity	The degree to which questionnaire scores are an adequate reflection of the dimensionality of the construct to be measured.
Hypothesis testing	The degree to which questionnaire scores are consistent with hypotheses (e.g., *regarding internal relationships, relationships to scores on other questionnaires, or differences between relevant groups*) based on the assumption that the questionnaire validly measures the relevant construct.
Cross-cultural validity	The degree to which performance of questions on a translated or culturally adapted questionnaire is an adequate reflection of the performance of the equivalent questions in the original language version of the questionnaire.
Criterion validity	The degree to which scores on the questionnaire are an adequate reflection of a "gold standard" (if there is one available).
Responsiveness	The ability of a questionnaire to detect change over time in the construct to be measured.
Interpretability	The degree to which one can assign qualitative meaning (i.e., clinical or commonly understood connotations) to quantitative scores or change scores on the questionnaire.

Source: Adapted from the COSMIN manual available at http://www.cosmin.nl/images/upload/files/COSMIN%20checklist%20manual%20v9.pdf [accessed 1506.14].

This is a key property of any instrument to be used as an outcome measure. The instrument will not be fit for purpose if it is not responsive to changes brought about by clinical interventions. But only 5 of 67 measures identified by Payne et al. were assessed for responsiveness.

Both test–retest reliability and responsiveness are harder to assess than internal consistency, because they require two separate administrations of the questionnaire, which is the most likely explanation for so many measures having been assessed for internal consistency but not for test–retest reliability or responsiveness.

A further concern identified in the review was that only 2 of 67 questionnaires identified had been assessed for interpretability. Although interpretability is not strictly a measurement property, it is a key attribute of a questionnaire intended to be used as a PROM. Interpretability refers to whether a specific change in questionnaire scores represents a trivial or important improvement or deterioration from the perspective of respondents. The usual method for assessing interpretability of a questionnaire is by establishing the minimal important change or minimal important difference (MID) in samples drawn from the target population [43,44].

It was clear from the findings of this systematic review that there is considerable lack of consensus about what PROs are important to measure from using CGS and furthermore, the psychometric quality of PROMs available in 2008 was very limited. This places limits on (i) what can be achieved in measuring PROs from genetic information because the validity of empirical findings depends on the quality of measures used and (ii) comparability between studies because so many different PROs are used.

Moving toward consensus on PROs

In an attempt to move toward some consensus over outcome domains (PROs) for CGS, the findings from the systematic review were used to design a two-round Delphi survey to explore genetics health professionals' and patients' views about which PROs are most useful [48]. Delphi surveys are useful tools to address issues that are not supported by a strong evidence base [49] and usually involve at least two rounds that collect views from a panel of experts. After each round an anonymous summary of the experts' views or judgments from the previous round is circulated to panel members who are encouraged to revise their previous answers in light of the responses of the whole panel. The theory is that the group will converge toward the "correct" (or consensus) answer. In this case, the survey contained 19 outcome domains and in the first round, respondents assessed the usefulness of each for CGS. The second-round survey comprised the same questions but was supplemented by bar charts summarizing the results from the first round. This approach enabled respondents to reformulate their views in light of feedback from the rest of the respondent panel.

At least 75% of the panel agreed that the following eight PROs were useful: knowledge of the condition, decision-making, perceived personal control, risk perception, satisfaction, meeting of expectations, coping, and quality of life. A ninth domain, accuracy of diagnosis, was also endorsed by 75% of the panel, but this is an objective rather than a subjective outcome, so accuracy of diagnosis would not be classed as a PRO. However, as argued earlier ("Educational impact of genetic counseling"), knowledge and risk perception are problematic as PROs because they may not represent outcomes that are either straightforward to measure or good representations of patient benefit. Satisfaction and meeting of expectations are also not without difficulties as PROs.

Satisfaction and expectations

Patient outcomes from health care are often thought to be influenced by whether patient expectations are met. There is some limited evidence of this in some areas of health care, for example, patients who expect to recover well from pain may achieve better pain relief outcomes [50]. However, the relationship between patient expectations and outcomes is not clear, and this is thought to be because conceptualizing "patient expectations" is difficult, leading to difficulties in measuring the construct [51–53].

In CGS there is some evidence that patient expectations are often met [54,55], but there is also evidence that clinical genetics patients are often unclear what to expect from the service prior to their first appointment [56,57]. A 1997 UK study showed that clinical genetics patients who had their expectations for reassurance and advice met were less anxious afterward even if their expectations for information, explanation, and decision-making support were not met [54]. However, the outcomes assessed were limited to psychological distress and satisfaction. Outcomes relating to decision-making, for example, decision quality or decisional control were not assessed. Furthermore, patient satisfaction with information provided could plausibly have been influenced by whether the patient had received good news or bad news.

The authors concluded that if patient expectations are judged by the genetic counselor to be unrealistic or inappropriate, that better patient outcomes may be achieved by negotiating expectations rather than simply not meeting them. These negotiations do take place in genetics clinics, for example, regarding availability or appropriateness of genetic testing in some circumstances, but do not necessarily lead to patient satisfaction in every case, for example, if genetic testing is perceived by the patient to be withheld for reasons of NHS rationing.

Although in the context of CGS, patient satisfaction has been shown to correlate with some PROs, for example, perceived personal control [45], patient satisfaction is, generally speaking, more closely linked to characteristics of health professionals and organizations that provide patient-centered care [58]. Satisfaction is higher among patients who perceive that they have been treated with care, dignity, and respect. Communication between patients and healthcare providers may, indeed, be the most important determinant of patient satisfaction [59]. Furthermore, patient satisfaction may, at least partly, reflect the type of information provided, for example, in clinical genetics, satisfaction may be partly determined by whether the patients is given a low or a high recurrence risk. But even that relationship may not be straightforward. For example, patients with a family history of breast cancer may come to clinical genetics seeking access to breast cancer screening at a younger than average age, but following risk assessment, may not qualify for early breast screening. They may still *perceive* that they are at high risk, even though their objective risk may not be substantially increased over the general population risk, and they may be very dissatisfied that they cannot gain access to early screening.

There is some evidence in other areas of health care that patient satisfaction is influenced by whether or not patient expectations have been met [51,60–62], but this has not been clearly demonstrated in clinical genetics. This may be due, at least in part, to aforementioned difficulties measuring patient expectations, with many studies collecting expectations data using questionnaires that have not been properly validated [54,63].

In summary, bearing in mind the difficulties outlined above with using knowledge, risk perception, satisfaction, and meeting expectations as PROs from clinical genetics, the findings to date would suggest that decision-making, perceived personal control, coping, and quality of life may be useful PROs.

What do key stakeholders think?

In parallel with the Nowgen systematic review and Delphi survey, qualitative research using focus groups and interviews was conducted to explore what key stakeholders value as PROs from CGS. Seven focus groups and 19 interviews were conducted with genetics health professionals, patients, and patient representatives [23,64]. Grounded theory methods were used with the aim to develop a theoretical framework describing patient benefits from using CGS. The resulting theoretical framework constructed from the data summarized patient benefits from using CGS using the term "patient empowerment" and comprised five dimensions:

1. Decisional control:
   a. Knowing what options are available for managing the genetic condition and/or the genetic risk.
   b. Feeling able to make informed decisions between available options.
2. Cognitive control (sense-making):
   a. Knowing that one has a clear scientific explanation (or the best one available) for what has happened in the family.
   b. Understanding the genetic risks to oneself and/or other relatives and future generations.
   c. Knowing what help and support is available.
3. Behavioral control:
   a. Feeling able to use health and social care systems effectively to reduce harm and/or improve life for oneself, one's children, at-risk relatives, and future descendants.
4. Hope:
   a. Having hope for a fulfilling family life, for oneself, relatives, and future descendants.

5. Emotional regulation:

    a. Feeling able to cope with the genetic condition in the family.

    b. Feeling able to manage feelings of distress, guilt, and anxiety about the genetic condition in the family, or about (potentially) having transmitted the condition to one's children.

There is considerable theoretical overlap between empowerment, as described in this model, and the PROs that seem most appropriate following the Delphi survey (decision-making, perceived personal control, coping, and quality of life). Three of the empowerment dimensions (cognitive, decisional, and behavioral control) together comprise the construct perceived personal control, which was previously applied to the genetic counseling context by Shiloh and colleagues in Israel [45,65]. The importance of perceptions of control is also recognized in chronic disease, where there is evidence that perceived control is adaptive and is associated with less mood disturbance [66,67]. Decision-making overlaps with decisional control, part of perceived personal control. Coping overlaps with emotional regulation, and, arguably, also with behavioral control. Coping has been described as comprising emotion-focused (emotional regulation) and problem-focused (behavioral control) activities [68]. Emotion-focused activities are those undertaken to reduce and manage the intensity of emotions associated with a problem—activities such as distraction, letting off steam, ignoring the problem, humor, physical exercise, expressing emotions creatively, for example, painting. Problem-focused activities are practical actions taken to solve the problem and can include taking control of the situation and acting to remove the stressor, evaluating options, and information seeking. Quality of life is a very broad construct that has been defined in many different ways. In health care, the term health-related quality of life (HRQoL) is often used and may include one or more of the following: general health (physical and/or mental), functioning (physical, emotional, cognitive, sexual), social well-being, coping, adjustment, and existential issues [69]. As defined above, patient empowerment could be considered an aspect of HRQoL.

The qualitative research also identified a set of interventions and a set of service attributes relating to clinical genetics that are valued by stakeholders. Five service attributes were highly valued by participating patients and genetics health professionals (Figure 3.1). Firstly, services that are local and accessible. Because genetic conditions are often rare condition that may affect multiple family members, and because CGS are specialized services, the financial and time burden upon families to attend appointments can be considerable. Patients value efforts made by services to ease this burden by offering outreach clinics.

Secondly, services that provide regular follow-up to patients with a single named point of contact. Patients value being recontacted after a genetic test result or a new diagnosis, when sufficient time has elapsed for the new information to "sink in." Patients also value a yearly contact letter reminding them that the service is available

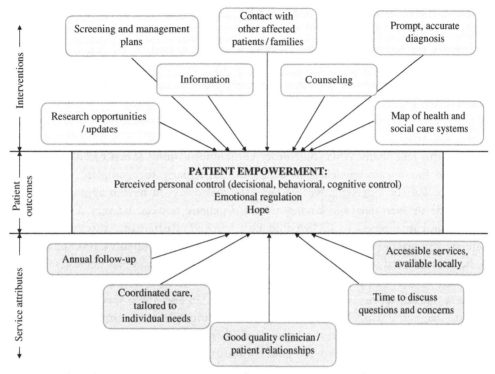

Figure 3.1 Clinical genetics services: service attributes, interventions, and outcomes.

to their family, should any family members wish to use the service, and patients with experience of a genetic register service find this approach very useful [70]. Different issues arise for family members at different life stages, and a long-term relationship between the family as a whole and the CGS enables family members to access the service as they need it over the life course. Most families do not need regular appointments, but may wish to access the service at times when genetic issues become salient for them, such as when their children reach reproductive age and might benefit from genetic counseling about their reproductive risks.

Thirdly, services that offer coordinated care, tailored to the specific needs of the family, that build long-term relationships with families through a named person to support family members to access the service when they need to. CGS that can orient their service toward whole families rather than being focused on single patients are highly valued.

Patients also value a good quality relationship with clinicians who have the social skills and flexibility to relate to people from many different backgrounds, who treat them with empathy and compassion, and who are perceived to be expert in the relevant genetic condition. Finally, having time to talk to the clinician is also valued—patients want enough time to ask questions and talk about their concerns without feeling rushed.

Seven interventions offered by CGS were highly valued by participating patients and genetics health professionals in the qualitative study. First and foremost, are the traditional medical interventions that can affect patient care. Participants in this study valued firstly, prompt and accurate diagnosis. In the case of genetic conditions, which are often very rare, diagnosis can be a considerable medical challenge, often requiring a multidisciplinary approach, sometimes requiring genetic testing, and in some instances can take many years to achieve. Difficulties gaining access to CGS appeared to be one factor causing delays in diagnosis, particularly for complex, multisystem conditions. Patient representatives commented that delays in having appropriate diagnoses made or even delays in having a wanted genetic test can be very distressing for families, and that access to CGS could be improved. Participants also reported that there are important psychological and social benefits to be gained by prompt diagnosis, because affected families are not left "in limbo," and even a bad news genetic test result can be reassuring for people because the psychological uncertainty has been removed. Health service providers who refer to CGS also value speed of service and accurate diagnosis, which may involve genetic testing and/or extensive family studies which are not possible within the context of other healthcare services.

Participants also valued advice and coordination of screening and management strategies, which for some genetic conditions, for example, inherited cancer syndromes and inherited cardiac conditions, can provide significant health improvements. These interventions are valued not only for the patients themselves, but also enable them to communicate the recommendations to other at-risk family members, thus empowering the wider family.

Without exception, all participants identified information as key to good patient outcomes. That could be information about a diagnosis; about how likely the condition in the family is to be genetic; about risks to the patient or their relatives; about treatment and management options; about options for genetic testing; or indeed, information provided by a genetic test. It also matters how information is given. Patients value information that is tailored according to their ability to understand, as well as information that is tailored to their specific circumstances.

Supportive counseling was also identified as contributing to patient empowerment. Patients value having someone to talk to who understands their situation, someone who is an expert and has specialist information, and is there to offer support, and who can help patients to understand complex information about the family condition. Simply talking to someone like this can reduce anxiety and feelings of guilt, and can help to relieve the tensions experienced by parents who may feel guilty for having transmitted a genetic condition to their child. Parents understand that the focus must be on the child who has problems, and yet the parent may have needs of their own, which, if not addressed, could potentially compromise their ability to parent their child effectively. The nondirective approach used in clinical genetics in relation to reproductive decision-making was praised as fostering effective decision-making and compared favorably to more "eugenic" approaches encountered in other healthcare specialties.

Feeling isolated with a condition in the family that is so rare, that there is often no lay knowledge about it, is very isolating. Participants valued being put in touch with other families affected with the same condition, or with a patient support organization. Sharing the experiences and challenges they face in their lives, with other people who have faced similar challenges is enormously helpful and empowering because it helps people to believe that they can cope, and that there is a future for them. Patients see CGS as ideally placed to take a key role in bringing people together.

Families who have been devastated by a diagnosis of a genetic condition, particularly in contexts where other family members are at risk for developing or transmitting the condition to the next generation, desperately need an anchor on which to hang some hope for a fulfilling or rewarding future of some kind, even if this is for their descendants and not for themselves. CGS can offer hope in a number of ways. Firstly, by providing families with information about, or providing opportunities to participate in clinical research. For individual patients, the opportunity to contribute to research that aims to better understand the condition, or that might lead to future treatment provides hope for their children and future descendants. This can be a considerable comfort for patients, even those who may be very ill or disabled and whose own future may be bleak. Associated with this, was the idea that clinical geneticists can act as powerful advocates for families, particularly those with academic and research reputations, because patients see them as well positioned to influence the progress of developments, winning research funding, and then feeding research findings back to families.

Hope can also be fostered through counseling interventions that support patients to adjust as positively as possible to the genetic condition in the family. Counseling interventions can help the patient and their family to put the condition in some sort of context, and to find an appropriate place for it in their lives that enables them to move on and engage with living again. Hope can also be fostered by providing information about variability in the effects of genetic conditions as well as variability in age at onset.

But medical information and interventions are not enough to empower patients and their families. They need guidance through the health and social care systems so that they can obtain practical help, social and respite care, and financial support. Because CGS are seen as the experts in these rare conditions, patients do see CGS as having some responsibility in providing guidance to families in these areas, even if this is limited to providing signposts toward third sector organizations who can help them. Genetics clinicians acknowledged that they can and do act as advocates by writing letters on behalf of patients in support of educational statementing or applications for rehousing or social security. Patient participants were, however, looking for more than this. They want information about how to navigate the health and social care system, about which they may know nothing, and about what help might be available to them to support life with the family condition, and in providing for the affected family member(s).

In summary, "triangulating" the three Nowgen studies provides support for the construct patient empowerment as a useful PRO from CGS. It is clear that empowering patients relies on more than simply providing them with genetic information. It is doubtful whether the information generated by genetic risk assessment and genetic testing would be empowering for patients without the counseling and other interventions and service attributes that help patients and families to make sense of genetic information and integrate it into the rich tapestry of family life.

MEASURING PATIENT EMPOWERMENT IN CGS

A 2008 systematic review of PROMs used to evaluate CGS demonstrated that the psychometric quality of measures available at that time was poor [42]. A model summarizing the patient benefits from using CGS specified patient empowerment as an overarching construct capturing five dimensions: cognitive, decisional and behavioral control (which together comprises the construct perceived personal control), hope, and emotional regulation [23,64]. Furthermore, the systematic review concluded that no single measure identified in the review captured all dimensions of empowerment, so it seemed the logical next step to develop a good quality measure to capture patient empowerment.

Funding was secured from the UK Medical Research Council to develop a robust new PROM for CGS, designed to capture patient empowerment as specified in the "theoretical framework" that emerged from the Nowgen qualitative research. The new measure is called the Genetic Counseling Outcome Scale (GCOS-24) and comprises 24 questions with seven Likert-style response categories (Figure 3.2).

Through collaboration with Genetic Alliance UK, a national UK charity comprising over 180 patient support groups for genetic conditions (see http://www.geneticalliance.org.uk/), GCOS-24 was tested using psychometric methods in large samples of people from families affected by genetic conditions who completed draft versions of the questionnaire [71]. These tests demonstrated that GCOS-24 has good internal consistency (Cronbach's $\alpha = 0.87$) and good test–retest reliability (intra-class correlation = 0.86). Hypothesis testing confirmed that people's GCOS-24 responses correlate with their responses on other questionnaires (measuring things like anxiety, depression, perceived personal control, and health locus of control) in the theoretically expected direction, demonstrating construct validity.

The Genetic Counselling Outcome Scale (GCOS-24)

Using the scale below, circle a number next to each statement to indicate how much you agree with the statement.
Please answer all the questions. For questions that are not applicable to you,
please choose option 4 (neither agree nor disagree).

	1 = strongly disagree 5 = slightly agree 2 = disagree 6 = agree 3 = slightly disagree 7 = strongly agree 4 = neither disagree nor agree	strongly disagree	disagree	slightly disagree	neither agree nor disagree	slightly agree	agree	strongly agree
1	I am clear in my own mind why I am attending the clinical genetics service.	1	2	3	4	5	6	7
2	I can explain what the condition means to people in my family who may need to know.	1	2	3	4	5	6	7
3	I understand the impact of the condition on my child(ren)/any child I may have.	1	2	3	4	5	6	7
4	When I think about the condition in my family, I get upset.	1	2	3	4	5	6	7
5	I don't know where to go to get the medical help I / my family need(s).	1	2	3	4	5	6	7
6	I can see that good things have come from having this condition in my family.	1	2	3	4	5	6	7
7	I can control how this condition affects my family.	1	2	3	4	5	6	7
8	I feel positive about the future.	1	2	3	4	5	6	7
9	I am able to cope with having this condition in my family.	1	2	3	4	5	6	7
10	I don't know what could be gained from each of the options available to me.	1	2	3	4	5	6	7
11	Having this condition in my family makes me feel anxious.	1	2	3	4	5	6	7
12	I don't know if this condition could affect my other relatives (brothers, sisters, aunts, uncles, cousins).	1	2	3	4	5	6	7
13	In relation to the condition in my family, nothing I decide will change the future for my children / any children I might have.	1	2	3	4	5	6	7
14	I understand the reasons why my doctor referred me to the clinical genetics service.	1	2	3	4	5	6	7
15	I know how to get the non-medical help I / my family needs (e.g. educational, financial, social support).	1	2	3	4	5	6	7
16	I can explain what the condition means to people outside my family who may need to know (e.g. teachers, social workers).	1	2	3	4	5	6	7
17	I don't know what I can do to change how this condition affects me / my children.	1	2	3	4	5	6	7
18	I don't know who else in my family might be at risk for this condition.	1	2	3	4	5	6	7
19	I am hopeful that my children can look forward to a rewarding family life.	1	2	3	4	5	6	7
20	I am able to make plans for the future.	1	2	3	4	5	6	7
21	I feel guilty because I (might have) passed this condition on to my children.	1	2	3	4	5	6	7
22	I am powerless to do anything about this condition in my family.	1	2	3	4	5	6	7
23	I understand what concerns brought me to the clinical genetics service.	1	2	3	4	5	6	7
24	I can make decisions about the condition that may change my child(ren)'s future / the future of any child(ren) I may have.	1	2	3	4	5	6	7

Figure 3.2 Genetic Counseling Outcome Scale (GCOS-24) [71].

GCOS-24 was also tested for sensitivity to change (responsiveness) in a sample of 241 patients who completed the PROM before and after attending an appointment at a CGS. This test showed that GCOS-24 can detect statistically significant improvement in GCOS-24 scores following clinic attendance with a medium-to-large effect size (Cohen's $d = 0.7$), demonstrating that empowerment levels are significantly higher after attendance at a genetics clinic, and that GCOS-24 can detect this change.

GCOS-24 is therefore a well-performing instrument that can be used to measure how much benefit patients derive from using CGS. Although empowerment is not a health outcome in the traditional sense of disease-associated mortality or morbidity, GCOS-24 does measure PROs that patients value from using CGS. GCOS-24 has potential as a useful PROM for CGS that could be used in research to evaluate new interventions in randomized controlled trials, and in clinical practice to assess whether CGS are providing benefits to the patients and families they serve, generating useful data for decision-makers.

I recently completed a structured literature review with a student [72] with the aim to summarize the published research on patient benefits from using CGS, to try and identify potential PROs. The review identified 39 studies reporting patient benefits. A qualitative approach was used to reinterpret the patient benefits identified by the included studies within the patient empowerment framework. Empowerment, and/or one or more of its dimensions, could be identified in 38 of the 39 included studies, although only 10 of the included studies explicitly identified empowerment or one of its dimensions. In the 19 studies where knowledge was identified as a patient outcome, this was reinterpreted as cognitive control. In the nine studies reporting reduction in psychological distress as an outcome, this was reinterpreted as emotional regulation. Eleven studies reported either intention or motivation to change health behavior, but no evidence of actual behavior change, except use of interventions to reduce breast cancer risk. This was reinterpreted as behavioral control, because of the lack of consistent evidence of actual behavior change. Three studies identified coping as a patient outcome and this was reinterpreted as behavioral control and emotional regulation, because coping activities can be either emotion-focused or problem-focused [68]. Benefits that could not be reinterpreted within the patient empowerment framework included family functioning, social functioning/support, risk perception, altruism, sense of purpose, and enabling development of future treatments, and research/research participation.

This review demonstrated that many patient benefits (PROs) previously identified in clinical genetics research can be reinterpreted within a patient empowerment framework, lending further support for patient empowerment, as measured in GCOS-24, as a useful approach to conceptualizing and measuring PROs from CGS.

Table 3.2 Service evaluation using GCOS-24 in five UK NHS clinical genetics centers

		Center 1	Center 2	Center 3	Center 4	Center 5
Before clinic	N =	42	44	54	54	74
	Mean score	105.16	109.71	110.84	116.8	104.53
	SD	17.2	18.39	17.47	8.93	17.55
After clinic	*Mean score*	118.98	115.07	118.83	121.78	113.59
	SD	25.75	19.51	15.46	8.99	22.05
Change	*Mean change score*	13.82	5.36	7.98	4.97	9.06
	SD	25.76	18.6	13.21	12.76	14.85
	p =	0.001	0.001	0.001	0.001	0.001

DOES ATTENDANCE AT CGS EMPOWER PATIENTS?

To answer this question, the GCOS-24 was used in service evaluation exercises in six UK NHS CGS in 2011—13. Patients were asked to complete GCOS-24 before and after clinic attendance, and group level before—after changes in GCOS-24 scores were calculated. Five centers demonstrated statistically significant improvement in GCOS-24 scores following clinic attendance even with relatively small samples sizes of 45, 44, 54, 54, and 74 ($p < 0.0001$). However, one center returned only nine matched preclinic and postclinic questionnaires, a sample that was too small to enable a useful analysis (Table 3.2).

Once center (Cardiff) also collected postclinic satisfaction data [73] and postclinic outcome data using an audit tool [74] designed to capture patient outcomes. The audit tool, because of item wording, can only be used after clinic attendance and not before, so it cannot be used to measure before—after *change*. Correlation analysis demonstrated that increases in patient empowerment scores (measured using GCOS-24) following clinic attendance correlate significantly with both patient satisfaction and patient outcomes, as captured in the audit tool.

These exercises were important because they demonstrated for the first time that NHS CGS can deliver measurable patient benefits. Furthermore, these exercises suggested that GCOS-24 has potential to evaluate routine NHS CGS, offering useful PRO data to supplement current approaches to service evaluation that could be useful to decision-makers, for example, commissioners. However, participating centers were not able to accurately report the patient response rates, so there is uncertainty about how representative these findings were. It is possible that only patients who experienced benefits completed and returned questionnaires.

PATIENT EMPOWERMENT: THE BROADER CONTEXT

The term "patient empowerment" has resonance for health care in a much wider context. The first part of the twenty-first century has seen patient empowerment, choice, and control taking a prominent place in healthcare policy, in the United Kingdom and elsewhere [75,76]. Alongside this, there are moves to link healthcare funding to performance against a range of quality measures including PROMs, reflecting the importance placed upon patients' subjective assessment of the health care they have received. The term "patient empowerment" is rooted in social reform and the civil rights movement, and is associated with a shift from paternalist models of health care toward more patient-centered models, and efforts to involve patients actively in decision-making about their own health and health care. These changes in health care reflect broader societal changes in which citizens, certainly in the West, increasingly expect to be treated as consumers who make their own informed choices between a range of healthcare products and services based upon their individual values and preferences.

The principle of patient empowerment is that patients are viewed as self-determining agents who have the right to some control over their own health and health care, rather than as passive recipients of health care who comply with decisions made in their best interests by healthcare providers. Although healthcare providers may hold medical authority, patients are seen as authoritative regarding their own or family experiences of the condition, and their own life and health goals, values, and preferences for treatment. The term "patient empowerment" has been used to describe both a state of being empowered and a process of becoming empowered [77,78]. The process of patient empowerment can be considered to have interactional as well as personal aspects. The interactional aspect involves some sharing of both information and power between healthcare provider and patient through processes such as collaborative deliberation and shared decision-making [79,80]. Patient empowerment has also been described as a process of personal transformation for the patient that involves the patient taking (or being provided with) some control and maintaining or regaining a sense of mastery over their condition and treatment [78,81]. The state of being empowered has been described to be indicated by constructs such as self-efficacy, decision-making capacity, and personal control in the health(care) domain, and these constructs have been included in questionnaires designed to measure patient empowerment [78,82]. Although patient empowerment is associated with patient-centered care, it is not limited to the healthcare context because patients increasingly empower themselves through use of the Internet to collect and share health information, and through involvement in patient support and advocacy activities [78]. However, interventions and approaches used in health care can foster patient empowerment, such as chronic disease self-management training, health literacy training, and shared decision-making [78,83−85].

PATIENT EMPOWERMENT AND GENETIC CONDITIONS

Much of the health policy rhetoric about patient empowerment focuses on patients affected by long-term conditions and assumes that empowered patients will manage their long-term condition by making rational decisions to maximize their long-term health and optimize their use of healthcare services. However, when it comes to genetic conditions, particularly single gene disorders, there may not be many options for treatment that will maximize long-term health. Although interventions to improve patient mortality and morbidity may be available for some people, for example, cancer screening and risk-reducing surgery for those at risk for HC syndromes, and pharmaceutical interventions and implantable devices for those at risk for inherited cardiac conditions, these are still many genetic conditions that can, as yet, be neither treated nor cured.

Yet despite this, patients of CGS value the empowerment benefits they derive from using a CGS, and this may be applicable to other long-term conditions. Measuring patient empowerment might be useful to capture nonhealth gains among patients affected by long-term conditions such as diabetes, asthma, and epilepsy [78,82]. Patients with long-term and even degenerative conditions who use healthcare services may not be able to derive significant measurable health gains. Measuring levels of empowerment in these patient groups could potentially capture gains that could offset losses in health.

GENOMIC INFORMATION, HEALTH LITERACY, HEALTH BEHAVIOR, AND PATIENT EMPOWERMENT

One group of stakeholders who have used the term "patient empowerment" are companies who market DTC genetic tests [86]. Most of the information generated by "DTC" tests is genomic susceptibility information that classifies individuals into population subgroups with differing risks of developing common chronic diseases (e.g., different types of common cancers). These risks are often not amenable to medical intervention, but are amenable to individual lifestyle and behavior modification. This "selling" of genetic testing as empowering has been criticized because of concerns that people who do not make the "right" health choices could be scapegoated, or even excluded from health care if they are judged to be irresponsible [86]. Furthermore, individuals' past lifestyle and environmental factors also contribute to their risk of developing some common chronic diseases, and these factors are not taken into account in generating their risk estimate. The individual's actual risk may be very different from that of the risk subgroup to which s/he has been assigned based on genetic information only [87].

So, interpreting results from DTC testing requires unusually high levels of health literacy. Health literacy includes critical as well as functional aspects. Critical health literacy includes the ability to critically evaluate health information and use it to gain control over life and health [88]. In the context of genomic susceptibility testing, to be empowered, the individual must understand the limitations of the information provided, and the risks and benefits of available options for risk management in order to make effective use of that information. Indeed, disempowerment may result from information overload or from information that people cannot understand or interpret.

What is the evidence that people will use genetic information to change their life-style and behavior to maximize their long-term health? There is reasonably good evidence that individuals who carry high-risk mutations for inherited cancer syndromes such as BRCA1/2 and Lynch syndrome, and who have benefited from intensive genetic counseling, are more likely to engage in cancer screening behavior than non-carriers [89]. However, these findings are specific to people from families with known high-risk cancer-causing mutations, in which carriers have lifetime risks of up to 85% of developing cancer, and noncarriers have average (general population) lifetime risks of developing cancer. In this context, genetic testing is generally considered to be "presymptomatic." It appears that "presymptomatic" genetic testing in the context of genetic counseling and testing for rare high-risk cancer-predisposing mutations, in people with a strong family history and personal experience of the condition, can positively influence cancer screening behavior.

These findings may not be applicable to genetic susceptibility testing for common chronic conditions where the increase in risk associated with specific gene variants is considerably smaller, more probabilistic and less predictive than in inherited cancer syndromes, and where people may have very little personal experience of the condition in close relatives.

In contrast to "presymptomatic" testing for rare high-risk cancer-predisposing mutations, there is actually very little evidence that people do, in fact, change their health behavior based upon genetic susceptibility information, even though researchers have looked hard to find this evidence [89,90]. Studies examining smoking cessation, diet, and exercise behavior in response to genetic susceptibility information have not been promising and show little added benefit of genetic information over general health promotion information. However, a limitation of many of these studies is that health literacy was not assessed. Furthermore, most of the studies that have been done have focused on one or a small number of genetic susceptibility tests, and there is even less evidence reported on behavioral outcomes from genomic tests that may generate risk information for numerous common gene variants simultaneously.

That is not to say that people will not want the information generated by genetic susceptibility tests. Juengst et al. [86] argue that without evidence that people will use genetic information to change their behavior, claims that genetic information is empowering are unfounded. However, the research conducted on patient outcomes from CGS indicates that people do value genetic information for its own sake. The concept of perceived personal control, a key component of patient empowerment as PRO from CGS, may be a useful way to think about benefits from genetic information despite lack of demonstrated health behavior change. Feeling that you have some control, including feeling that you have enough knowledge and understanding of genetic information, may be valued for its own sake, even if that control is not or cannot be translated into health behavior change with long-term health benefits.

Indeed, this may explain why people have been willing to pay for genomic testing through companies such as 23andMe offering DTC genomic tests.

Juengst et al. [86] also raise the issue that genetic susceptibility information may unfairly inflate patients' responsibilities for their own long-term health by raising social and health service expectations of health behavior change, and that institutions may abandon regulatory and other approaches that aim to promote health and reduce risk. As a result, they warn that individuals, rather than being empowered, may be constrained by the limits of their own abilities to change their health behavior, as well as by the limits of their control over the healthcare systems upon which they depend. However, in the absence of good evidence for health behavior change in response to genetic susceptibility information, it is (surely?) unlikely that this testing will ever be adopted by public health programs or by national health services, and will remain in the private sector.

What about genomic testing (whole genome sequencing and whole exome sequencing) in the context of health services? Traditional approaches limited genetic testing to single genes, but genomic testing presents new problems, because of the sheer volume of data generated, and current limitations in our ability and capacity to interpret those data. It is unclear what the outcomes will be of current initiatives such as the 100,000 Genomes Project, which focuses initially on genetic testing for diagnosis of rare disorders, cancers, and infectious diseases (see http://www.1000genomes.org/home).

However, genomic testing is being offered to parents of children affected by developmental disorders in the United Kingdom, to try and identify a cause for the child's condition, because genomic tests are faster and provide a better diagnostic yield than testing individual single genes [91,92]. It is worth noting that current approaches involve testing parental samples as well as the child's sample, where potential causative genetic variants are identified in the child because this can help to identify whether the variant is indeed the cause of the child's problems. The American College of Medical Genetics (ACMG) recently recommended communicating IFs relating to a predetermined set of genes [93]. Others have been more cautious, including the European Society for Human Genetics, and advocate limiting detailed analysis of genomic data only to those genes likely to be relevant to the original clinical question, arguing that analyzing the whole genome is no different to opportunistic screening [94,95]. Part of the rationale for the latter approach is that the population prevalence of known disease-causing variants is unknown. Although we can (generally speaking) say that a known disease-causing variant found in an individual affected by a condition associated with that gene is the likely cause of their condition, it is not necessarily true that the same disease-causing variant will cause the same condition in an unaffected person with no relevant family history. Research is under way in the United Kingdom to collect views and attitudes toward sharing of genomic IFs among families who have had this kind of genomic testing [95].

IFs are, by definition, unexpected, and their unexpected nature could mean that families need to take on information about which they have no personal experience, which is rather different to obtaining a result from a genetic test driven by a clinical question. As outlined above, there is evidence that both genetic risk perception and psychological responses to genetic risk information are strongly influenced by personal experience. Probabilistic risk information about IFs may be significantly less salient to people, because there is unlikely to be any relevant family history and may have considerably less emotional resonance for recipients than genetic information about a condition already known to have caused problems in the family. This could have potential implications for psychological responses and for health behavior.

Many of the genes that the ACMG recommends for return of IFs are genes that predispose to inherited cancer syndromes, and this is deliberate because there are screening interventions that can be offered to those who carry mutations in those genes—the conditions are considered medically actionable. The assumption is that individuals identified to be carriers of gene variants predicted to be disease causing will take full advantage of screening opportunities to maximize their long-term health. However, the evidence on health behavior in response to genetic information could be interpreted as suggesting that in the absence of a strong family history, there is a risk that even information about a high risk of cancer (e.g., BRCA1/2) might not result in the same positive health (screening) behavior reported in families where BRCA1/2 testing was done *because* of a strong family history.

Furthermore, the sheer volume of information that accompanies the offer of genomic testing in the context of childhood developmental delay challenges traditional genetic counseling models of informed consent for genetic testing and limits the capacity of genetic counselors to offer appropriate support. A recent Canadian study exploring decision-making about genomic testing in this context found that participants were often focused on the benefits of genomic testing, for example, obtaining a diagnosis for their child, but later realized that they lacked relevant information about IFs [96]. It emerged that not all participants' informational needs had been met at the time they made a decision to proceed with testing, but at the same time, participants reported that the large volume of information presented to them at the time they gave informed consent was difficult to take in and digest. Staged consent has been advocated, whereby patients consent to testing for the primary clinical question, and provide consent at a later date for IF information [97].

Web-based decision support tools may provide a good solution where large volumes of information about genomic tests are required for informed consent, and where different individuals may be seeking different amounts of information at different times [37,98]. Participants in Li's study showed interest in web-based tools of this kind as an adjunct to discussions with healthcare professionals about genomic testing, although they did not consider these as appropriate replacements for face-to-face discussions. Work is under way to develop and evaluate web-based tools to support decision-making about genomic tests for parents making a decision about these tests for their child (Patricia Birch, University of British Columbia Vancouver, personal communication). However, one of the risks of overreliance on web-based decision support tools, in this busy twenty-first century when time is such a rare commodity, is that people may not use them and simply request the information anyway. It may be helpful for these tools to include functions that enable patients themselves to determine when they would like to have information about IFs, as they explore the information about these within the tool.

CONCLUSION

In conclusion, there is evidence that genetic information can be empowering for people. Most of this evidence comes from CGS, which provide information and testing to families where a genetic condition may be present, and where risks to relatives and future descendants may be high. Indeed, patient empowerment may be the most important PRO from using CGS. Many of the benefits from genetic information, in the context of high-risk genetic conditions, relate to feelings of personal control and empowerment, along with emotional benefits such as hope for the future. There is some evidence of positive behavior change, likely to lead to important long-term health benefits, in the context of high-risk inherited cancer syndromes. There is less evidence of clear benefit or of patient empowerment in the context of genetic susceptibility testing, and there is still considerable uncertainty about genomic testing and the value or otherwise of information about IFs and VUSs.

The benefits accrued by patients of CGS are achieved through provision of genetic information in the context of genetic counseling. This is often provided against a backdrop of (sometimes repeated) disease-related family tragedies, such as pregnancy losses, suffering, and death of close relatives that considerably heighten emotional responses to genetic information. These experiences are known to influence how people perceive genetic risks, and this may have implications for risk perception regarding IFs identified through genomic testing. Genetic risks associated with IFs identified through genomic testing, in the absence of a suggestive family history, will be difficult for both clinicians and patients to interpret, with potential implications for associated health behavior. It is reassuring that providers of genomic sequencing are cautious about how to proceed regarding IFs in particular. If genomic testing is to realize its potential for patient empowerment, it will be important to conduct research to investigate people's responses to genomic testing. It will also be vital to conduct research to evaluate ways in which informed consent might best be achieved. Use of web-based decision support tools may be useful, and evaluation of these could include measures of empowerment, psychological distress, health behavior change, as well as measures of decision quality. Another useful avenue for research will be to assess behavioral responses to IFs in the absence of a relevant family history, particularly for IFs that predict significant lifetime risks of developing cancer where health behavior change can significantly influence long-term health.

REFERENCES

[1] Stern AM. Telling genes: the story of genetic counseling in America. Baltimore, MD: The Johns Hopkins University Press; 2012.

[2] Coventry PA, Pickstone JV. From what and why did genetics emerge as a medical specialism in the 1970s in the UK? A case-history of research, policy and services in the Manchester region of the NHS. Soc Sci Med 1999;49(9):1227—38.

[3] Department of Health. Specialised Services National Definition Set: 20 Medical genetic services (all ages). Available from: <http://webarchive.nationalarchives.gov.uk/+/www.dh.gov.uk/en/Managingyourorganisation/Commissioning/Commissioningspecialisedservices/Specialisedservicesdefinit-ion/DH_4001694>; 2007 [accessed 18.06.13].

[4] Raffan E, Semple RK. Next generation sequencing—implications for clinical practice. Br Med Bull 2011;99:53—71.

[5] Devlin NJ, and Appleby J. Getting the most out of PROMs: putting health outcomes at the heart of NHS decision-making. London, United Kingdom: The King's Fund. Available from: <http://www.kingsfund.org.uk/publications/getting-most-outproms>; 2010 [accessed on 24.06.13].

[6] Snyder CF, Aaronson NK, Choucair AK, et al. Implementing patient-reported outcomes assessment in clinical practice: a review of the options and considerations. Qual Life Res 2012;21:1305—14.

[7] Royal College of Physicians. Commissioning clinical genetics services: activity, outcome, effective-ness and quality. London: RCP Clinical Genetics Committee; 1998.

[8] Clarke AJ. Outcomes and process in genetic counselling. Clin Genet 1997;50:462—9.

[9] MacLeod R. The genetic counselling process: an interpretative phenomenological analysis of patient—counsellor interactions. PhD Thesis, The University of Manchester; 2003.

[10] Resta R, Bowles Biesecker B, Bennett RL, et al. A new definition of genetic counseling: national society of genetic counselors' task force report. J Genet Couns 2006;15:77—83.

[11] Somer M, Mustonen H, Norio R. Evaluation of genetic counselling: recall of information, post-counselling reproduction, and attitude of the counselees. Clin Genet 1988;34:352—65.

[12] Evers-Kiebooms G, van den Berghe H. Impact of genetic counseling: a review of published follow-up studies. Clin Genet 1979;15:465—74.

[13] Evans DGR, Blair V, Greenhalgh R, Hopwood P, Howell A. The impact of genetic counselling on risk perception in women with a family history of breast cancer. Br J Cancer 1994;70:934—8.

[14] Tversky A, Kahneman D. Judgments under uncertainty: heuristics and biases. Science 1974;185:1124—31.

[15] Katapodi MC, Dodd MJ, Facione NC, Humphreys JC, Lee KA. Why some women have an opti-mistic or a pessimistic bias about their breast cancer risk: experiences, heuristics, and knowledge of risk factors. Cancer Nurs 2010;33:64—73.

[16] Dieng M, Watts CG, Kasparian NA, Morton RL, Mann GJ, Cust AE. Improving subjective per-ception of personal cancer risk: systematic review and meta-analysis of educational interventions for people with cancer or at high risk of cancer. Psychooncology 2014;23:613—25.

[17] Gigerenzer G, Todd P, Group AR. Simple heuristics that make us smart. New York, NY: Oxford University Press; 1999.

[18] Simon HA. Models of bounded rationality: empirically grounded economic reason. Boston, MA: MIT Press; 1982.

[19] Austin JC. Re-conceptualizing risk in genetic counseling: implications for clinical practice. J Genet Couns 2010;19:228—34.

[20] McAllister M. Personal theories of inheritance, coping strategies, risk perception and engagement in hereditary non-polyposis colon cancer families offered genetic testing. Clin Genet 2003;64:179−89.

[21] McAllister M. Predictive genetic testing and beyond: a theory of engagement. J Health Psychol 2002;7:491−508.

[22] Edwards A, Gray J, Clarke A, Dundon J, Elwyn G, Gaff C, et al. Interventions to improve risk communication in clinical genetics: systematic review. Patient Educ Couns 2008;71:4−25.

[23] McAllister M, Dunn G, Todd C. Empowerment: qualitative underpinning of a new patient reported outcome for clinical genetics services. Eur J Hum Genet 2011;19:125−30.

[24] Livneh H, Antonak RF. Psychosocial adaptation to chronic illness and disability: a primer for counselors. J Couns Dev 2005;83:12−20.

[25] Meiser B, Dunn S. Psychological impact of genetic testing for Huntington's disease: an update of the literature. J Neurol Neurosurg Psychiatry 2000;69:74−578.

[26] Almqvist EW, Brinkman RR, Wiggins S, Hayden MR, Canadian Collaborative Study of Predictive Testing. Psychological consequences and predictors of adverse events in the first 5 years after predictive testing for Huntington's disease. Clin Genet 2003;64:300−9.

[27] Duisterhof M, Trijsburg RW, Niermeijer MF, Roos RA, Tibben A. Psychological studies in Huntington's disease: making up the balance. J Med Genet 2001;38:852−61.

[28] Timman R, Roos R, Maat-Kievit A, Tibben A. Adverse effects of predictive testing for Huntington disease underestimated: long-term effects 7−10 years after the test. Health Psychol 2004;23:189−97.

[29] Butow PN, Lobb EA, Meiser B, Barratt A, Tucker KM. Psychological outcomes and risk perception after genetic testing and counselling in breast cancer: a systematic review. MJA 2003;178:77−81.

[30] Codori AM, Hanson R, Brandt J. Self-selection in predictive testing for Huntington's disease. Am J Med Genet 1994;54(3):167−73.

[31] Marteau TM, Dormandy E. Facilitating informed choice in prenatal testing: how well are we doing? Am J Med Genet 2001;106:185−90.

[32] Kaiser AS, Pastuszak AL, Llewellyn-Thomas H, Johnson J, Conacher S, Shaw BF. The effects of prenatal group genetic counselling on knowledge, anxiety and decisional conflict: issues for nuchal translucency screening. J Obstet Gynaecol 2002;22:246−55.

[33] Hunter AGW, Cappelli M, Humphreys L, et al. A randomized trial comparing alternative approaches to prenatal diagnosis counseling in advanced maternal age patients. Clin Genet 2005;67:303−13.

[34] van den Berg M, Timmermans DRM, ten Kate LP, van Vugt JMG, van der Wal G. Are pregnant women making informed choices about prenatal screening? Genet Med 2005;7:332−8.

[35] Rowe HJ, Fisher JR, Quinlivan JA. Are pregnant Australian women well informed about prenatal genetic screening? A systematic investigation using the multidimensional measure of informed choice. Aust N Z J Obstet Gynaecol 2006;46:433−9.

[36] Dormandy E, Bryan S, Gulliford MC, et al. Antenatal screening for haemoglobinopathies in primary care: a cohort study and cluster randomised trial to inform a simulation model. The Screening for Haemoglobinopathies in First Trimester (SHIFT) trial. Health Technol Assess 2010;14:1−160.

[37] Birch P. Interactive e-counselling for genetics pre-test decisions: where are we now? Clin Genet 2014; EPub 14 May. Available from: http://dx.doi.org/10.1111/cge.12430.

[38] Kessler S. Psychological aspects of genetic counseling: VI. A critical review of the literature dealing with education and reproduction. Am J Med Genet 1989;34:340−53.

[39] Phipps S, Zinn AB. Psychological response to amniocentesis. Mood state and adaptation to pregnancy. Am J Med Genet 1986;25:131−42.

[40] van Zuuren FJ. Coping style and anxiety during prenatal diagnosis. J Reprod Infant Psychol 1993;11:57−9.

[41] Tercyak KP, Bennett Johnson S, Roberts SB, Cruz AC. Psychological response to prenatal genetic counseling and amniocentesis. Patient Educ Couns 2001;43:73−84.

[42] Payne K, Nicholls S, McAllister M, MacLeod R, Donnai D, Davies LM. Outcome measurement in clinical genetics services: a systematic review of validated measures. Value Health 2008;11:497−508.

[43] Mokkink LB, Terwee CB, Patrick DL, et al. The COSMIN checklist for assessing the methodological quality of studies on measurement properties of health status measurement instruments: an international Delphi study. Qual Life Res 2010;19:539−49.

[44] Terwee CB, Mokkink LB, Knol DL, et al. Rating the methodological quality in systematic reviews of studies on measurement properties: a scoring system for the COSMIN checklist. Qual Life Res 2012;21:651−7.

[45] Berkenstadt M, Shiloh S, Barkai G, Bat-Miriam-Katznelson M, Goldman B. Perceived personal control (PPC): a new concept in measuring outcome of genetic counseling. Am J Med Genet 1999;82:53−9.

[46] McAllister M, Wood AM, Dunn G, Shiloh S, Todd C. The perceived personal control (PPC) questionnaire: reliability and validity in a sample from the United Kingdom. Am J Med Genet A 2012;158A:367−72.

[47] Streiner DL, Norman GR. Health measurement scales: a practical guide to their development and use. Oxford: Oxford University Press; 2008.

[48] Payne K, Nicholls S, McAllister M, MacLeod R, Middleton-Price H, Ellis I, et al. Outcome measures for clinical genetics services: a comparison of genetics healthcare professionals and patients' views. Health Policy 2007;84:112−22.

[49] Pill J. The Delphi method: substance, context, a critique and annotated bibliography. Socioecon Plann Sci 1971;5:57−71.

[50] Myers SS, Phillips RS, Davis RB, et al. Patient expectations as predictors of outcome in patients with acute low back pain. J Gen Intern Med 2008;23(2):148−53.

[51] Bowling A, Rowe G, Lambert N, et al. The measurement of patients' expectations for health care: a review and psychometric testing of a measure of patients' expectations. Health Technol Assess J 2012;16(30):i−xii, 1−509.

[52] Haanstra TM, Berg T, Ostello RW, et al. Systematic review: do patient expectations influence treatment outcomes in total knee and hip athroplasty? Health Qual Life Outcomes 2012;10:152.

[53] Younger J, Gandhi V, Hubbard E, Mackey S. Development of the Stanford Expectations of Treatment Scale (SETS): a tool for measuring patient outcome expectancy in clinical trials. Clin Trials 2012;9:767−76.

[54] Michie S, Marteau TM, Bobrow M. Genetic counselling: the psychological impact of meeting patients' expectations. J Med Genet 1997;34(3):237−41.

[55] Davey A, Rostant K, Harrop K, Goldblatt J, O'Leary P. Evaluating genetic counselling: client expectations, psychological adjustment and satisfaction with service. J Genet Couns 2005;14 (3):197−206.

[56] Bernhardt BA, Biesecker BB, Mastromarino CL. Goals, benefits, and outcomes of genetic counseling: client and genetic counselor assessment. Am J Med Genet Part A 2000;94(3):189−97.

[57] MacLeod R, Craufurd D, Booth K. Patients' perceptions of what makes genetic counseling effective: an interpretive phenomenological analysis. J Health Psychol 2002;7(2):145−56.

[58] Cleary PD, McNeil BJ. Patient satisfaction as an indicator of quality care. Inquiry 1988;25 (1):25−36.

[59] Paddison CAM, Abel GA, Roland MO, Elliott MN, Campbell JL. Drivers of overall satisfaction with primary care: evidence from the English General Practice Patient Survey. Health Expect 2013. Available from: http://dx.doi.org/10.1111/hex.12081. [Epub ahead of print].

[60] Rao JK, Weinberger M, Kroenke K. Visit-specific expectations and patient-centered outcomes. Arch Fam Med 2000;9:1148−55.

[61] Noble PC, Conditt MA, Cook KF, Mathis KB. Patient expectations affect satisfaction with total knee arthroplasty. Clin Orthop Relat Res 2006;452:35−43.

[62] Verbeek J, Sengers M-J, Riemens L, Haafkens J. Patient expectations of treatment for back pain: a systematic review of qualitative and quantitative studies. Spine 2004;29(20):2309−18.

[63] Platten U, Rantala J, Lindblom A, et al. The use of telephone in genetic counseling versus in-person counseling: a randomized study on counselees' outcome. Fam Cancer 2012;11(3):371−9.

[64] McAllister M, Payne K, Nicholls S, et al. Patient empowerment in clinical genetics services. J Health Psychol 2008;13:887−97.

[65] Shiloh S, Berkenstadt M, Meiran N, Bat-Miriam-Katznelson M, Goldman B. Mediating effects of perceived personal control in coping with a health threat: the case of genetic counselling. J Appl Soc Psychol 1997;27:1146−74.

[66] Affleck G, Tennen H, Pfeiffer C, Fifield J. Appraisals of control and predictability in adapting to a chronic disease. J Pers Soc Psychol 1987;53(2):273−9.

[67] Scharloo M, Kaptein AA, Weinman J, Hazes JM, Willems LNA, Bergman W, et al. Illness perceptions, coping and functioning in patients with rheumatoid arthritis, chronic obstructive pulmonary disease and psoriasis. J Psychosom Res 1998;44:573−85.

[68] Lazarus RS, Folkman S. Stress, appraisal, and coping. New York, NY: Springer; 1984.

[69] Fayers PM, Machin D. Quality of life: the assessment, analysis and interpretation of patient-reported outcomes. 2nd ed. Chichester: John Wiley & Sons; 2007.

[70] Kerzin-Storrar L. Genetic registers. In: eLS [Online]. Available from: <http://www.els.net/> [accessed 14.10.14.] Available from: http://dx.doi.org/10.1002/9780470015902.a0005619.

[71] McAllister M, Wood A, Dunn G, Shiloh S, Todd C. The genetic counseling outcome scale: a new patient-reported outcome measure for clinical genetics services. Clin Genet 2011;79:413−24.

[72] McAllister M, Dearing A. Patient reported outcomes and patient empowerment in clinical genetics services. Clin Genet. Available from: <http://dx.doi.org/10.1111/cge.12520>. [Epub ahead of print].

[73] Zellerino B, Milligan SA, Brooks R, Freedenberg DL, Collingridge DS, Williams MS. Development, testing, and validation of a patient satisfaction questionnaire for use in the clinical genetics setting. Am J Med Genet C Semin Med Genet 2009;151C(3):191−9.

[74] Skirton H, Parsons E, Ewings P. Development of an audit tool for genetic services. Am J Med Genet A 2005;136(2):122−7.

[75] Darzi A. High quality care for all: NHS next stage review (final report). Department of Health London. Available from: <https://www.gov.uk/government/uploads/system/uploads/attachment_data/file/228836/7432.pdf>; 2008 [accessed 17.03.14].

[76] United States Congress. Patient Protection and Affordable Care Act, USA. Available from: <http://www.hhs.gov/healthcare/rights/law/patient-protection.pdf>; 2010 [accessed 14.03.201].

[77] Aujoulat A, d'Hoore W, Deccache A. Patient empowerment in theory and practice: polysemy or cacophony? Patient Educ Couns 2007;66:13−20.

[78] Bravo P, Edwards A, Barr PJ, Scholl I, Elwyn G, McAllister M. Conceptualising patient empowerment: a mixed methods study. BMC Health Serv Res 2015;15:252. Available from: http://dx.doi.org/10.1186/s12913-015-0907-z.

[79] Makoul G, Clayman M. An integrative model of shared decision making in medical encounters. Patient Educ Couns 2006;60:301−12.

[80] Elwyn G, Lloyd A, May C, et al. Collaborative deliberation: a model for patient care. Patient Educ Couns 2014. Available from: http://dx.doi.org/10.1016/j.pec.2014.07.027>. [Epub ahead of print].

[81] Aujoulat I, Marcolongo R, Bonadiman L, Deccache A. Reconsidering patient empowerment in chronic illness: a critique of models of self-efficacy and bodily control. Soc Sci Med 2008;66:1228−39.

[82] Barr PJ, Scholl I, Bravo P, Faber M, Elwyn G, McAllister M. Assessment of patient empowerment: a systematic review of measures. *PLOS ONE* 2015;10(5):e0126553.

[83] Edwards M, Davies M, Edwards A. What are the external influences on information exchange and shared decision making in healthcare consultations: a meta-synthesis of the literature. Patient Educ Couns 2009;75:37−52.

[84] Edwards M, Wood F, Davies M, Edwards A. The development of health literacy in patients with a long-term health condition: the health literacy pathway model. BMC Public Health 2012;12:130.

[85] Funnell MM, Anderson RM. Empowerment and self-management of diabetes. Clin Diabetes 2004;22(3):123−7.

[86] Juengst E, Flatt MA, Settersten RA. Personalized genomic medicine and the rhetoric of empowerment. Hastings Cent Rep 2012;42(5):34−40.

[87] Covolo L, Rubinelli S, Orizio G, Gelatti U. Misuse (and abuse?) of the concept of empowerment. The case of online offer of predictive direct-to-consumer genetic tests. J Public Health Res 2012;1:e3.

[88] Nutbeam D. Health literacy as a public health goal: a challenge for contemporary health education and communication strategies into the 21st century. Health Promot Int 2000;15(3):259−67.

[89] McBride CM, Koehly LM, Sanderson SC, Kaphingst KA. The behavioral response to personalized genetic information: will genetic risk profiles motivate individuals and families to choose more healthful behaviors? Annu Rev Public Health 2010;31:89−103.

[90] Marteau TM, French DP, Griffin SJ, et al. Effects of communicating DNA-based disease risk estimates on risk-reducing behaviours. Cochrane Database Syst Rev 2010;10. Available from: http://dx.doi.org/10.1002/14651858.CD007275.pub2.

[91] Firth HV, Wright C, DDD Study. The deciphering developmental disorders (DDD) study. Dev Med Child Neurol 2011;53:702−3.

[92] Wright CF, Middleton A, Burton H, Cunningham F, Humphries SE, Hurst J, et al. Policy challenges of clinical genome sequencing. BMJ 2013;22:347.

[93] Green RC, Berg JS, Grody WW, Kalia SS, Korf BR, Martin CL, et al. ACMG recommendations for reporting of incidental findings in clinical exome sequencing and genome sequencing. Genet Med 2013;15:565−74.

[94] Wolf SM, Annas GJ, Elias S. Patient autonomy and incidental findings in clinical genomics. Science 2013;340:1049−50.

[95] Middleton A, Parker M, Wright CW, Bragin E, Hurles M. Empirical research on the ethics of genomic research. Am J Med Genet A 2013;161(8):2099−101.

[96] Li K. Supporting decision-making in whole genome/exome sequencing: parents' perspectives. MSc dissertation. University of British Columbia, Vancouver; 2014.

[97] Appelbaum PS, Waldman CR, Fyer A, Klitzman R, Parens E, Martinez J, et al. Informed consent for return of incidental findings in genomic research. Genet Med 2013;1−7. Available from: http://dx.doi.org/10.1038/gim.2013.145.

[98] Overby CL, Kohane I, Kannry JL, Williams MS, Starren J, Bottinger E, et al. Opportunities for genomic clinical decision support interventions. Genet Med 2013;15(10):817−23.

CHAPTER 4

The Expanding Scope of Gen-Ethics

Ruth Chadwick
Healthcare Ethics and Law, School of Law, The University of Manchester, Manchester, UK

Contents

INTRODUCTION—GENOMICS AND BIOETHICS

The history of bioethics in relation to genetics and genomics, sometimes called gen-ethics—as branches of applied ethics become increasingly specialized—has been marked by varying emphases, at different times, on environmental and biological explanations of human health and behavior, and by changing perceptions of the relationship between the individual's interests and the public good. In the early years of the twentieth century, in relation to the sterilization of Carrie Buck, it was claimed that "three generations of imbeciles are enough" by the judge in the case of *Buck v Bell* [1]. The history of eugenics in the early twentieth century, and responses to that, had a significant influence on how debates became framed in the second half of the twentieth century, in particular in relation to the emergence of the nondirective ethos in genetic counseling. Reaction against forced sterilization of those deemed "unfit," such as Buck, led not only to concerns in succeeding decades about the improper interpretation of scientific information, but also to worries about abuse of genetic

Genomics and Society
DOI: http://dx.doi.org/10.1016/B978-0-12-420195-8.00004-5

science; to an emphasis on free choice in the genetics clinic; and a preference for environmental explanations and interventions at a social level. In the second half of the twentieth century, however, in the light of the success and high profile of genetic and genomic research, genetic explanations and promises regained popularity.

Toward the end of the twentieth century, rapid developments in genomics, and the initiation and progress of the Human Genome Project (HGP), saw the allocation of a dedicated budget to address ethical, legal, and social issues (ELSI). This was followed by an ELSA (ethical, legal, and social *aspects*) program in Europe. Although ELSI and ELSA have become recognized as brands in themselves (subsequently succeeded by responsible research and innovation), the broader field of bioethics was also itself greatly affected by the huge explosion in work on the ethical dimensions of genomics.

Over time, the issues giving rise to the most scrutiny in bioethics and genomics have changed, as genomics has moved out of the clinic and into society at large. What have also been noticeable are the ways in which scientific developments have posed challenges for ways of thinking about bioethical issues such as informed consent, leading to a considerable amount of work on ethical frameworks [2]. This chapter aims to demonstrate the coevolution of bioethics and genomics in gen-ethics as the science expanded beyond the clinic to the potential transformation of the practice of medicine itself, along with implications for other areas of social life.

ISSUES IN CLINICAL GENETICS: GENETIC TESTING AND COUNSELING

When the focus of discussion was within the genetics clinic, and concerned with diseases recognized as having an identifiable genetic component—for example, Huntington's disease, the ethical issues turned largely on the autonomy and confidentiality of the client in relation to testing, which may be diagnostic of a current condition or predictive of a late onset one, and relating not only to adults but also to their children and fetuses. In the clinical context a genetic *test* (as opposed to *screening*) is typically carried out where there is some reason to think there may be a problem. A good example is the situation where there is a strong family history of breast cancer. For adults, it has been argued that there is a right not to know (see later) the result of a predictive test which reveals a risk factor but not certainty about onset or severity of a disease [3].

Where children are concerned, for parents with a child with an apparent disorder, the result of a *diagnostic* test may bring relief in knowing what is wrong. In the case of prenatal testing however, there might be difficult questions concerning termination decisions, in cases where a fetus is found to be suffering from a genetic or chromosomal abnormality (relevant factors also include the context of social support, or lack of it, for persons with disabilities which might affect a mother's decision). Preimplantation genetic diagnosis and embryo selection avoid the issues of termination of a fetus but the issues of the criteria for selection of embryos, resource issues, and what should be done with "spare" embryos remain controversial.

Nondirectiveness

As mentioned earlier, following the abuses of genetics in the early twentieth century and the rise in individualism in medical ethics and bioethics, the prevailing ethos in counseling was that of nondirectiveness, facilitating and sharing decision-making but not directing the decision of the client in relation to decisions such as termination [4]. Nondirectiveness may of course be difficult to achieve, in situations where there is a power differential between counselor and client; and some clients may actually want to be directed or at least advised, but for present purposes the main point is that the ethos reflects the primacy of autonomy.

In some situations there is a clear tension between the interests of the client and those of third parties. Some of the most discussed questions in the early 1990s concerned possibilities of disclosure to family members of the results of genetic tests on blood relatives, and the conflict between the interests of those family members and the confidentiality of the person tested—for example, is it right to disclose a finding of nonpaternity where a genetic test on the fetus reveals that the mother's partner could not actually be the genetic father? The balance between competing interests here is normally struck by the position that individuals should be encouraged to share their information, but that there may be some cases at the extreme, where disclosure is permissible.

Children

The genetic testing and screening of children has long been acknowledged to be a special case [5]. In relation to children, the issues of autonomy are more complicated, where it might be the parents' wish to have a child tested, because *they* want to know. The possibilities of stigmatization and self-fulfilling prophecy are real issues for a child who is symptom free but identified as being at risk of a late onset disorder. Hence there are strong arguments for the view that children should not be tested for late onset disorders until they are capable of deciding for themselves whether or not to be tested. The situation is different where there are potentially life-threatening or life-limiting disorders which can be treated in childhood, where the expectation would be that parents should opt for treatment as they would with any other illness.

Genetic Exceptionalism

The fact that blood relatives may share genetic predispositions to particular diseases was one of the principal factors driving discussion of the thesis of "*genetic exceptionalism*," the thesis that there is something special or different about the ethical issues in genetics, the other factors being that genetic information is predictive and independent of time: hence a test in childhood has the potential to reveal an individual's risk status with regard to certain late onset diseases. The thesis of genetic exceptionalism, however, has been criticized on the grounds that other types of medical information may share these features in varying degrees [6]: although the ethical issues in genetics may be particularly complicated, they are not in a wholly different category from those relevant in other areas of medicine, so that there is no need to develop a bioethics specific to genetics *per se*. Different issues need to be considered on their own merits and in specific contexts.

While the issues in clinical genetics and genetic counseling remain live, they are no longer the primary focus of concern or ethical debate. What has changed is that genetic information has become relevant not only in the genetics clinic but throughout medical practice, as the biological factors underlying not only genetic disorders but also the common diseases such as heart disease and the cancers are identified. This is perhaps one of the most marked changes in relation to genetics—the "geneticization" of health and health care.

SCREENING

In contrast to genetic testing, where there are factors relating to an individual which suggest that testing is indicated, a screening program is carried out on a population where there is no reason to think, for any given individual, that he or she is at higher risk. In the 1990s there was considerable discussion about the principles underlying the introduction of a genetic screening program—adapting the 1968 Wilson and Jungner principles, for screening in general, to the particular context of genetic screening [7]. The first criterion is that the condition screened for must be important, or serious; another is that there must be an acceptable treatment—which was changed in the genetic context to "scope for action," as the intervention in question may be a termination or change of lifestyle. There was considerable but to some extent inconclusive discussion about what constituted a condition sufficiently "serious" to warrant a screening program. A classic example is phenylketonuria, for which newborns are screened and which can be controlled by a diet low in phenylalanine.

There have been cases of interventions at the population level to counteract high incidence of genetic disease, for example, in the case of thalassemia in Cyprus. The prevention program changed over time with developing technologies, but in 1980 premarital screening was made compulsory by law. The incidence of thalassemia has declined and patients with the condition have a better life expectancy as treatment improves. This example makes clear, however, that genetics retains its connection with reproduction decisions even for those who may be unaffected themselves, but carriers. In other words it is not only concerned with individuals who are diagnosed either as having a condition or being at risk of developing one.

GENE THERAPY

One of the reasons why issues in genetics have been so high profile lies in the nature of genetic disorder and the options for treatment. It is in the nature of genetic disorders that they cannot typically be cured in the way that infectious diseases can, for example, by prescribing drugs (notwithstanding the current anxieties about antibiotic resistance). The boundary between genetic and nongenetic disease is not, of course, clear-cut: infectious diseases have a genetic component, not only in relation to the genome of the infecting organism but in its interaction with the genome of the persons affected. People will differ in their susceptibility and response to infectious disease in accordance with their own genomic variation.

The early prospects of gene therapy, of introducing a functioning gene into the body of a patient who lacked a functioning one, by using a vector such as a virus, led to disappointment arising partly from the results of tests and partly because of ethical concerns about some trials. The focus of interest turned away somewhat to the new promise of stem cell therapy. In the twenty-first century, however, some forms of gene therapy are seeing encouraging results. In 2014, for example, positive news was reported in relation to a trial using gene therapy to treat eye disease [8]. There remain concerns about the difference between somatic gene therapy (where the new genetic information affects the individual patient only) and germline therapy (where the germline is also affected). In the latter case, future generations will be affected who have no opportunity to consent. To some, however, the latter argument is overstated, as we cannot avoid decisions that will affect future generations in any case.

Debates about clinical genetic testing, screening programs, and therapy, however, were overtaken by genomics moving center stage with the advent of the HGP.

THE HGP AND GENOMICS

The aim of the HGP was to produce a complete map and sequence of all the genes in the human genome, so genomics marked a shift of attention from genes to genomes. Until quite recently most of the bioethical discussion related to genomics concerned the nuclear genome, which was what was referred to in the HGP. The sequencing of the mitochondrial genome, completed in the 1980s, passed almost unattended by any ethical comment, presumably because of how the role of the mitochondrial genome, which exists within the cell outside of the nucleus, was perceived. The analogy of a battery has been used, suggesting that the role of mitochondria is just to provide energy to the cell, in contrast to the nuclear genome which, together with other factors, influences the characteristics of the phenotype.

It is nevertheless the case that a defect in the mitochondria, which are inherited through the female line, can have a devastating effect on the phenotype. The battery analogy, however, has been influential in making the positive case for mitochondrial replacement therapy, which involves removing the diseased mitochondria from a woman's egg and replacing them with those from donor. The egg can then be fertilized by the woman's husband or partner. This has been described in some sections of the press as producing babies with three parents, but the supporters of the technique argue that it prevents potentially very great suffering by the prevention of mitochondrial disease which cannot otherwise be cured.

With regard to the nuclear genome, it was thought by some that to know the genome of an individual would enable us to know everything about that person—that a human being was identified with the sum total of his genome. Genetic determinism, the thesis that a person's genes determine the outcome in the phenotype, had considerable support. It therefore came as a shock to some when the HGP was completed and revealed that there are far fewer genes in the human genome than had previously been thought—about 22,000 rather than 100,000. This led to the suggestion that genetic determinism was dead. If there are so few genes in the human genome, genes cannot possibly explain the huge variety and complexity that we see in human beings. Genetic *factors*, as opposed to whole genes, are clearly influential in relation to development and phenotypic expression, but so are many other factors, such as gene—environment interaction, copy number variation—variation in the number of copies of a genetic factor that an individual has—and epigenetics, or factors over and above the genome that affect gene expression.

BIOBANKS

In addition to trying to make sense of the information provided by the outcome of the HGP itself, attention turned in the last decade of the twentieth century to population-wide genomic research [9]. Biobanks were established in a number of countries, such as Iceland and Estonia. The purpose of a biobank was to collect both DNA samples and phenotypic information from participants, in an effort to establish links between genetic factors and the development of disease, and not only diseases classifiable as "genetic diseases." For example, the UK Biobank aimed to clarify the factors at work in the common diseases that typically manifest themselves in midlife, such as heart disease and the cancers.

Attention turned to the study of variation in the genome—the ways in which individuals differ from each other at the genetic level, leading to effects which may be due to a single base pair in a particular part of the genome. There had been earlier interest in variation, in the Human Genome Diversity project, but that had attracted considerable amount of hostility among different ethnic groups who had concerns about the motivation behind the study of diversity.

Biobanks, particularly in Iceland, were also initially very controversial and gave rise once more to discussions about whether there is anything special about the ethical issues involved. One prominent example concerned informed consent. The doctrine of informed consent had become a cornerstone of biomedical ethics in the twentieth century, following not only the concerns about eugenics mentioned earlier but also malpractice in relation to research subjects, such as had occurred in Europe in the Second World War and in the United States.

Participants asked to donate DNA samples to biobanks were asked to give their informed consent. The problem that emerged was that it was not always clear what they were being asked to consent to. Biobank operators were proposing to collect and store their samples, but for what? It might not be clear even to the scientists and researchers what they might want to do with the samples a few years down the line. New scientific discoveries would almost certainly reveal new lines of inquiry. Thus it was argued by some that genuine informed consent could not be provided in relation to a donation to a biobank. Again, this did not support a thesis of genetic exceptionalism, because there is a view that fully informed consent is an ideal that can never be fully reached in any branch of medicine. The new context, however, gave rise to the bioethical debate about the difference between "narrow" and "broad consent." A person who gives narrow informed consent agrees to specific, detailed research purposes, whereas broad consent involves agreement to research for purposes that may not be specified or specifiable at the time. UK Biobank adopted a form of broad consent, asking participants to consent to being in the Biobank—or not.

There are different arguments in favor of broad consent in the biobank genomics context. First, the purposes for which biobanks are established are clearly, to some extent, undermined if there are restrictions on the research that may be done on the samples. This is not just an argument about scientific freedom: it is argued that this kind of population research is essential for future population health, in addressing the incidence of the common diseases. Another argument is that the kind of harm which may befall contributors to a biobank is very different from that involved in conventional medical research such as a drug trial. The physical intervention involved is limited to the taking of a blood sample: the potential harm arises largely from the potential from access to and misuse of a participant's information, especially if the samples are linked to phenotypic information of various kinds.

As noted earlier, in the clinical genetics context, confidentiality and disclosure had been an issue. Biobank research however gave rise to a considerable amount of new debate about the handling of genomic information, both as regards storage and access. Storage issues included to what extent and in what ways samples were linked, or not, to phenotypic information, by techniques of coding and anonymization. If samples could not be linked at all, then the value of the research was decreased. On the other hand, the greater the possibilities of identifiability, the greater the risk to the privacy of the individual participant. Increasingly it was realized, in any case, that where DNA samples are collected and stored, the possibility of maintaining complete privacy is undermined, and it was argued by some that the better course would be to acknowledge this fact. This led to a call for "open consent" where participants would donate in the knowledge that there was no guarantee of privacy, in the interests of research [10].

The development of biobank research and population-wide genomic research has been associated with a move toward an emphasis on public health in bioethical debate. It would be a mistake to characterize the issues here as a trade-off between a personal interest in privacy on the one hand and a public interest in health one the other: there is also a public interest in the maintenance of privacy protections, and the balance of interest has to be negotiated.

Nevertheless, given the predominance of the public good argument in relation to biobanks, it is not surprising that there has been a shift in the bioethical framing of the issues. In addition to the arguments for broad consent, there has been a move toward a greater emphasis on principles other than the individualistic ones at play in the earlier days of bioethics, including equity and justice. Another obvious example is the principle of solidarity, which has been appealed to in this context. The principle of solidarity, which has been defined in the report prepared by Prainsack and Buyx for the Nuffield Council of Bioethics on solidarity [11] as the willingness to bear costs for another's good, can be used, for example, to provide a reason for contributing to a biobank. Autonomy nevertheless remains important—the individual must be free to give consent, whether broad or narrow, but other principles are increasingly emphasized.

Feedback and the right (not) to know

Much of the discussion over genetic testing and screening concerned an individual's right to access to and control over their genetic data. The ways in which this played out in the clinical context of genetic counseling have already been alluded to.

In the biobank context, these issues took on a particular focus in relation to feedback of results of biobank research. Opposing arguments have been advanced in relation to this; on the one hand, it is argued that participants should not be given information other than the general results of the research. In many cases the raw data that emerges will not be meaningful to individuals in any case. On the other hand, it is argued that if information specific to an individual (sometimes called incidental findings) emerges which could be relevant to their health, even potentially lifesaving, then there is an obligation to tell them. At the time of writing this debate is far from resolved, and there are ongoing debates about what information participants in biobanks and genomic research should have access to.

Debates in the 1990s had already considered whether there could be a right not to know genetic information [3] The historical predominance of autonomy, choice, and informed consent in the second half of the twentieth century perhaps produced a default position that individuals had a right to information: what had to be justified was withholding it. Different understandings of autonomy, however, suggested that an individual can make an autonomous decision to live their life without certain information which might not only cause distress but undermine their conception of themselves and their life plan.

PERSONALIZED MEDICINE

The potential irony in the situation where principles such as solidarity and concepts of the common good are appealed to in the ethical discussions of population-wide genomic research is that the interest in the variations in the genome between individuals, the discovery of which was an aim of this research, was to drive the quest for a new form of individual-centered medicine, the so-called personalized medicine, whereby treatment could be prescribed in accordance with the particular genome of the patient.

The argument for personalized medicine made much of the fact that traditional medical practice and drug prescribing had a considerable problem of disease caused by medical examination or treatment, or iatrogenic disease. There was a large burden of not only morbidity but also mortality arising from the side effects of pharmaceutical products. While it was known that any given drug would not be appropriate for every individual, in the past for the most part, the only way to find out would be for that individual to take the drug—a trial-and-error approach predominated in prescribing. The theory behind pharmacogenomics was that it should be possible through genomic research to discover the associations between genetic factors and drug response. Implementation in clinical practice would of course have to have access to genetic information about the individual patient, and this would then inform the choice of drug, or the appropriate dosage. The promise was safer and more effective medicine—and also a reduction in time wasted through trying drugs that did not work for a given patient [12].

The label "personalized medicine" has not proved totally beneficial and is now being challenged by terms such as "precision medicine" [13,14]. First, it is not always clear what "personalization" means. Initially personalized medicine might have meant assigning patients to groups of good or poor responders to a particular drug on the basis of a targeted test—a fairly limited sense of personalization, which might more appropriately be called patient stratification. Developments in science and technology, from genome-wide association studies to next generation sequencing, have given a new meaning to personalization, ultimately involving investigating the whole genome sequence of an individual. (Some approaches to personalized medicine also include wider ranges of nongenetic data; [13,14]). The metaphor of "tailoring" has been used to describe the fit between the individual's genome sequence and the prescription of medical and lifestyle advice. The cost for such sequencing has also fallen rapidly and continues to fall.

There are also new challenges, however. While on the one hand it fits into an individualistic paradigm in society that employs the rhetoric of choice, this has also led to one of the main criticisms: that on a wide scale it is only feasible in wealthy societies (despite the falling costs). Indeed, it has been called "boutique" medicine by Daar and Singer [15]. They claimed, that the underlying research could be beneficial for other purposes—what has been described as "drug resuscitation" of products taken off the market because of adverse events in the west, to help underserved populations in developing countries, where differences in genomic factors might avoid the adverse effects. This discussion is part of a much wider debate about the sharing of the benefits of genomic research. Benefit sharing, to be understood in a wide sense beyond commercial benefit, has become a standard ethical dimension of proposals for genetic and genomic studies, one that points to the greater consideration of the equity and justice [16].

Direct-to-consumer testing

Following the HGP companies emerged offering testing direct to consumers, where the findings could include information about genetic ancestry, and health-related information. As regards the latter they gave rise to many concerns about the bypassing of professional counseling and advice in the delivery of the information, as well as criticisms regarding the accuracy and usefulness of the information provided. Whereas some people regarded it as an aspect of the "democratization" of science to facilitate this kind of accessing of their genomic information, there were worries about the possible adverse effects of individuals acting on misleading information about themselves and their families. In December 2013 in the United States, for example, the Food and Drug Administration (FDA) ordered the company 23andMe to halt its disclosure of health-relevant results to new direct-to-consumer testing clients. Early in 2015, however, the FDA authorized marketing of a direct-to-consumer genetic carrier test for Bloom syndrome and announced the exemption of carrier screening tests from premarket review.

"RECREATIONAL" AND BEHAVIORAL GENOMICS

While the main argument for genetic testing, screening, and now sequencing has always been the medical benefits, other uses have increasingly become apparent, some of which are regarded as recreational. Ancestry tracing has already been mentioned. Behavioral and forensic genomics raise ethical issues about responsibility, freedom, and punishment—could there be a legitimate defense in court relating to one's genetic predisposition, for example. Also giving rise to concern are the promotion of tests related to character and natural talents of children, and to pick out budding sports stars so that they cans be given appropriate training. In some cases this overlaps with medical matters, as in the case of a predisposition to sudden cardiac death which is clearly important for someone considering a sporting career. The main issues here, however, remain the quality of the scientific basis of the information on offer and the tension between the marketplace and the professions as the medium through which genetic information should be conveyed or bought and sold.

An area which also crosses boundaries is that of nutrigenomics. At first sight it seems analogous to pharmacogenomics. Just as the latter facilitates personalized prescribing on the basis of genetic information, so the former, in principle, makes possible personalized dietary advice which can also be extremely important in the healthcare setting, especially in the light of a purported obesity epidemic. However, nutrigenomics is very different from pharmacogenomics in that foodstuffs, with multiple ingredients are far more complex and work on more aspects of the body than do drugs, which are designed to act on specific targets and pathways in the body. Establishing the associations on the basis of which useful advice can be given is therefore more challenging.

GENOMICS AND IDENTITY

The ethical debates about genetics and genomics have raised identity issues in a number of ways. First, there has been considerable controversy about what counts as a disorder. This is of course a live issue even absent the genetic element: mental disorders are a case in point, and testing for predisposition to such disorders is particularly sensitive. In the genetics clinic the question of what counts as a disorder has been a particular issue because of the fact that termination was one potential outcome of prenatal genetic counseling, giving the impression that the birth of babies with disabilities was regarded as an outcome to be avoided. Disability rights groups objected to the identity and worth of people being reduced to a disorder, questioned the very label of disorder, and feared the return of a form of eugenics. Arguments on the other side attempted to draw a distinction between the conditions and the people who experienced them: between discrimination against a disorder and discrimination against persons with disabilities, with varying degrees of success.

For a given individual or community, the results of genomic research can disturb long held beliefs about their ancestry or relatedness, leading to considerable distress and upset in some well-publicized cases, such as that of research on the Havasupai Indians, who gave consent to the use of their samples in research on diabetes, which was a serious problem in their community. However, they subsequently learned that their samples had been used to study other things, and that publications on the results contradicted their history as they saw it and a lawsuit led to the payment of significant compensation [17].

Another identity issue concerns that of the human species as a whole. Although the majority of the ethical writing concerns human genomics, the genomes of other species have been and continue to be sequenced, with results about the number of genes in some of those species that have been surprising to some. Information about both the complexity of the genomes of other species and their degree of relatedness to the human has the potential to undermine some long held beliefs about the identity of the human species in relation to those others.

INTERNATIONAL PERSPECTIVES AND CULTURAL ISSUES

International issues in genomics may take different forms—first there are questions both ethical and practical about international research and the transfer of samples and data across borders—issues of ownership are at stake here in both subnational research projects and national initiatives. Whereas biobanks were initially set up in different countries such as Iceland and Estonia, as mentioned earlier, arguments for linking between different national efforts appeal to the need for large amounts of data to underpin statistical validity.

The second type of international issue, which is related to the first, concerns cultural differences in various parts of the world where not only ethical values but also the cultural meaning of genetic material may be different. In these situations ethical principles may at least need to be revisited, and informed consent may need to be approached in different ways. Public engagement has become increasingly important, partly as a necessary condition of a process of benefit sharing. The Human Heredity and Health in Africa initiative, for example, which includes a number of different genomics projects funded by the Wellcome Trust and the National Institutes of Health in the United States, is addressing issues of public engagement and informed consent in multiple populations in different social contexts, both rural and urban.

In terms of international approaches to the issues in genomics, these include declarations such as the Universal Declaration on the Human Genome and Human Rights [18], which states in Article 1 that "The human genome underlies the fundamental unity of all members of the human family, as well as the recognition of their inherent dignity and diversity. In a symbolic sense, it is the heritage of humanity." While this Declaration lays down universal principles, other international instruments incorporate rules. At the level of the European Union, for example, the proposed new Regulation on Data Protection lays down parameters for transfer of data across the Union and beyond.

CONCLUSION

Despite the close interaction between bioethics and genomics, and the ways in which bioethical principles have been subject to reinterpretation in the light of new and emerging challenges, issues are far from settled. The rapid advances in technology and prospects for implementation offer the promise of considerable development in the relationship between science and bioethical debate in the years to come. Genomics as applied in the quest for precision medicine, for example, is now joined not only by epigenomics but by multi-omics approaches, which promise further complexity to gen-ethics as the science coevolves with ethics.

REFERENCES

[1] U.S. Buck v Bell 274 U.S. 200; 1927.
[2] Knoppers BM, Chadwick R. Human genetic research: emerging trends in ethics. Nat Rev Genet 2005;6:75—9.
[3] Chadwick R, Levitt M, Shickle D, editors. The right to know and the right not to know. Cambridge: Cambridge University Press; 2014.
[4] Elwyn G, Gray J, Clarke A. Shared decision making and non-directiveness in genetic counselling. J Med Genet 2000;37(2):135—8.
[5] British Society for Human Genetics (BSHG) Report on the Genetic Testing of Children Birmingham; 2010.
[6] Green MJ, Botkin JR. Genetic exceptionalism" in medicine: clarifying the differences between genetic and nongenetic tests. Ann Intern Med 2003;138(7):571—5.
[7] Nuffield Council on Bioethics. Genetic screening: ethical issues. London: Nuffield Council on Bioethics; 1993.
[8] Scholl HPN, Sahel JA. Gene therapy arrives at the macula. Lancet 2014;383(9923):1105—7.
[9] Gottweis H, Petersen A, editors. Biobanks: governance in comparative perspective. London: Routledge; 2008.
[10] Lunshof JE, Chadwick R, Vorhaus DB, Church GM. From genetic privacy to open consent. Nat Rev Genet 2008;9:406—11.
[11] Prainsack B, Buyx A. Solidarity: reflections on an emerging concept in bioethics. London: Nuffield Council on Bioethics; 2011.
[12] Nuffield Council on Bioethics. Pharmacogenomics: ethical issues. London: Nuffield Council on Bioethics; 2003.
[13] [US] National Academy of Sciences (NAS). Toward precision medicine: building a knowledge network for biomedical research and a new taxonomy of disease. Washington, DC: NAS; 2011.
[14] European Science Foundation (ESF). Personalised Medicine for the European Citizen—towards more precise medicine for the diagnosis, treatment and prevention of disease. Strasbourg: ESF; 2012.
[15] Daar AS, Singer PA. Pharmacogenomics and geographical ancestry: implications for drug development and global health. Nat Rev Genet 2005;6:241—6.
[16] Human Genome Organisation (HUGO). Ethics Committee. Statement on Benefit-Sharing <www.hugo-international.org> [accessed 08.09.15].
[17] Lunshof JE, Chadwick R. Editorial: genetic and genomic research—changing patterns of accountability. Account Res 2011;18(3):121—31.
[18] United Nations Educational, Scientific and Cultural Organization (UNESCO). Universal Declaration on the Human Genome and Human Rights. <www.unesco.org>; 1997 [accessed 08.09.15].

CHAPTER 5

Health Economic Perspectives of Genomics

Sarah Wordsworth[1], James Buchanan[1] and Adrian Towse[2]
[1]Health Economics Research Centre, Nuffield Department of Population Health, University of Oxford, Headington, Oxford
[2]Office of Health Economics, London, UK

Contents

Genomics and Society
DOI: http://dx.doi.org/10.1016/B978-0-12-420195-8.00005-7

INTRODUCTION

Alongside the excitement and perceived opportunities for health improvement from using genomics in health care, there is also concern that healthcare funders will not be able to afford all the emerging technological advancements, such as whole-genome sequencing (WGS) and uncertainty regarding whether these technologies will actually deliver anticipated improvements in patient health. While there are several perspectives on the impact of genomics on health care, those relating to health economics tend to fall into two related schools of thought: that genomic advancements increase healthcare costs, but save lives, or that they could actually lead to a reduction in costs.

The first school of thought considers genomics as a range of very new advanced technologies which will escalate healthcare costs. For example, identification of the *human epidermal growth factor receptor 2* (*HER2/neu*) gene and the development of trastuzumab (Herceptin) drug therapy have increased the cost of breast cancer treatment, partly as a result of the additional cost of *HER2/neu* expression testing, but mainly because of the cost of the associated drug (trastuzumab) for the 25—30% of women who test *HER2/neu*-positive [1]. The total cost of this genomic advance, with substantial clinical benefit, has been estimated at more than $750 million per year in the United States [2]. This is a clear example of an advance in genomics which has improved patient survival and quality of life, but has resulted in placing significant pressure on limited healthcare budgets.

The second school of thought that genomics is likely to be cost saving in health care relates to the potential of genomic technologies to identify individuals who will experience little to no benefit from an intervention such as a drug. These individuals may be at low risk of getting ill without the intervention, or they may be unlikely to respond to the intervention, or it could be the case that a genomic test provides information which allows clinicians to rule out a disease. An example of a cost-saving application of genomic technologies can be found in colorectal cancer. This type of cancer is closely related to the epidermal growth factor receptor (EGFR) pathway [3], and studies of this pathway have led to the development of targeted drug therapies (anti-EGFR monoclonal antibodies) such as cetuximab (Erbitux) and panitumumab (Vectibix). Although both drugs demonstrated antitumor properties in initial studies, patient response rates were unfortunately poor (~10%) [4,5]. Research efforts were therefore focused on trying to find out why this was the case, with scientists searching for genetic aberrations downstream in the EGFR pathway that may be responsible for resistance to anti-EGFR antibodies. KRAS (Kirsten Ras) forms a vital part of the EGFR mediated pathway and mutations in this gene have now been established as a mechanism for the development of resistance to these drugs [6—8]. KRAS mutational testing in metastatic colorectal cancer is now routinely used to identify patients unlikely to benefit from treatment with expensive anti-EGFR monoclonal antibodies. Reducing the use of interventions in patients who will gain little or no benefit could have important implications for healthcare funders, especially if a treatment is used frequently or is very expensive. Such possibilities are highlighted in a recent editorial in the *Journal of the American Medical Association*, which notes that genomics has the potential to "bend the cost curve" by ensuring that the most effective treatment is used in the most appropriate patients [9].

As yet, however, outside of a small number of examples in breast and colorectal cancer, there is little evidence of the impact of using genomics in routine clinical care, in terms of patient health and healthcare costs. It is therefore unclear what the true benefits of genomics to patients are likely to be across a range of diseases and treatments. Furthermore, the financial implications for healthcare funders of using genomic information on a wide scale are also unclear. Health economists can help to fill this evidence gap and should be at the forefront of discussions concerning whether genomics represents good value for money for healthcare providers.

This chapter introduces the discipline of health economics and explains how it can contribute toward discussions about the "true" costs and consequences of using genomics in health care. The chapter begins by defining what health economics is and describing the concept of economic evaluation, a decision-making tool used to compare the costs and outcomes of different healthcare interventions. The chapter then sets out where health economic approaches can be applied to assess the efficiency of using genomic information to diagnose, treat, and assess the risk of disease. This is followed by a section on the practical challenges faced by health economists in this context, including the assessment of new and fast paced technologies, and the recent development of next-generation sequencing (NGS) and WGS approaches. We then consider the evidence required by healthcare decision-makers in genomics and discuss the methodological challenges that health economist's face when generating this evidence, considering whether commonly used outcome measures such as quality-adjusted life years (QALYs) are relevant. The chapter then explores several policy issues which will influence the nature and pace of the introduction of genomics into health care. These include regulatory challenges and value-based pricing and assessment. We conclude by summarizing the main points raised in the chapter and suggesting future research priorities for health economists working in this area.

WHAT IS HEALTH ECONOMICS

The mother discipline to health economics, economics, is commonly defined as the study of how individuals and societies choose to allocate scarce resources (e.g., land, labor, capital) among competing alternative uses (e.g., the production of different goods and services). Given that resources are limited, it is argued that they should be used in as efficient a manner as possible to maximize the gains to society. Standard economic theory suggests that the best way to achieve this goal is to allow consumers and producers to interact in a free market, using price as a rationing tool to equalize demand and supply. However, this requires several assumptions to hold, including the existence of fully informed consumers, the absence of externalities (consequences of economic activity that are experienced by unrelated third parties [10]), and a willingness to accept the distribution of income (and therefore purchasing power) that results. If these assumptions do not hold in a specific sector of the economy, the free market may not be the most efficient way to allocate scarce resources, and alternative rationing tools are required.

Health economics uses the tools and methods of economics to consider how to allocate scarce healthcare resources in order to achieve the objectives of a healthcare system. Health care is a clear example of a sector of the economy in which many of the assumptions that underlie the free market approach do not hold. For instance, it is difficult for individuals to predict when they are going to require health care. In a free market, firms will respond to this uncertainty by offering healthcare insurance. However, healthcare insurance markets suffer from two fundamental problems: moral hazard (individuals are likely to change their behavior if they possess insurance) and adverse selection (individuals know more about their health risk status—such as preexisting conditions—than those offering healthcare insurance). Both effects can have a negative impact on population health. For example, moral hazard may encourage individuals to take more risks with their health, such as adopting unhealthy lifestyles, if individuals are not totally financially responsible for the full healthcare costs associated with being overweight (or smoking or drinking too much alcohol). Adverse selection may force poorer individuals (often the sickest members of society) to opt out of healthcare insurance systems altogether if premiums increase. Healthcare markets also suffer from the presence of externalities. A classic example is vaccination programs: these benefit both the participating individual and also the rest of society. A third problem is that physicians act on behalf of patients to demand and provide health care, which can lead to a conflict of interests and to physicians potentially supplying more health care than is actually necessary. Finally, a system of professional licensure must be in place as mistakes in health care can have serious consequences. However, this gives clinicians a degree of market power, which could be exploited.

Given that a free market system can fail in multiple ways in health care, most economists believe that some degree of government intervention is both necessary and inevitable to ensure the efficient allocation of scarce healthcare resources [10]. The exact level of intervention depends to some extent on the aims of a healthcare system. It has been previously suggested that healthcare systems have three basic aspirations: to provide "comprehensive, high quality medical care [...] to all citizens on the basis of professionally judged medical need without financial barriers to access" [11]. However, if healthcare resources are limited, these aims are said to be an "inconsistent triad"—it is not possible to achieve all three simultaneously and stay within a budget [11]. For instance in the UK National Health Service (NHS), a key aim is that health care should be available to all. As a consequence, the NHS sometimes compromises on the range and quality of services offered to patients, using limited funds to provide a "basic" level of health care to as many people as possible. However, this can result in the most clinically effective healthcare interventions not always being provided. Healthcare resource allocation decisions such as these can be made by using waiting lists or on a case-by-case basis, but these approaches could lead to random, implicit, or subjective decisions. An alternative approach is to make these decisions using supporting evidence from systematic and explicit decision tools: economic evaluation is such a decision support tool.

Economic evaluation

Economic evaluation is commonly defined as "the comparative analysis of alternative courses of action in terms of both their costs and consequences" and it can be used to compute and compare the costs and outcomes of different healthcare interventions with a view to identifying those which represent the best value for money [12]. Economic evaluation can take several forms, each with different aims. We focus in this chapter on the three most common forms: cost-effectiveness analysis, cost–utility analysis, and cost–benefit analysis. All three approaches measure healthcare costs in monetary terms, but each differs in the way in which the outcomes of healthcare interventions are measured.

Cost-effectiveness analysis and cost–utility analysis

Cost-effectiveness and cost–utility analyses are often considered together since both measure outcomes in health-related terms. Cost-effectiveness analysis uses natural units, which are often (but not always) disease specific, such as number of cases detected (e.g., cancer) or life years gained. Cost–utility analysis uses a broader measure—the QALY—which combines information on longevity and quality of life and facilitates comparisons between healthcare interventions in different disease areas. Although the two approaches differ in their treatment of outcomes, the term "cost–utility analysis" is rarely used in the literature, with "cost-effectiveness analysis" used generically to reflect the measurement of outcomes using any health-related metric (including QALYs).

In both cases, when an economic evaluation is conducted, the difference in costs and outcomes between a new intervention and its comparator are calculated [12]. If the new intervention is less costly and more effective, it is said to dominate the comparator and is likely to be introduced into clinical practice. If the new intervention is more costly and less effective, the comparator dominates and the new intervention is unlikely to be adopted. When the new intervention is both more costly and more effective than its comparator, the cost and outcome differences are expressed as a ratio (the incremental cost-effectiveness ratio, or ICER). This is then compared to a threshold (the maximum acceptable ICER) representing the maximum amount that society is willing to pay for an additional unit of health outcome. This threshold may vary depending on the patient group affected, the severity of the disease, and other factors that reflect a willingness to give greater priority to some uses of money than others. If the ICER falls below this threshold, the new intervention is said to be cost-effective and is likely to be adopted [12].

Incremental costs and effects can be expressed visually using the cost-effectiveness plane (Figure 5.1) [13]. This plane consists of four quadrants, reflecting different combinations of positive and negative incremental costs and effects. These four quadrants are commonly referenced using compass points, and each represents a different adoption decision. The most common scenario is that the ICER for a new intervention versus its comparator (C) falls in the northeast quadrant, with the adoption decision depending on the maximum acceptable ICER.

Cost—benefit analysis

In cost—benefit analysis, health outcomes are valued in monetary terms. There are two principal approaches to monetizing health outcomes. The human capital

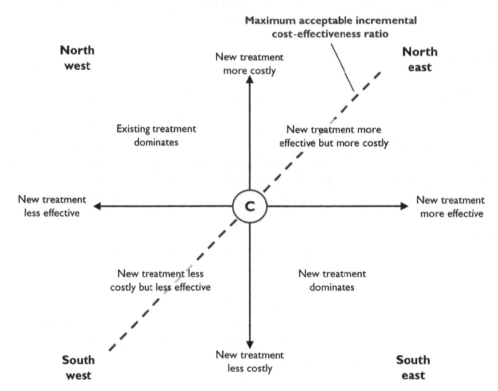

Figure 5.1 The cost-effectiveness plane.

approach considers the value of future earnings of those benefiting from an intervention, thus placing a value on health outcomes that reflects one source of benefits to the patient and the rest of society. The willingness-to-pay approach monetizes health outcomes by considering the preferences of the affected individual(s). These preferences can either be revealed (by observing the behavior of individuals in relation to health-related activities or purchasing habits) or stated (using choice modeling approaches such as discrete choice experiments, DCEs). Whichever approach is used, cost—benefit analysis results are usually expressed as either a ratio of benefits to costs or a simple sum representing the net benefit of one intervention compared to another. The decision rule used when deciding which intervention to adopt is therefore straightforward: if a new intervention generates a positive net benefit (or a ratio of benefits to costs greater than one), it is worthwhile to adopt this intervention. The use of a common unit (money) to value outcomes means that healthcare interventions in a wide variety of disease areas can be easily compared in terms of their net benefit.

APPLYING ECONOMIC EVALUATION APPROACHES TO GENOMICS

In the previous section, we summarized a range of economic evaluation methods that could be used to evaluate genomic interventions. In this section we explore some of the areas where these methods have been used in genetics and genomics to date, notably the diagnosis and treatment of disease and in the risk assessment of developing a disease/condition.

Genomic diagnostics

One of the main areas where economic evaluation is being used in genomics is in diagnostics. Diagnosis is the process of identifying whether a patient has a specific disease, condition, or syndrome at the time of testing. It is usually performed for patients with specific complaints or in whom signs or symptoms have been noted that may indicate a disease. As well as making or excluding a particular diagnosis, diagnostic tests can also be used to monitor diseases and provide prognostic information. A wide range of diagnostic tests are used in health care generally, ranging from tests which take physiological measurements (e.g., blood pressure), to imaging tests such as X-rays, magnetic resonance imaging, and ultrasound scans, which produce images of part or all of the body. More invasive tests such as endoscopic examinations, which include tests such as colonoscopies, are also used for diagnostic purposes and either produce images or can be viewed directly through lenses. Laboratory and pathology tests involve taking samples of body fluids (saliva, blood, etc.) or tissues and subjecting them to some form of analysis. These latter tests are an area of diagnostics where the use of genomic information is becoming especially prolific.

Much of the earlier health economics work in disease diagnosis using genomic information from laboratory diagnostic processes has been in the area of prenatal testing. The existing literature has demonstrated the cost-effectiveness of screening and diagnosis for conditions such as Down syndrome in the general population [14]. In the arena of prenatal diagnosis, particular health economic methodological and ethical concerns include whether the effects of such testing on individuals other than the patient are included, how termination of pregnancy is included in the models, and how screening may reassure or cause anxiety in patients depending on their results. The use of QALYs in this context may not be relevant, appropriate, or even useful.

The primary motivation for most screening tests is to detect a condition at a stage when treatment is more effective than waiting for the appearance of signs and symptoms. Screening can take place at the population level or be limited to patients in known risk categories. General population screening is a public health service in which members of a defined population, who may not be aware that they are at risk of, or already affected by a disease, are offered a test. This test seeks to identify those people for whom further tests or treatments to reduce the risk of the disease, or its complications, are likely to be beneficial. Newborn screening is one area in which health economists have provided evidence to support general population screening. For example, neonatal screening for phenylketonuria and Medium-chain acyl-CoA dehydrogenase deficiency (two inherited metabolic disorders) can improve patient outcomes and is cost-effective, because screening and disease management costs (fairly simple dietary measures) are less than the costs of treating children if they become ill. Many countries now include these two conditions in their newborn screening testing programs [15].

The health outcomes arising from diagnostic tests usually result from the treatment that follows the use of a diagnostic technology, rather than from the diagnostic test itself. These outcomes include benefits and also potential harms from using the diagnostic technology. However, there are situations where no treatment is available following a diagnosis, for example, using genomic tests to diagnose learning disability, where the "value" might be that a family finally having a diagnosis for their child, which could facilitate access to educational and other support, and also avoid a series of nongenomic clinical tests [16].

To date, most of the well-known genomic diagnostic tests have been in oncology, which is the area with the most development activity and clinically available applications to date. The prominence of cancer diagnostics reflects the importance of genomic variation in the genesis of cancer and the role that specific variations play as therapeutic targets. However, genomic diagnosis can be helpful in other disease areas. Below we present an example of an economic evaluation that was undertaken to evaluate the cost-effectiveness of using genomics to diagnose an inherited heart disease: hypertrophic cardiomyopathy (HCM) [17].

Economic evaluation of using genomic diagnostics in heart disease

HCM is the most common monogenic cardiac disorder [18] and the most frequent cause of sudden cardiovascular death (SCD) in young people and trained competitive athletes [19]. HCM is defined by unexplained hypertrophy of the ventricular myocardium (enlarged heart) and prevalence among adults is around 0.2% (1:500) [20]. In recent studies, the annual SCD rate from HCM varies between 0.1% and 1.7% [21] with a subset of patients having an estimated annual SCD rate between 4% and 5% [22]. HCM is caused by mutations in at least 10 genes and is transmitted in an autosomal dominant fashion, with the child of an affected parent having a 50% chance of inheriting the disease-causing allele [18,20,23,24]. Most individuals with HCM are asymptomatic and SCD can be the first manifestation of disease [21,24−26].

Some patients with HCM can present with left ventricular hypertrophy by echocardiography or abnormal electrocardiogram (ECG) during any phase of life [24]. However, not everyone with a disease-causing mutation will manifest HCM and this presents a problem for physicians assessing families with a history of HCM. With the advent of genomic approaches, it has become possible to use DNA testing to help identify asymptomatic individuals with HCM. However, the existence of such tests does not necessarily mean that DNA testing for those at risk of HCM is a worthwhile use of healthcare funding. We therefore undertook an economic evaluation which examined the long-term costs and effects of diagnosing asymptomatic members of families with HCM using both the traditional clinical approach and a molecular genetic approach, with a view to enabling the clinical management of those at high risk of SCD. The study addressed whether DNA testing could be used routinely in the UK NHS to identify genetically at-risk individuals and to discharge those not at risk.

Case study methods

An economic decision model (Figure 5.2) was developed to examine the cost-effectiveness of genetic and clinical testing strategies for identifying individuals at risk of sudden cardiac death due to HCM in the United Kingdom. The model estimated the lifetime resource costs and health outcomes (life years gained) of alternative strategies for assessing first-degree (e.g., parent, sibling, or child) asymptomatic family members (children) of an individual diagnosed with HCM, the proband (first affected family member who is found to have the genetic disorder). We used the children of a diagnosed proband (aged 18, 20, and 22) in our analysis as an example. The clinical diagnostic tests in the model comprised clinical history and physical examination (12-lead ECG and echocardiogram). Genetic diagnostics involved the taking of a clinical history followed by comprehensive screening of the four genes most commonly associated with HCM: MYH7, MYBPC3, TNNT2, and TNNI3 (which cover around 80% of individuals).

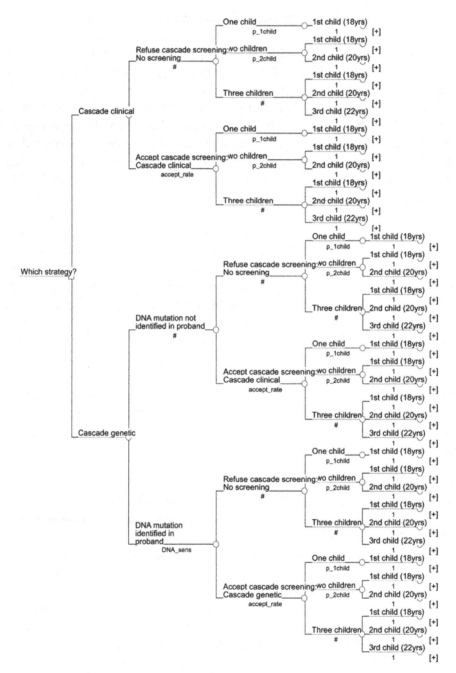

Figure 5.2 Economic decision model for diagnosing hypertrophic cardiomyopathy. *Published on behalf of the European Society of Cardiology. All rights reserved.* © *The Author 2010. For permissions email journals.permissions@oxfordjournals.org.*

The decision tree in Figure 5.2 simulates the impact of the clinical and genetic diagnostic strategies upon the first-degree asymptomatic family members. Cascade testing depends on the probability of identifying an HCM mutation in the proband (DNA_sens) and the acceptance rate for cascade testing by the family members (accept_rate). The successful identification of individuals with HCM is determined by the sensitivity of the testing approach. The lifetime treatment/surveillance resource costs and life years gained for each child are simulated in the natural history Markov model.

Case study results

The expected costs and life years for patients undergoing the different diagnosis strategies (at baseline) are presented in Table 5.1.

Although the cascade clinical strategy is the least expensive option (€14,872), this strategy also provides the fewest life years saved (43.61). Overall, the cascade genetic strategies generate higher costs and effects compared with the clinical strategies due to the higher rate of implantable cardioverter defibrillators being used resulting from the greater number of high-risk patients detected. However, the incremental (difference between strategies) cost per life year saved of the genetic cascade strategies was well below the threshold of £30,000 (€41,000) per life year saved used by the National Institute for Health and Clinical Excellence (NICE) in England and Wales [27]. Above this figure, NICE needs to identify a strong case for supporting the technology as an effective use of NHS resources [28]. In this context, the findings from the HCM study suggested that the genetic strategies were highly cost-effective options. More specifically, the incremental cost per life year saved for the cascade genetic compared with the cascade clinical approach was €14,397. Under the scenario of a single child undergoing screening (rather than three children), the cascade genetic strategies were still more cost-effective when compared with the clinical strategies (€16,185 and €21,561 per life years saved) given the defined cost-effectiveness threshold of £30,000.

Table 5.1 Expected years of life and costs by strategy

Diagnostic strategy	Discounted lifetime cost per patient (€)	Discounted life expectancy per patient (LY)	Incremental cost per life year (€/LY)
Cascade clinical	€14,872	43.61	
Cascade genetic	€19,459	43.93	€14,397
Cascade clinical (5-year surveillance)	€21,208	43.91	Dominated by genetic strategies
Cascade genetic (5-year surveillance)	€21,803	44.04	€21,561

Overall, these results show that genetic diagnostic strategies are more cost-effective than clinical tests alone. An important feature highlighted by this study is the potential for more accurate molecular genetic testing to discharge individuals who do not have a disease-causing mutation that may have had an ambiguous, or even false-positive clinical result requiring ongoing clinical surveillance. The genetic test had a higher rate of both sensitivity and specificity than clinical testing, which made the genetic approach more accurate.

Treatment

A second area where economic evaluation is being used in genomics is in treatment. There are four main areas where genomics could be useful in the context of treatment.

Reducing or avoiding treatment adverse events

Depending on the severity of side effects, genomic testing can have several benefits. For example, it could allow a treatment to receive marketing authorization by improving the risk—benefit ratio associated with the treatment. This could increase treatment adoption rates, in cases where a treatment is licensed but not widely used because of its perceived unfavorable average risk—benefit balance when considered across a wide patient population. For example, HLA-B*5701 is an allele associated with hypersensitivity to abacavir, which is part of a multidrug regimen for HIV-1. Screening for HLA-B*5701 reduces the risk of an adverse reaction. As a result, identification of the HLA-B*5701 marker has increased prescribing of abacavir, which is now recommended for HLA-B*5701-negative patients in European and US guidelines. Besides health benefits to patients, reducing adverse events could also lessen the medical and nonmedical costs of managing these events, such as hospital admissions.

Reducing or avoiding time delays in selecting the most appropriate treatment

This could provide a notable health gain for patients. For instance, if a disease is at an advanced stage (e.g., metastatic cancer), identifying nonresponders and switching them to an alternative dosage, treatment, or care at the right time may have a significant impact on patients' length and/or quality of life. A further benefit to patients is that results from a genomic diagnostic could be generated more quickly than standard testing, especially if high-throughput sequencing is used. This could avoid or reduce inconvenience for patients who no longer have to experience a lengthy diagnostic process, or who no longer have to try different therapies to identify the most suitable. For health services, genomic testing could lead to cost savings if nonresponders could be predicted in advance, because this avoids or reduces the cost of treating these patients (including drug costs). An example here is the BCR-ABL test which measures minimal residual disease in chronic myelogenous leukemia (CML). The test can identify CML patients who are not responding to current treatment, generating health gains for patients by preventing the disease from progressing to blast crisis and death. It can also lead to cost savings to the healthcare system and nonhealth benefits to patients (experience of care) as it enables first-line treatment to stop when it is no longer effective.

Enabling a treatment to become available which is effective only in a small fraction of the population

There are several ways in which a treatment which is only effective in a very small number of people could be made available due to using genomic information. First, treatments could be "rescued" when they might otherwise have been refused a license or withdrawn because of the limited treatment effect across the overall population (favorable clinical effects in a subgroup go unnoticed when the majority of the group are nonresponders). An example here is the EGFR mutation test, which predicts response to tyrosine kinase inhibitors (TKIs) treatments, such as gefitinib, for nonsmall cell lung cancer (NSCLC). Gefitinib was initially approved based on positive Phase II trial results but subsequently withdrawn when Phase III studies failed to show a survival benefit. After the identification of the EGFR mutation and its association with a positive response rate to TKIs, gefitinib received regulatory approval in the European Union and other markets in combination with EGFR mutation testing. This has enabled responder patients to benefit from access to an intervention which is effective for them but not for most patients.

More generally, targeting can improve the chances of a treatment meeting reimbursement criteria (if a diagnostic test targeting responders improves cost-effectiveness) or being included in clinical guidelines (if the evidence provided is deemed sufficient to change treatment protocols). An example here is the use of the HER2/neu test in breast cancer to predict patients' response to trastuzumab. NICE recommends trastuzumab for advanced breast cancer and for adjuvant treatment of early-stage HER2/neu-positive breast cancer in England and Wales, as cost per QALY estimates for the test—treatment combination were found to be below the standard threshold. The presence of a test to select responders therefore means that treatments are more likely to deliver good value for money from a payer perspective (because of the higher health gains per patient due to an increase in drug effectiveness from using the test) and also improves the health outcomes of patients identified as eligible for treatment.

A further potential gain is that treatment research and development (R&D) processes can be accelerated when a biomarker (or another genetic characteristic allowing for patient stratification) is discovered in the early stages of treatment development. One potential benefit of such improved patient stratification in oncology clinical trials could be that clinical development attrition rates are reduced, in particular attrition rates from Phase II to Phase III [29]. An example here is the

ALK fluorescence in situ hybridization (FISH) test, used in combination with crizotinib, a treatment licensed in the US targeting a small subset—between 3% and 8%—of NSCLC patients with an ALK-positive molecular abnormality. Research on crizotinib began before the discovery that a fusion of two genes (ALK and EML4) could cause some lung cancers. However, the subsequent development of the ALK FISH test accelerated the development process and increased the likelihood of a successful market launch for crizotinib, therefore delivering health benefits to a subset of patients.

Reducing uncertainty about the value of potential new treatments and likely effectiveness of available treatments

In the first case, a test could provide better information on the prevalence of a particular untreatable condition, which helps to direct R&D toward that unmet need. In the case of available treatments, there is often uncertainty surrounding expected health effects and costs. This relates to knowledge of the disease (diagnosis, prognosis, casual explanation) and to the clinical and cost-effectiveness of treatments. This uncertainty influences the risk of poor value for money for payers, that is, the likelihood that treatments are not cost-effective. For example, criteria currently used to predict risk of recurrence in breast cancer patients following surgery are not very accurate. As a result, many patients are either over- or undertreated with adjuvant chemotherapy. Oncotype DX and MammaPrint are multigene assays which can identify patients with a high risk of recurrence, guide intervention decisions, and reduce the risk of dispensing unnecessary chemotherapy. This can lead to cost savings to the healthcare system and improvements in the care experience of patients, who can avoid therapies with limited benefits and potential side effects. These assays also provide patients and clinicians with information about the benefits and risks of treatment which are of value independently of any changes in treatment that result from the use of the information.

Another type of uncertainty is the perceived value of information (VOI) to patients of reduced uncertainty as to their medical condition, independent of the expected health outcomes [30]. The literature has defined this as the "value of knowing" or, as Ash et al. put it, "knowing for the sake of knowing," even if the condition under examination is untreatable [31,32]. Patients may value information from a test regardless of the impact on their treatment strategy for the following reasons:

(i) First, genomic information could decrease the level of "ambiguity," which is a situation where probabilities of certain outcomes are highly uncertain [33]. There is evidence that people dislike ambiguous situations and prefer to receive information regardless of its nature ("bad" and "good" news) [32,34]. In some cases, test results could produce disutility as patients are not always indifferent to the outcome of the test [31]. In the case of degenerative diseases such as Alzheimer's, which has limited treatment options and a high level of emotional burden, fear of living with the possibility of developing the disease can be very distressing. If the disutility associated with "bad news" is higher than the utility gain of "good news," then testing may not be of benefit to the patient. In such situations, the choice as to whether or not to test should ultimately be left to the individual patient as part of an informed decision-making process with their clinicians.

(ii) Information could also provide reassurance to patients (value of "rule-out"), particularly to those already identified as "at-risk." A person with family history of a certain genetic disease may value a predictive test which provides proof of the absence (or lower chance) of contracting or already having the disease in the future [34]. Genetic testing for the inherited heart condition HCM, described earlier in this chapter, is an example of information from a genomic test leading to some individuals being ruled out of having the disease, avoiding the need for long-term clinical surveillance [17].

(iii) Individuals might want to undertake a test with no treatment options because the results will impact on their family/life planning, including choices related to personal finances, work, and leisure time [35]. In these cases, diagnostics can impact on patient utility beyond QALY-oriented health outcomes, and these effects are not necessarily reflected in payers' decisions, which are usually based only on health-related quality of life measures.

Disease prevention

A third area where economic evaluation is being used in genomics is in disease prevention. One possibility is that genomic testing could increase patient willingness to undertake preventative measures, including changes in behavior. For example, a test estimating the risk of developing Type 2 diabetes over the next 5 years may encourage patients to follow a healthy lifestyle. This could improve health outcomes for these patients, who will be less likely to develop the disease, and yield cost savings to the health system if net costs are reduced in the long run (when the cost of treating patients who develop diabetes exceeds the cost of testing and preventative interventions).

Type 2 diabetes is a good example of a genetically complex chronic disease, in which the use of genetic information could alter individual disease profiles [36]. Over 20 million people suffer from diabetes in the United States alone, and more than a million new cases are diagnosed each year [37]. There is strong evidence that lifestyle changes can delay progression, yet uptake of recommended behavior changes, such as weight loss for obesity, are low, as shown by increasing rates of obesity globally. To date, over 40 common DNA variants, or single nucleotide polymorphisms, have been found to be associated with increased risk for Type 2 diabetes [38,39]. One factor limiting the broader application of genetic testing is that the magnitude of increased risk conferred by each variant is relatively small. Thus, genetic risk testing for these variants may not yet offer significant predictive value for individual patients sufficient to alter providers' screening and treatment recommendations.

Genetic risk information could, however, provide clinical utility (net benefit in improving health outcomes) in other ways. This could be demonstrated through increasing patient activation or positively influencing patient attitudes, beliefs, and health-related behaviors. The relatively scarce research into clinical utility of genetic testing has produced mixed results [40,41]; however, some studies have found evidence that providing results of genetic tests for chronic diseases increases patients' preventive behavior [42,43]. A recent systematic review of the impact of genetic risk information on chronic adult diseases found some psychological benefits of including genetic information in treatment of chronic diseases, but concluded that many gaps in knowledge must be addressed before genetic science can be effectively translated into clinical practice [40].

CHALLENGES FOR ECONOMIC EVALUATION

The past decade has seen a remarkable evolution in genetic and genomic testing. This began with a move away from standard single gene tests (such as Sanger sequencing) to targeted panels, which explore sets of genes. These panels greatly increased the number of genes that could be tested compared to standard single gene tests, but for some time they were still constrained by the extent of our current knowledge. However, the advent of NGS technologies (which include targeted panels, whole exome sequencing (WES) and WGS), has now enabled parallel testing of multiple genes. In contrast with targeted panels, WGS can provide a mechanism to interrogate a patient's entire genome in a hypothesis-free manner, with the increased probability of a diagnostic outcome. It is widely believed that WGS in particular could be a revolutionary technology for clinical diagnostics, providing a fast and comprehensive platform applicable to many areas of medicine, such as rare diseases, cancer, and infectious disease, and facilitating improved patient treatment stratification.

However, NGS technologies are considered to be expensive and a lack of evidence on their effectiveness and cost-effectiveness is slowing their adoption into mainstream clinical care. Some of the challenges for health economists evaluating these technologies include questions surrounding how they should be valued, especially where one test can be revisited and used for different clinical purposes and is no longer tied to a one-off purpose. The costs of NGS technologies are also unclear. In some ways NGS technologies are a disruptive innovation because of the significant decrease in the cost of analyzing many gene sequences in parallel, compared to the use of a range of "traditional" tests. However, of the three drivers of the cost of NGS (the preanalytics and assay, the bioinformatics pipeline, and the evidence base for clinical utility), only the costs of the preanalytics and assay are currently declining, with the costs of the bioinformatics pipeline (which provides much of the "value" of NGS) in particular not reducing at all [44]. This raises interesting questions concerning how health systems should assess and value platform testing technologies. Furthermore, NGS technologies could replace a suite of alternative tests, but some of these tests may not have actually been entirely necessary in the first place.

An additional consideration is that while NGS technologies are likely to prove cost-effective on a simple cost per mutation basis (although test costs are generally higher, NGS technologies can test for more genes and are more effective/accurate); it is harder to demonstrate that these technologies are cost-effective when treatment information is added to analyses. For example, if NGS results suggest that a patient should receive an expensive targeted drug therapy, an economic evaluation might indicate that the NGS approach is not cost-effective. In this instance, should NGS technologies effectively "get the blame" for the high costs of some targeted therapies?

Finally, WGS can reveal information on genetic variants that have known or unknown effects on an individual's risk for other, often unrelated, diseases. These incidental findings (IFs) may have unanticipated costs and consequences which should feed into economic evaluations of these technologies and are therefore a new and growing topic of concern.

The remainder of this section discusses these challenges in detail. We begin by considering whether existing economic evaluation methods are suitable for genomics.

Methods—are existing economic evaluation methods suitable for genomics?

Health technology assessment (HTA) agencies currently conduct economic evaluations of genomic technologies using the same guidelines as used for nongenomic technologies, emphasizing a health service perspective and the use of QALYs. However, health economists have recently started to question whether this approach is entirely appropriate in genomics [45,46]. For example, QALYs do not, by definition, capture information on nonhealth outcomes (e.g., the VOI from genomic test results to patients). Other methodological challenges include the choice of analytical perspective, the heterogeneity of cost data, and how to incorporate complex effectiveness data into economic analyses. How these issues are tackled is important because the use of different methodologies will produce different results which may change decisions as to whether particular genomic technologies are made available in routine clinical practice.

So are current economic evaluation methods sufficient in genomics? To answer this question, we need to identify the methodological challenges that arise when conducting economic evaluations of genomic technologies, and consider to what extent these challenges can be managed within current economic evaluation guidelines. These challenges can be divided into several broad categories, including the analytical approach, estimation of costs and resource use, measurement of outcomes, and measurement of effectiveness. Issues surrounding the measurement of test effectiveness (how does a test result in health gain or other benefits?) have been considered in earlier sections, so we focus here on the remaining challenges. We consider each in turn highlighting where current methods may not be fit for purpose and suggesting ways in which we might be able to overcome methodological issues.

Analytical approach

Prior to undertaking an economic evaluation, health economists must make several decisions concerning the appropriate analytical approach. Four challenges arise when making these decisions in the context of a genomic intervention.

Perspective

Most economic evaluations of nongenomic interventions consider the direct impact of an intervention on the health service within which it is used, as recommended by most HTA agencies. In contrast, a societal perspective considers indirect impacts such as the financial costs incurred by families as a result of an intervention. Almost all HTA agencies do, however, permit the submission of additional analyses to accommodate a wider societal perspective (provided that these additional results are presented in a disaggregated manner and combined with the healthcare perspective analysis results). This wider perspective may be beneficial in genomics as testing can affect both healthcare choices and broader life decisions (e.g., regarding family planning or schooling). However, at present, only 27% of all genomic economic evaluations adopt this perspective [47,48].

More generally, health economists should carefully consider how the genomic intervention that is being evaluated operationalizes the personalization of medical treatment as this can impact on the choice of evaluative framework. Rogowski et al. suggest that if physiological biomarkers are being used to stratify healthcare interventions to patient subpopulations ("personalization by physiology"), conventional (extra-welfarist) evaluative frameworks will generally be appropriate. However, if individuals' preferences for elements of health and related well-being are being used to decide the health care they get ("personalization by preferences"), then evaluative frameworks may need to be adapted [49].

Time frame of analysis

The information provided by genomic interventions can have long-term implications for patients, for example, identifying the presence of mutations linked to Lynch syndrome in asymptomatic patients [50]. Consequently, studies focusing on short-term costs and consequences risk underestimating the cumulative costs and effects of genomic interventions realized over the entire time frame of the technologies

[46,48,51−54]. The time horizon needs to reflect all of the benefits and costs which may occur in a patient's lifetime rather than just the stage at which they might undergo diagnostic testing with a genomic technology.

Timing of analysis

Genomic testing practice changes continually, and the costs, effectiveness, and outcomes of new genomic technologies are particularly poorly defined [48,52,55−59]. For example, the cost of sequencing (using older approaches) has fallen dramatically in just a decade [60]. Patient categorization can also vary over time, as additional subgroups are identified with implications for individualized treatment (e.g., Type 2 diabetes risk assessment using genotyping) [61−64]. Uncertainty concerning the appropriate timing of an economic evaluation does not necessarily mean that an evaluation should not be undertaken. Indeed, evidence from economic evaluations is valued by decision-makers at all stages in the translational process, particularly in the early stages of product development when decisions about prioritization of research funding are being made [52]. However, allowing for the impact of a learning curve (which typically occurs when a test first starts to be used) on both unit costs and effects could be useful. This would mean modeling the potential impact of a learning curve and then validating (or otherwise) those estimates by collecting resource use and cost data toward the end of a study period, once any reductions in price have occurred and earlier higher rates of testing errors (which increase costs) have reduced.

Given the difficulties associated with timing a genomic economic evaluation in the face of near constant changes in testing practice, health economists may need to be prepared to work in a more iterative manner in genomics, quickly incorporating new evidence into analyses and rerunning simulations regularly [48,55,56,59,61,63,64]. However, ensuring that health economic evidence is available to decision-makers at all stages of the translational process is only a good use of limited health economist time if HTA agency processes allow technology adoption decisions to be reviewed far more regularly than is currently the case. For example, in England and Wales, NICE commonly considers published guidance for review 3 years after publication. This practice could be at odds with the pace of technological development in genomics.

Analytical context

Genomic tests often have multiple applications in varied contexts and it is rarely possible to synthesize results from economic evaluations undertaken in these differing settings [54,65]. Furthermore, different applications of a specific technology may not be similarly cost-effective (e.g., Oncotype DX cost-effectiveness differs between breast and colon cancer [66,67]). This means that study design must be carefully considered, with particular attention paid to three elements.

First, standard genomic testing practice rarely exists; it varies considerably both between and within different clinical specialties [68]. It is therefore important to carefully choose the comparators in a genomic economic evaluation. It may be appropriate to evaluate genomic testing both against, and in addition to, nongenetic testing. For example, expensive yet specific genomic tests have been shown to be more cost-effective in colorectal cancer if combined with preliminary screening using cheaper less specific tests [56,65,69–72].

Second, study designs should take into account the therapeutic decisions that follow from the use of a genomic intervention [57]. For example, genetic testing for Duchenne muscular dystrophy can detect mutations across 79 genetic regions, with 30 different therapies required, each targeted at specific faulty regions. In this scenario, it is insufficient to conduct an economic evaluation of just one test specification: costs and effects will vary depending on mutation prevalence and treatment efficacy [73]. Multiple potential test designs exist, and analysts must carefully consider which comparators are appropriate prior to undertaking an economic evaluation.

Finally, genomic testing outcomes are often influenced by multiple genes. Each genetic variant can influence multiple outcomes, and the influence of a variant on a given outcome can vary across individuals [56]. This makes it challenging to structure a model-based economic evaluation which evaluates a test for multiple biomarkers across several contexts. This may necessitate a move away from commonly used Markov state transition cohort model approaches to microsimulation or discrete-event simulation approaches [49]. Given the computational and data requirements of such approaches, such a decision should be considered at an early point in the study design process.

Next steps

Some of the issues related to the appropriate analytical approach in genomics can be addressed by undertaking realistic simulations and comprehensive parameter and structural sensitivity analyses, such as considering alternative discount rates and time frames [46,48]. In a genomic context, it may be necessary to widen the sensitivity analyses to consider multiple comparators and different combinations of testing and/or treatment. These considerations should not necessarily present a problem to current guidelines, although the end result may be that economic evaluations of genomic technologies could be larger undertakings than those conducted for nongenomic technologies.

Estimating costs and resource use

All economic evaluations require health economists to collect information on the costs of an intervention. In genomics, there is uncertainty concerning which costs should be collected and when, and how costs vary between laboratories and countries.

Which costs should be included?

The wide range of potential costs in genomic economic evaluations requires health economists to carefully plan data collection to ensure that all relevant costs are included. Potentially relevant costs include those associated with patient recruitment (e.g., publicity and patient education) [48,51], sample collection, laboratory testing [48], data analysis (e.g., informatics solutions, data libraries) [55,74], and the reporting of test results to patients [48]. The costs associated with actions taken on the basis of test results are also relevant, including the costs of treatment and follow-up testing (or treatment and tests avoided), management of adverse drug reactions, and genetic counseling [48,51,75–78]. These costs can be significant. A study comparing microarray testing with karyotyping in idiopathic learning disability found that including all the costs of follow-up testing after karyotyping changed the study conclusions in favor of microarray testing (the more expensive test on a standalone basis) [16]. The time and cost burden of monitoring disease progression and drug response may also be important: the most cost-effective genomic tests are likely to be those for conditions where monitoring is difficult and expensive [78].

Depending on the analytical perspective chosen, the indirect costs accruing to patients may also require estimation (e.g., time spent seeking treatment, time lost from work due to poor health). For example, if genomic technologies can identify patients who will not benefit from treatment, these patients may be spared the costs associated with inappropriate treatment, including work absenteeism. However, few studies have considered these costs: a French study evaluating the cost-effectiveness of Oncotype DX in breast cancer is one exception [79].

Finally, it may be appropriate to account for the wider infrastructure costs associated with genomic technologies (e.g., staff training costs). However, this can be difficult to achieve in practice as genomic platform diagnostics often have multiple applications: a piece of equipment may initially provide one specific test, but ultimately develops into a "one stop shop" for genomic testing in a variety of clinical areas [55,59]. In this scenario, the appropriate allocation of infrastructure costs is not clear: initial applications of a genomic technology may be burdened by a disproportionate share of overall infrastructure costs, while later applications benefit from economies of scale realized through more intensive use. These distortions may bias economic evaluation results.

How much do genomic interventions cost?

National guidelines and agreed reimbursement rates (e.g., NHS tariffs or Medicare payment rates) for genomic tests rarely exist, hence costs can vary considerably between laboratories [44,46,55,77,80,81]. Cost differences between tests developed by accredited laboratories and commercially marketed test kits can be particularly acute [46,75]. Costs also vary between countries due to local price differences, variations in testing practice, and differences in patient contributions to test costs [82]. Estimating test costs will become increasingly problematic as more tests consider multiple genetic changes in a single assay.

When should cost data be collected?

Frequency of data collection depends on the type of genomic intervention being evaluated. For some diseases (e.g., chronic lymphocytic leukemia) multiple genomic tests may be required to identify changes in tumor genomes which are acquired over time [83]. In addition, genomic tests are often packaged with informatics which are regularly updated to incorporate new research. This means that results may need to be retrieved and reanalyzed multiple times, incurring additional costs [51,84]. Furthermore, as WGS becomes more widespread, it may become cheaper and easier to sequence a patient's entire genome multiple times, rather than retrieving and reanalyzing data for new indications, presenting an additional cost complication.

Next steps

Health economists need to be aware of the issues surrounding costing in this context, including how to apportion the infrastructure and training costs associated with platform diagnostics with multiple applications. HTA agencies should consider updating guidelines to make specific recommendations regarding such costs so that evaluations of genomic technologies are undertaken in a consistent manner. Cost differences between laboratories can be handled by comprehensive sensitivity analysis at present, but the development of national tariffs, which are available in most nongenomic clinical areas for tests, would be valuable in the longer term. Finally, guidelines should specify that economic evaluations of genomic technologies need to reflect all future uses of the technology by a patient to capture all costs associated with repeat testing.

Measurement of test outcomes

Several methodological issues arise when measuring the outcomes of a genomic intervention, relating to the type of outcome measure used, whether to include information on personal utility, the validity of the cost—benefit analysis approach, and the importance of individual outcomes.

Disease-specific and preference-based outcome measures

Disease-specific outcome measures are occasionally informative in genomics [75], but in general they limit comparability and do not capture all relevant dimensions, in particular nonhealth outcomes. This is a particular problem for genomic interventions which provide only diagnostic information, which can reduce anxiety and help patients to make future plans. Generic preference-based measures such as the EQ-5D improve comparability by collecting data across a broad range of health-related quality of life domains. However, QALYs measured using such metrics do not capture nonhealth outcomes (e.g., the value of possessing genomic information) [48,51,53,55,56,64,75,76,85−89], or family spillover effects [87], and thus cannot reflect all possible health states for genomic interventions [70]. As some genomic interventions do not improve health or extend life (e.g., microarray testing in learning disability), preference-based outcome measures may not pick up differences that matter, resulting in high ICERs which suggest that these tests are poor value [90].

Personal utility

Genomic interventions do not always provide clinical utility but may impact on patient well-being via nonclinical routes [46,64,74,90−92]. Such effects can be captured in the term personal utility (defined as "benefits or harms that are manifested primarily outside medical contexts") [89,91]. Factors that can have a positive effect on personal utility include the acquisition of prognostic or diagnostic information that can enhance individuals' sense of control, improving personal accountability for health-related choices, informing a sense of self-identity and autonomy, and potentially improving uptake and adherence [51,59,91,93]. Genomic information also provides reassurance and relieves anxiety [52,65,94,95], allowing patients and relatives to make lifestyle modifications (e.g., APOE testing in Alzheimer's disease) [52,65,90,96], and reducing the frequency of diagnostic odysseys (e.g., idiopathic learning disability [16]). Finally, genomic interventions can provide healthcare process-related benefits such as improvements in waiting time [92].

Genomic interventions can also have a negative impact on personal utility. Anxiety may be increased if a test suggests treatment nonresponse [56,97] leaving the patient without the prospect of amelioration, or if IFs of unknown significance are reported [51,85,90,91]. Genomic test results can also disrupt family dynamics and complicate reproductive decisions [51,85,91]. They can lead to fear of discrimination and stigmatization [54,87] if an adverse result is found, or, conversely, lead to complacency, encouraging unhealthy behaviors [87], if a test result appears favorable.

Unfortunately, the metrics for measuring personal utility are not well established [57,65,96]. Work is currently under way in England and Wales to consider how the current definition of value used by NICE could be expanded to consider personal utility and other factors [59], but these efforts have not been replicated in other countries. Some studies have developed outcome measures based on particular concepts of personal utility (e.g., empowerment [90]), but these approaches have limitations [63]. The capabilities approach—which suggests that well-being should be measured by what people have the potential to do (their capabilities, including their ability to make choices), not what they actually do (their functioning in daily life)—may also be useful for capturing information on personal utility, but this has not, to date, been applied widely in this context [90].

Cost—benefit analysis

The area where there is the most disagreement among health economists in the genomics area is how to measure the health outcomes associated with genomic technologies, with many analysts now advocating the more widespread use of cost—benefit analysis in order to incorporate information on personal utility into economic evaluations [48,87]. Given the challenges associated with frequently used outcome measures and the likely importance of information on personal utility, cost—benefit analysis might have a greater role to play in genomics. Several of the approaches used to value outcomes in monetary terms in cost—benefit analysis may be well equipped to capture the nonhealth outcomes of genomic testing [48,51,54,55,59,64,75,85,87,90]. Stated preference approaches such as DCEs may be particularly appropriate. DCEs present participants with choices between hypothetical genomic testing scenarios, with their responses revealing their preferences for different test attributes. These valuations can feed directly into cost—benefit analysis, providing information on whether patients value genomic testing [59]. DCEs have, however, rarely been used in genomics, or to inform cost-benefit analyses, possibly because people struggle to make hypothetical choices in unfamiliar contexts [48,87,96].

An alternative approach which potentially avoids some of the problems associated with DCEs is best—worst scaling (BWS). In BWS, participants are presented with a series of scenarios which each describe the attributes associated with one test, from which they are asked to select their most and least preferred attribute. This approach places a lower cognitive burden on patients compared to DCEs, which is important when respondents are making choices in an unfamiliar context such as genomics [98]. However, only one study has applied BWS in a genomic context to date, evaluating physician preferences for personalized medicine [99].

Given these drawbacks, and the fact that cost—benefit analyses are rarely accepted by HTA agencies, analysts may also wish to consider alternative approaches such as cost-consequence or comparative effectiveness analysis [55]. However, these have also been applied infrequently to date and may not be accepted by HTA agencies either.

Individual versus population outcomes

It is important to consider patient-level outcomes when evaluating genomic interventions. Response to treatment informed by genomic testing will vary by patient [56,91,94,100]. Patient subgroups will have different opinions on acceptable levels of test sensitivity and specificity [101]. These individual differences in risk perceptions and attitudes are to be expected given the complexity of genomic test information [56]: one study found that the value placed on genetic testing for colorectal cancer varied significantly according to patient gender, income, and education [56,102]. Consequently, adopting a narrow "population" approach will likely obscure differences in cost-effectiveness across patient subgroups. However, converting the evidence base from individual tests into an understanding of the value of obtaining the same information across patient subgroups is difficult [74]. Prescreening activities such as counseling to identify the likely impact on patients may improve the cost-effectiveness of these interventions and so need to be incorporated into economic evaluations [69,91].

Next steps

As already mentioned, health economists are somewhat constrained in their approach to outcome measurement by the regulatory context within which economic evaluations of genomic technologies are conducted. Cost–benefit analyses are occasionally permitted as supplementary analyses by agencies such as Canadian Agency for Drugs and Technologies in Health (CADTH), and the Pharmaceutical Benefits Advisory Committee (PBAC) in Australia, but these analyses rarely influence HTA decisions. The outcome measures that are currently commonly used in HTA submissions are unlikely to be able to capture all of the outcomes associated with genomic technologies that are important to patients, so agencies should consider building some flexibility into guidance to permit the use of alternative approaches for a carefully defined subset of interventions. Examples of such interventions include those that do not improve health or extend life (i.e., only provide prognostic or diagnostic information). Ultimately, however, cost–benefit analysis and alternative outcome measures are unlikely to be permitted to be used more frequently until the evidence base for their use in genomics is improved. Studies which compare different economic evaluation approaches in genomics will be helpful, as will studies applying DCE and BWS methods in this context.

Other issues

Three further issues relating to service delivery, research prioritization decisions, and IFs do not fit neatly into any of the above categories, and are considered in the following sections.

Service delivery

Performance standards for genomic interventions vary considerably between laboratories; hence the choice of service delivery model is likely to be an important driver of economic evaluation results. For some disorders (e.g., HER2 testing in breast cancer), laboratories with lower testing volumes are more likely to report incorrect findings, suggesting that use of specialist regional testing centers might yield more favorable economic evaluation results [57,59,103,104].

Research prioritization

VOI analysis is now widely used to inform research prioritization decisions in health care. VOI techniques allow researchers to compare the expected costs and benefits of future research, highlighting areas where new evidence is likely to be highly valued by decision-makers. These techniques may be particularly useful in genomics, where evidence of efficacy is commonly weak [54,56,57,76,89,105]. The costs and consequences of collecting additional data to better inform translational research must therefore be carefully considered, particularly if robust methods of data collection such as randomized controlled trials are not feasible [53,106]. However, only one study has applied these techniques in this context to date, noting the high cost associated with improving the evidence base for population-based genetic screening for hemochromatosis in Germany [105].

There are several issues that must be addressed to enable these techniques to be more extensively applied in genomics in the future. A key problem is that the populations that are likely to benefit from these interventions are small, so low cost approaches to evidence generation are required to ensure that the benefits derived from new evidence are greater than the costs incurred when undertaking the research. VOI analysis also incorporates an assumed time horizon for the research prioritization decision. This is challenging to estimate in genomics. Although the pace of technological progress is rapid, inferring a relatively short product life cycle, this contrasts with the fact that when new biomarkers are discovered they are often complements to existing knowledge rather than substitutes, and full validation of these biomarkers will take time, inferring a longer time horizon. Overall, these two effects add a great deal of complexity when selecting an appropriate time horizon [49].

IFs from WGS

WGS is able to reveal information that is unrelated to the original research/clinical question but which may be of clinical value to the patient. Such findings are referred to as IFs. Effectively responding to these IFs could incur significant initial financial costs, but may lead to a reduction in morbidity, mortality, and overall cost if it results in diseases being caught at an early stage. However, there is no current evidence on the additional health benefits produced by the extra information from IFs, or whether analysis plus clinical action based on IFs provides value for money for healthcare funders. The American College of Medical Genetics and Genomics (ACMG) has recommended that clinical laboratories performing NGS analyze and return pathogenic variants for 56 specific genes it considers as medically actionable. Veenstra and colleagues [107] aimed to evaluate the clinical and economic impact of returning these results. They developed a decision-analytic policy model to project the QALYs and lifetime costs associated with returning ACMG-recommended IFs in three hypothetical cohorts of 10,000 patients: patients with hypertrophic or dilated cardiomyopathy; patients with colorectal cancer or polyposis; and generally healthy individuals receiving testing because family members possess genomic risk factors (or family history indicates a specific disease risk). The authors concluded that returning IFs to patients could be cost-effective for certain populations. However, screening of generally healthy individuals is unlikely to be cost-effective based on current data, unless sequencing costs are less than $500 per patient. The UK 100,000 Genomes initiative, led by Genomics England, will be a useful source of actual data to explore this issue in much more detail, for a greater range of diseases. Clearly there are also many ethical issues associated with returning IFs to patients. This is another large component of work in the 100,000 Genomes Project and this work, which we do not discuss here, is more advanced to date than the economics component.

Evidence

Alongside the consideration of challenges in health economics and genomics, it is also important to reflect on how the results from genomic economic evaluations are currently used by decision-makers when deciding which healthcare interventions to introduce into clinical practice. HTA agencies now exist in many countries around the world, including NICE in England and Wales, CADTH, and PBAC in Australia. In most of these organizations, evidence from economic evaluations plays a role in the decision-making process. Given this, strict guidelines for undertaking economic evaluations are set out by each agency, and for almost every assessment these guidelines require the use of cost-effectiveness or cost–utility analysis. One exception is the evaluation of public health programs by NICE: the use of cost–benefit analyses is permitted in this context if it has been judged that standard approaches are not suitable [108].

To date, most genomic technologies have been evaluated by HTA organizations according to the same economic evaluation guidelines as used for nongenomic technologies (e.g., testing for familial hypercholesterolemia [109]). It is unlikely at present that entirely different guidelines will be developed for genomic technologies. In the United Kingdom, the Diagnostics Assessment Programme (DAP) has been created to assess diagnostic technologies (including genomic technologies) within NICE's remit, and this agency has recently completed an appraisal of the use of genomic testing to guide adjuvant chemotherapy decisions in early breast cancer management [110]. However, current assessment methods for this program closely follow those used to evaluate medicines, measuring patient benefits using QALYs [80]. The UK Genetic Testing Network (UKGTN) also acts as a gatekeeper for genetic tests in the NHS, requiring that laboratories who wish to introduce a new genetic test submit a "gene dossier," containing both clinical and nonclinical evidence in support of this test. If a genetic test is replacing another test, then this must include an economic evaluation of the new test compared to the old test. However, UKGTN also states that his should take the form of cost-effectiveness analysis [111].

There are also some genomic tests (such as the use of microarrays for the genetic analysis of children with learning disabilities in the United Kingdom) which gradually emerge into clinical practice after previously being offered on a research basis (instead of formally entering the healthcare system following assessment by HTA agencies) [46]. However, adoption decisions concerning most genomic technologies are made in exactly the same way as decisions for nongenomic technologies.

Obtaining the evidence required by decision-makers is not, however, straightforward. One paper explored nine case studies where diagnostic tests had successfully brought personalized medicine into clinical practice with health and economic impacts for patients, healthcare systems, and manufacturers [112]. It examined the availability of evidence of clinical utility, important not only for clinicians but also for payers and budget holders. It found that that the predominant funders of evidence on the clinical utility of diagnostic tests are drug developers (as part of a codevelopment) and public research bodies. Diagnostic companies play a limited role. Payers funded one study, but this was terminated when ownership changed. The authors concluded that demonstrating diagnostic clinical utility and the development of economic evidence is currently feasible (i) through drug-diagnostic codevelopment and (ii) when the research is sponsored by payers and public bodies. It is less clear whether the diagnostic industry can routinely undertake the work necessary to provide evidence as to the clinical utility and economic value of its products. This is because of limited resources, limited intellectual property protection, and the low prices that diagnostic tests obtain in many health systems. The authors concluded that it would be good public policy to increase the economic incentives to produce evidence of clinical utility: otherwise, opportunities to generate value from personalized medicine—in terms of both cost savings and health gains—may be lost.

Policy

Many healthcare payers do not have in place institutional arrangements to assess the cost-effectiveness of tests when evidence of clinical utility and of relative test performance is available. As we have noted, NICE has established its Diagnostic Assessment Programme to focus on the cost-effectiveness of diagnostics, while "companion diagnostics will be directed, via a Department of Health led scoping process, through the NICE Technology Appraisal route." Garau et al. [92] have argued that this is an optimal institutional structure for all payers to introduce, combined with a willingness to pay for genetic and other diagnostic tests at prices that reflect value. Other health systems may benefit from introducing arrangements of this sort.

We have already noted that diagnostic companies do not have incentives to collect evidence. Addressing this requires consideration of (i) pricing, (ii) intellectual property protection—for example a form of data exclusivity for clinical utility evidence, and (iii) evidentiary standards and the costs of collecting evidence. Drug manufacturers have incentives to collect evidence in some circumstances as noted above, as do payers, although in a competitive insurance market payers will have problems appropriating the benefits of evidence generation for themselves. As Towse et al. [112] argue, public funding of some research will be essential.

Promoting a robust evidence base may require reforms to intellectual property rights. Competition between genetic and other molecular diagnostic tests is already being seen: for example, Oncotype DX now faces several biomarker-based test competitors that raise fascinating and as yet unaddressed questions about market dynamics and efficient competition, as well as about study design to collect good evidence—especially how to conduct indirect comparisons among them. There are also incentive issues around the use of In Vitro Diagnostic (IVD) kits from manufacturers versus "homebrews" (in-house tests (IHTs) in the United Kingdom and laboratory developed tests (LDTs) in the United States). IVDs require marketing authorization from regulatory authorities and can be subject to some evidence requirements. IHTs/LDTs have some quality regulation, but evidentiary requirements are often less onerous. This tends to undermine the incentives for the generation of data needed for companion diagnostics.

The use of economic evaluations of genomic applications in health care therefore gives rise to a number of policy issues, especially around evidence generation. Appropriate incentives for evidence generation make it much more likely that there is evidence to identify cost-effective genomic technologies that provide both benefits to patients and good value to the healthcare system. These could include:

(i) greater willingness on the part of payers to accept prices for genomic technologies that reflect value. This will involve a need for price flexibility for targeted drugs as evidence of their value for different groups of patients emerges with evidence of heterogeneity of treatment effect over time, and the need for value-based rather than cost-based pricing for diagnostics;

(ii) consideration of some form of intellectual property protection (e.g., data exclusivity) for diagnostics, so enabling diagnostic companies to obtain a return on investment if they generate good evidence of the clinical utility and cost-effectiveness of their tests. Such a measure would also ensure that there are incentives to bring higher quality tests to the market;

(iii) realistic expectations around the standards for evidence. This involves the use of coverage with evidence development and real-world evidence collection for both drugs and diagnostics;

(iv) public investment to complement the efforts of payers and manufacturers to collect evidence, recognizing the limitations on the incentives for both to invest in evidence collection on all of the questions that matter.

CONCLUSIONS

In this chapter we have defined health economics as a discipline which studies how individuals and societies choose to allocate scarce health care among competing alternative uses, introduced economic evaluation as a decision-making tool used to compare the costs and outcomes of different healthcare interventions and described how it can be applied to different areas of genomics. We have outlined several challenges in evaluating the economics of genomic technologies, but also recognized that in many respects these issues are not unique to genomics. In the future, it is likely that health economists will expand their research from addressing questions about whether using a genomic technology will be cost-effective in one specific disease context, to evaluating whether technologies such as WGS are cost-effective across multiple disease contexts. This will no doubt involve a shift from testing being considered as part of specialized genetics services to more mainstream clinical areas. To prepare for this change, several developments will be required. First, reliable information on the costs and effects of different genomic technologies, updated on a regular basis, will be essential. To achieve this, more investment will be needed to generate evidence that will inform evaluations of both the clinical utility and cost-effectiveness of genomic technologies. This is likely to require policy changes to increase the incentives for technology developers to finance or undertake studies. Second, health economists need to be prepared to consider a wide range of both health and nonhealth outcomes in genomic economic evaluations. In addition, policy makers need to be prepared to allow a variety of outcomes to be considered, and also better understand where evaluations might be straightforward and where they might be more complex. This all suggests that there are exciting times ahead for the discipline of health economics in the era of genomic medicine.

REFERENCES

[1] Slamon DJ, Leyland-Jones B, Shak S, Fuchs H, Paton V, Bajamonde A, et al. Use of chemotherapy plus a monoclonal antibody against HER2 for metastatic breast cancer that overexpresses HER2. N Engl J Med 2001;344(11):783–92.
[2] Hillner BE, Smith TJ. Do the large benefits justify the large costs of adjuvant breast cancer trastuzumab? J Clin Oncol 2007;25(6):611–13.
[3] Martinelli E, De Palma R, Orditura M, De Vita F, Ciardiello F. Anti-epidermal growth factor receptor monoclonal antibodies in cancer therapy. Clin Exp Immunol 2009;158(1):1–9.
[4] Jonker DJ, O'Callaghan CJ, Karapetis CS, Zalcberg JR, Tu D, Au H-J, et al. Cetuximab for the treatment of colorectal cancer. N Engl J Med 2007;357(20):2040–8.
[5] Amado RG, Wolf M, Peeters M, Van Cutsem E, Siena S, Freeman DJ, et al. Wild-type KRAS is required for panitumumab efficacy in patients with metastatic colorectal cancer. J Clin Oncol 2008;26(10):1626–34.
[6] Tabin CJ, Bradley SM, Bargmann CI, Weinberg RA, Papageorge AG, Scolnick EM, et al. Mechanism of activation of a human oncogene. Nature 1982;300(5888):143–9.

[7] Egan SE, Giddings BW, Brooks MW, Buday L, Sizeland AM, Weinberg RA. Association of Sos Ras exchange protein with Grb2 is implicated in tyrosine kinase signal transduction and transformation. Nature 1993;363(6424):45−51.

[8] Rozakis-Adcock M, Fernley R, Wade J, Pawson T, Bowtell D. The SH2 and SH3 domains of mammalian Grb2 couple the EGF receptor to the Ras activator mSos1. Nature 1993;363 (6424):83−5.

[9] Armstrong K. Can genomics bend the cost curve? JAMA 2012;307(10):1031−2.

[10] Donaldson C, Gerard K, Jan S, Milton C, Wiseman V. Economics of health care financing: the visible hand. 2nd ed. Palgrave Macmillan; 2004.

[11] Weale A. Rationing health care. BMJ 1998;316(7129):410.

[12] Drummond MF, Sculpher MJ, Torrance GW, O'Brien BJ, Stoddart GL. Methods for the economic evaluation of health care programmes. 3rd ed Oxford University Press; 2005.

[13] Gray A, Clarke P, Wolstenholme J, Wordsworth S. Applied methods of cost-effectiveness analysis in healthcare. Oxford University Press; 2010.

[14] Gekas J, Gagne G, Bujold E, Douillard D, Forest JC, Reinharz D, et al. Comparison of different strategies in prenatal screening for Down's syndrome: cost effectiveness analysis of computer simulation. BMJ 2009;338:b138.

[15] Bodamer OA, Hoffmann GF, Lindner M. Expanded newborn screening in Europe 2007. J Inherit Metab Dis 2007;30(4):439−44.

[16] Wordsworth S, Buchanan J, Regan R, Davison V, Smith K, Dyer S, et al. Diagnosing idiopathic learning disability: a cost-effectiveness analysis of microarray technology in the National Health Service of the United Kingdom. Genomic Med 2007;1(1−2):35−45. [Epub 2008/10/17].

[17] Wordsworth S, Leal J, Blair E, Legood R, Thomson K, Seller A, et al. DNA testing for hypertrophic cardiomyopathy: a cost-effectiveness model. Eur Heart J 2010;31(8):926−35.

[18] Maron BJ, McKenna WJ, Danielson GK, Kappenberger LJ, Kuhn HJ, Seidman CE, et al. American college of cardiology/European Society of Cardiology Clinical Expert Consensus Document on Hypertrophic Cardiomyopathy. pp. 1965−91; 2003.

[19] Maron BJ, Shirani J, Poliac LC, Mathenge R, Roberts WC, Mueller FO. Sudden death in young competitive athletes. Clinical, demographic, and pathological profiles. JAMA 1996;276(3): 199−204.

[20] Elliott P, McKenna WJ. Hypertrophic cardiomyopathy. Lancet 2004;363(9424):1881−91.

[21] Elliott PM, Gimeno JR, Thaman R, Shah J, Ward D, Dickie S, et al. Historical trends in reported survival rates in patients with hypertrophic cardiomyopathy. Heart 2006;92(6):785−91.

[22] Elliott PM, Poloniecki J, Dickie S, Sharma S, Monserrat L, Varnava A, et al. Sudden death in hypertrophic cardiomyopathy: identification of high risk patients. J Am Coll Cardiol 2000;36 (7):2212−18.

[23] Ashrafian H, Watkins H. Reviews of translational medicine and genomics in cardiovascular disease: new disease taxonomy and therapeutic implications cardiomyopathies: therapeutics based on molecular phenotype. J Am Coll Cardiol 2007;49(12):1251−64.

[24] Maron BJ. Hypertrophic cardiomyopathy: a systematic review. JAMA 2002;287(10):1308−20.

[25] Maron BJ, Olivotto I, Spirito P, Casey SA, Bellone P, Gohman TE, et al. Epidemiology of hypertrophic cardiomyopathy-related death: revisited in a large non-referral-based patient population. Circulation 2000;102(8):858−64.

[26] McKenna WJ, Behr ER. Hypertrophic cardiomyopathy: management, risk stratification, and prevention of sudden death. Heart 2002;87(2):169−76.

[27] National Institute for Health and Care Excellence. Guide to the methods of technology appraisal; 2013.

[28] National Institute for Health and Clinical Excellence. Social value judgments: principles for the development of NICE guidance; 2005.

[29] Walker I, Newell H. Do molecularly targeted agents in oncology have reduced attrition rates? Nat Rev Drug Discov 2009;8(1):15−16.

[30] Han PK, Klein WM, Arora NK. Varieties of uncertainty in health care: a conceptual taxonomy. Med Decis Making 2011;31(6):828−38.

[31] Asch DA, Patton JP, Hershey JC. Knowing for the sake of knowing: the value of prognostic information. Med Decis Making 1990;10(1):47–57.

[32] Neumann PJ, Cohen JT, Hammitt JK, Concannon TW, Auerbach HR, Fang C, et al. Willingness-to-pay for predictive tests with no immediate treatment implications: a survey of US residents. Health Econ 2012;21(3):238–51.

[33] Ellsberg D. Risk, ambiguity, and the savage axioms. Q J Econ 1961;75(4):643–69.

[34] Kenen RH. The at-risk health status and technology: a diagnostic invitation and the "gift" of knowing. Soc Sci Med 1996;42(11):1545–53.

[35] Lee DW, Neumann PJ, Rizzo JA. Understanding the medical and nonmedical value of diagnostic testing. Value Health 2010;13(2):310–14.

[36] Cho AH, Killeya-Jones LA, O'Daniel JM, Kawamoto K, Gallagher P, Haga S, et al. Effect of genetic testing for risk of type 2 diabetes mellitus on health behaviors and outcomes: study rationale, development and design. BMC Health Serv Res 2012;12:16.

[37] Centers for Disease Control and Prevention. National diabetes fact sheet: national estimates and general information on diabetes and prediabetes in the United States; 2011.

[38] McCarthy MI. Genomics, type 2 diabetes, and obesity. N Engl J Med 2010;363(24):2339–50.

[39] de Miguel-Yanes JM, Shrader P, Pencina MJ, Fox CS, Manning AK, Grant RW, et al. Genetic risk reclassification for type 2 diabetes by age below or above 50 years using 40 type 2 diabetes risk single nucleotide polymorphisms. Diabetes Care 2011;34(1):121–5.

[40] Scheuner MT, Sieverding P, Shekelle PG. Delivery of genomic medicine for common chronic adult diseases: a systematic review. JAMA 2008;299(11):1320–34.

[41] McBride CM, Koehly LM, Sanderson SC, Kaphingst KA. The behavioral response to personalized genetic information: will genetic risk profiles motivate individuals and families to choose more healthful behaviors? Annu Rev Public Health 2010;31:89–103.

[42] Taylor JY, Wu CY. Effects of genetic counseling for hypertension on changes in lifestyle behaviors among African-American women. J Natl Black Nurses Assoc 2009;20(1):1–10.

[43] Christensen KD, Roberts JS, Uhlmann WR, Whitehouse PJ, Obisesan T, Bhatt DL, et al. How does pleiotropic information affect health behavior changes? Initial results from the REVEAL Study, a randomized trial of genetic testing for Alzheimer's disease. Oral presentation at the 2010 annual meeting of the American College of Medical Genetics; 2010.

[44] Deverka PA, Schully SD, Ishibe N, Carlson JJ, Freedman A, Goddard KAB, et al. Stakeholder assessment of the evidence for cancer genomic tests: insights from three case studies. Genet Med 2012;14(7):656–62.

[45] Buchanan J, Wordsworth S, Schuh A. Issues surrounding the health economic evaluation of genomic technologies. Pharmacogenomics 2013;14(15):1833–47.

[46] Payne K. Fish and chips all round? Regulation of DNA-based genetic diagnostics. Health Econ 2009;18(11):1233–6. [Epub 2009/10/13].

[47] Djalalov S, Musa Z, Mendelson M, Siminovitch K, Hoch J. A review of economic evaluations of genetic testing services and interventions (2004–2009). Genet Med 2011;13(2):89–94.

[48] Giacomini M, Miller F, O'Brien BJ. Economic considerations for health insurance coverage of emerging genetic tests. Community Genet 2003;6(2):61–73.

[49] Rogowski W, Payne K, Schnell-Inderst P, Manca A, Rochau U, Jahn B, et al. Concepts of "personalization" in personalized medicine: implications for economic evaluation. PharmacoEconomics 2015;33(1):49–59.

[50] Evans JP, Berg JS, Olshan AF, Magnuson T, Rimer BK. We screen newborns, don't we? Realizing the promise of public health genomics. Genet Med 2013;15(5):332–4.

[51] Rogowski WH, Grosse SD, John J, Kaariainen H, Kent A, Kristofferson U, et al. Points to consider in assessing and appraising predictive genetic tests. J Community Genet 2010;1(4):185–94.

[52] Rogowski W. Current impact of gene technology on healthcare. A map of economic assessments. Health Policy 2007;80(2):340–57. [Epub 2006/05/09].

[53] Rogowski WH. The cost-effectiveness of screening for hereditary hemochromatosis in Germany: a remodeling study. Med Decis Making 2009;29(2):224–38.

[54] Rogowski WH, Grosse SD, Khoury MJ. Challenges of translating genetic tests into clinical and public health practice. Nat Rev Genet 2009;10(7):489−95. [Epub 2009/06/10].

[55] PHG Foundation. Next steps in the sequence: the implications of whole genome sequencing for health in the UK; 2011.

[56] Conti R, Veenstra DL, Armstrong K, Lesko LJ, Grosse SD. Personalized medicine and genomics: challenges and opportunities in assessing effectiveness, cost-effectiveness, and future research priorities. Med Decis Making 2010;30(3):328−40. [Epub 2010/01/21].

[57] Goddard KAB, Knaus WA, Whitlock E, Lyman GH, Feigelson HS, Schully SD, et al. Building the evidence base for decision making in cancer genomic medicine using comparative effectiveness research. Genet Med 2012;14(7):633−42.

[58] Lin X, Tang W, Ahmad S, Lu J, Colby CC, Zhu J, et al. Applications of targeted gene capture and next-generation sequencing technologies in studies of human deafness and other genetic disabilities. Hear Res 2012;288(1−2):67−76.

[59] The Academy of Medical Sciences. Realising the potential of stratified medicine; 2013.

[60] Wetterstrand KA. DNA sequencing costs: data from the NHGRI Genome Sequencing Program (GSP). Available from <www.genome.gov/sequencingcosts2013> [cited 2013].

[61] Rogowski W. Genetic screening by DNA technology: a systematic review of health economic evidence. Int J Technol Assess Health Care 2006;22(3):327−37. [Epub 2006/09/21].

[62] Mihaescu R, van Hoek M, Sijbrands EJ, Uitterlinden AG, Witteman JC, Hofman A, et al. Evaluation of risk prediction updates from commercial genome-wide scans. Genet Med 2009;11(8):588−94. [Epub 2009/07/29].

[63] Ransohoff DF, Khoury MJ. Personal genomics: information can be harmful. Eur J Clin Invest 2010;40(1):64−8.

[64] Roth JA, Garrison LP, Burke W, Ramsey SD, Carlson R, Veenstra DL. Stakeholder perspectives on a risk-benefit framework for genetic testing. Public Health Genomics 2011;14 (2):59−67.

[65] Grosse SD, Kalman L, Khoury MJ. Evaluation of the validity and utility of genetic testing for rare diseases. Adv Exp Med Biol 2010;686:115−31.

[66] Alberts SR, Yu T, Behrens RJ, Renfro LA, Srivastava G, Soori GS, et al. Real-world comparative economics of a 12-gene assay for prognosis in stage II colon cancer. J Clin Oncol 2013;31(4):391.

[67] Holt S, Bertelli G, Humphreys I, Valentine W, Durrani S, Pudney D, et al. A decision impact, decision conflict and economic assessment of routine Oncotype DX testing of 146 women with node-negative or pNImi, ER-positive breast cancer in the UK. Br J Cancer 2013;108 (11):2250−8.

[68] Wordsworth S, Buchanan J, Papanicolas I, Taylor J, Frayling I, Tomlinson I. Molecular testing for somatic mutations in common cancers: the views of UK oncologists. J Clin Pathol 2008;61 (6):761−5.

[69] Hall PS, McCabe C, Stein RC, Cameron D. Economic evaluation of genomic test-directed chemotherapy for early-stage lymph node-positive breast cancer. J Natl Cancer Inst 2012;104(1):56−66.

[70] Ladabaum U, Wang G, Terdiman J, Blanco A, Kuppermann M, Richard Boland C, et al. Strategies to identify the Lynch syndrome among patients with colorectal cancer. Ann Intern Med [Internet] 2011;(2):69−79.

[71] Mvundura M, Grosse SD, Hampel H, Palomaki GE. The cost-effectiveness of genetic testing strategies for Lynch syndrome among newly diagnosed patients with colorectal cancer. Genet Med 2010;12(2):93−104. [Epub 2010/01/20].

[72] Sanderson S, Zimmern R, Kroese M, Higgins J, Patch C, Emery J. How can the evaluation of genetic tests be enhanced? Lessons learned from the ACCE framework and evaluating genetic tests in the United Kingdom. Genet Med 2005;7(7):495−500.

[73] McCormick P. Editor. Development of personalised medicines for rare genetic conditions: some problems. Society for Genomics, Policy and Population Health, Spring Conference. London, UK; 2012.

[74] Feero W, Wicklund C, Veenstra DL. The economics of genomic medicine: insights from the IOM roundtable on translating genomic-based research for health. JAMA 2013;309(12):1235−6.

[75] Payne K. Towards an economic evidence base for pharmacogenetics: consideration of outcomes is key. Pharmacogenomics 2007;9(1):1−4.

[76] Faulkner E, Annemans L, Garrison L, Helfand M, Holtorf A-P, Hornberger J, et al. Challenges in the development and reimbursement of personalized medicine—payer and manufacturer perspectives and implications for health economics and outcomes research: a report of the ISPOR personalized medicine special interest group. Value in Health 2012;15(8):1162−71.

[77] Singer DRJ, Watkins J. Using companion and coupled diagnostics within strategy to personalize targeted medicines. Pers Med 2012;9(7):751−61.

[78] Veenstra DL, Higashi MK, Phillips KA. Assessing the cost-effectiveness of pharmacogenomics. AAPS PharmSci 2000;2(3):E29. [Epub 2001/12/14].

[79] Vataire AL, Laas E, Aballea S, Gligorov J, Rouzier R, Chereau E. [Cost-effectiveness of a chemotherapy predictive test]. Bulletin du cancer. 2012;99(10):907−14. [Epub 2012/10/09]. Analyse cout-efficacite d'un test predictif de la chimiotherapie dans le cancer du sein (Oncotype DX((R))) en France.

[80] van Rooij T, Wilson DM, Marsh S. Personalized medicine policy challenges: measuring clinical utility at point of care. Expert Rev Pharmacoecon Outcomes Res 2012;12(3):289−95.

[81] Malhotra AK, Zhang JP, Lencz T. Pharmacogenetics in psychiatry: translating research into clinical practice. Mol Psychiatry 2012;17(8):760−9.

[82] Van Den Akker-Van Marle ME, Gurwitz D, Detmar SB, Enzing CM, Hopkins MM, Gutierrez De Mesa E, et al. Cost-effectiveness of pharmacogenomics in clinical practice: a case study of thiopurine methyltransferase genotyping in acute lymphoblastic leukemia in Europe. Pharmacogenomics 2006;7(5):783−92.

[83] Knight SJL, Yau C, Clifford R, Timbs AT, Sadighi Akha E, Dreau HM, et al. Quantification of subclonal distributions of recurrent genomic aberrations in paired pre-treatment and relapse samples from patients with B-cell chronic lymphocytic leukemia. Leukemia 2012;26(7):1564−75.

[84] Gurwitz D, Zika E, Hopkins MM, Gaisser S, Ibarreta D. Pharmacogenetics in Europe: barriers and opportunities. Public Health Genomics 2009;12(3):134−41.

[85] Jarrett J, Mugford M. Genetic health technology and economic evaluation: a critical review. Appl Health Econ Health Policy 2006;5(1):27−35. [Epub 2006/06/16].

[86] Bajaj PS, Veenstra DL. A risk-benefit analysis of factor V Leiden testing to improve pregnancy outcomes: a case study of the capabilities of decision modeling in genomics. Genet Med 2012;15(5):374−81.

[87] Grosse SD, Wordsworth S, Payne K. Economic methods for valuing the outcomes of genetic testing: beyond cost-effectiveness analysis. Genet Med 2008;10(9):648−54. [Epub 2008/11/04].

[88] Guzauskas GF, Garrison LP, Stock J, Au S, Doyle DL, Veenstra DL. Stakeholder perspectives on decision-analytic modeling frameworks to assess genetic services policy. Genet Med 2013;15 (1):84−7.

[89] Veenstra DL, Piper M, Haddow JE, Pauker SG, Klein R, Richards CS, et al. Improving the efficiency and relevance of evidence-based recommendations in the era of whole-genome sequencing: an EGAPP methods update. Genet Med 2013;15(1):14−24.

[90] Payne K, McAllister M, Davies LM. Valuing the economic benefits of complex interventions: when maximising health is not sufficient. Health Econ 2012;22(3):258−71.

[91] Foster MW, Mulvihill JJ, Sharp RR. Evaluating the utility of personal genomic information. Genet Med 2009;11(8):570−4. [Epub 2009/05/30].

[92] Garau M, Towse A, Garrison L, Housman L, Ossa D. Can and should value-based pricing be applied to molecular diagnostics? Pers Med 2012;10(1):61−72.

[93] Grosse SD, Khoury MJ. What is the clinical utility of genetic testing? Genet Med 2006;8 (7):448−50. [Epub 2006/07/18].

[94] Asch DA, Hershey JC, Pauly MV, Patton JP, Jedrziewski MK, Mennuti MT. Genetic screening for reproductive planning: methodological and conceptual issues in policy analysis. Am J Public Health 1996;86(5):684−90.

[95]   Caughey AB, Kuppermann M, Norton ME, Washington AE. Nuchal translucency and first trimester biochemical markers for Down syndrome screening: a cost-effectiveness analysis. Am J Obstet Gynecol 2002;187(5):1239−45.

[96]   Grosse SD, McBride CM, Evans JP, Khoury MJ. Personal utility and genomic information: look before you leap. Genet Med 2009;11(8):575−6. [Epub 2009/07/23].

[97]   Khoury MJ, Gwinn M, Dotson WD, Bowen MS. Is there a need for PGxceptionalism? Genet Med 2011;13(10):866−7.

[98]   Flynn TN. Valuing citizen and patient preferences in health: recent developments in three types of best-worst scaling. Expert Rev Pharmacoecon Outcomes Res 2010;10(3):259−67.

[99]   Najafzadeh M, Lynd LD, Davis JC, Bryan S, Anis A, Marra M, et al. Barriers to integrating personalized medicine into clinical practice: a best-worst scaling choice experiment. Genet Med 2012;14(5):520−6.

[100] Issa AM. Evaluating the value of genomic diagnostics: implications for clinical practice and public policy. Adv Health Econ Health Serv Res 2008;19:191−206.

[101] Issa A, Tufail W, Ajike R, Tenorio J, McKeever J. Breast and colorectal cancer patients' preferences for pharmacogenomics: a discrete choice experiment. 27th International conference on pharmacoepidemiology and therapeutic risk management; Chicago, IL. Pharmacoepidemiology and Drug Safety; 2011.

[102] Van Bebber SL, Liang S-Y, Phillips KA, Marshall D, Walsh J, Kulin N. Valuing personalized medicine: willingness to pay for genetic testing for colorectal cancer risk. Pers Med 2007;4(3):341−50.

[103] Phillips KA, Liang S-Y, Van Bebber S, The CRG, Afable-Munsuz A, Elkin E, et al. Challenges to the translation of genomic information into clinical practice and health policy: utilization, preferences, and economic value. Curr Opin Mol Ther 2008;10(3):260−6.

[104] Thariani R, Veenstra DL, Carlson JJ, Garrison LP, Ramsey S. Paying for personalized care: cancer biomarkers and comparative effectiveness. Mol Oncol 2012;6(2):260−6.

[105] Rogowski WH, Grosse SD, Meyer E, John J, Palmer S. Using value of information analysis in decision making about applied research : the case of genetic screening for hemochromatosis in Germany. Bundesgesundheitsblatt—Gesundheitsforschung—Gesundheitsschutz 2012;55(5):700−9.

[106] Flowers CR, Veenstra D. The role of cost-effectiveness analysis in the era of pharmacogenomics. Pharmacoeconomics 2004;22(8):481−93. [Epub 2004/06/26].

[107] Bennette CS, Gallego CJ, Burke W, Jarvik GP, Veenstra DL. The cost-effectiveness of returning incidental findings from next-generation genomic sequencing. Genet Med 2014;17(7):587−95.

[108] National Institute for Health and Clinical Excellence. Methods for the development of NICE public health guidance. 3rd ed.; 2012.

[109] National Institute for Health and Care Excellence. Identification and management of familial hypercholesterolaemia; 2008.

[110] National Institute for Health and Care Excellence. Gene expression profiling and expanded immunohistochemistry tests for guiding adjuvant chemotherapy decisions in early breast cancer management: MammaPrint, Oncotype DX, IHC4 and Mammostrat; 2013.

[111] Zimmern RL, Kroese M. The evaluation of genetic tests. J Public Health 2007;29(3):246−50.

[112] Towse A, Ossa D, Veenstra D, Carlson J, Garrison L. Understanding the economic value of molecular diagnostic tests: case studies and lessons learned. J Pers Med 2013;3(4):288−305.

CHAPTER 6

Legal Aspects of Health Applications of Genomics

Ellen Wright Clayton[1] and Ma'n Zawati[2]
[1]Center for Biomedical Ethics and Society, Vanderbilt University, Nashville, TN, USA
[2]Centre of Genomics and Policy, Department of Human Genetics, Faculty of Medicine, McGill University, Montreal, Canada

Contents

INTRODUCTION

Much has been written about the ethical issues raised by the incorporation of new genomic technologies into clinical practice, issues that have been addressed elsewhere in this book (see Chapter 5). But the limits of what is permissible in practice are defined by law, which delineates what people can and cannot do at the risk of other penalties or liability. The law, however, is complex and comes from a variety of sources. In much of the world, a country's constitution defines the outer limits of what the government can do. Within those limits, governing bodies enact laws. In many countries, legislative bodies are the primary entities that have the power to pass laws. In other cases, the executive branch, such as the president, can directly set policy as well, albeit often in a more limited way. In either case, laws and executive directives frequently do not spell out all the details necessary for their enforcement but rather delegate authority to regulatory bodies to create final rules or regulations. The courts play a role as well. In some countries, the role of the courts is simply to interpret the laws, directives, and regulations either for enforcement or to adjudicate claims for relief by plaintiffs, people who claim that they were injured because the laws and regulations were not followed. In federal systems that have governments at both the national, state, or more local levels, law, regulations, and adjudication occur at multiple levels, at times coming into conflict. In common law countries, moreover, some courts may also apply judge-made law to analyze claims that plaintiffs were inappropriately injured and so were entitled to damages.

FRAMING THE ISSUES

The discussion that follows is organized in the order in which legal issues arise in the process of implementing clinical genomic testing. The first is what clinicians are permitted to order and use, since clinical practice is highly regulated in most countries, especially in civil law jurisdictions. The second is what genomic information patients can access. This involves two parallel inquiries: (i) what medical records they can obtain and (ii) what genomic information they can acquire outside the clinical setting, for example, through direct-to-consumer genetic testing, and what they can expect their clinicians to do with these results. The third is reimbursement or coverage, what tests payers will reimburse. The last is the role of potential liability in shaping physician behavior regarding the use of genomic testing. We will not, however, address legal issues posed by reproductive genetic testing or state-run newborn screening.

Because constitutions, statutes, regulations, and case law vary from country to country, this chapter will apply this framework primarily in two different settings, the United States, which is a federal and common law country, and Quebec, which is a civil law province in another federal country, Canada. References will be made to other jurisdictions, where appropriate. Our purpose is to illustrate some of the legal issues that must be explored in order to implement genomic testing effectively in the clinic. This discussion is not intended to be interpreted as legal advice nor does it purport to be comprehensive, especially given the rapidly changing legal landscape around the world, so that the reader is strongly advised to seek advice from local counsel for any legal concerns.

THE UNITED STATES
What genomic information clinicians can order and use

The practice of medicine is heavily regulated in the United States, which can be challenging in an environment where genomic technologies are rapidly changing [1]. Probably the most important law in the area of genomic testing is the Clinical Laboratory Improvements Amendments of 1988 (CLIA) [2]. This law was enacted after a scandal revealed that poor laboratory procedures in reading Pap smears caused many women to receive inaccurate results, including results that were not theirs. This law requires detailed procedures to ensure that samples are not mixed up and that test results are accurate and interpreted appropriately. Clinicians are mandated to use CLIA-compliant laboratory results in caring for patients. Recently, however, some scholars have argued that these regulations impermissibly violate the clinician's right to free speech under the US Constitution's protection of commercial speech [3]. How the courts would respond to such a claim were it ever to be litigated remains to be seen, especially since such an interpretation would substantially undermine the basis for regulation in the United States. In any event, however, as will be demonstrated later in the discussion about patients, the impact of CLIA has been substantially reduced by new regulation in another area.

The US Food and Drug Administration (FDA) have long asserted their ability to regulate genomic technologies, but until recently, has not chosen to do so. Recently, however, the FDA has proposed to require that new laboratory-developed tests (LDTs), which are commonly used in genomics, require an Investigational Device Exemption (IDE) "if test results are returned to the patients without confirmation by a medically accepted diagnostic product or procedure" [4]. At this early stage of genomics, confirmatory tests are often not available. The process of obtaining an IDE is an onerous one and would be a major disincentive to innovation. Here, some argue that this proposal is impermissible regulation of the practice of medicine, which is precluded by the FDA's authorizing statute, as well as a violation of commercial speech rights [5], how the courts would address these arguments were they to be raised remains to be seen.

Some states impose additional requirements on genetic testing as well. New York, for example, requires that DNA testing services for its residents must take place in a laboratory accredited by its Department of Health [6]. California imposes additional requirements as well [7].

Notably, all of these laws in the United States address issues of process that must be met before results can be returned. Some other countries, by contrast, decide centrally which specific tests will be available for clinical use and under what circumstances.

Returning to the United States, another issue is to whom clinicians can disclose patient information. Physicians are bound ethically by norms of protecting patient's confidentiality. A long-standing ethical consensus holds that clinicians should not inform patients' relatives about particular genetic risks they may share except in the most unusual situations and then only after the clinician has counseled the patient about the importance of sharing the information and has offered to help the patient to do so [8]. The major law about protecting patient information on the national level in the United States is the Health Insurance Portability and Accountability Act (HIPAA), which has recently adopted a different, and more permissive, position. The so-called privacy rule, which generally prevents clinicians from disclosing patients' personal health information, has a number of exceptions for such purposes as payment for clinical care, public health, and criminal prosecution. Recently, the Office of Civil Rights (OCR), which enforces HIPAA, stated that "*[p]roviders may share genetic information...with providers treating family members...who are seeking to identify their own genetic health risks,... provided the individual has not requested...a restriction on...disclosure*" [9]. The OCR then went on to say that clinicians need not agree to their patients' request for confidentiality. Thus, patients whose information was disclosed are left to state law for protection of their privacy, a topic addressed in the section on liability later.

Notably, countries differ widely in their protection of medical records and their willingness to permit physicians to disclose their patients' genetic information to relatives [10]. Thus, this is an issue for which local legal advice is particularly important.

What genomic information patients can obtain

The trend in the United States is to expand patients' rights to review information in their medical records. Driven in part by HIPAA, which mandates that patients have broad access to their records, many healthcare institutions now provide patient portals to enable patients easily to obtain laboratory and imaging results. New regulations issued under the Health Information Technology for Economic and Clinical Health (HITECH) Act expand this access dramatically [11], by providing patients with access directly from the laboratory to any information in a "designated record set," which is defined as "(1) A group of records maintained by or for a covered entity that is: (i) The medical records...about individuals maintained by or for a covered health care provider;...or (iii) Used, in whole or in part, by or for the covered entity to make decisions about individuals" [12]. The OCR explicitly stated that "[t]his final rule is intended to remove barriers in the HIPAA Privacy and CLIA regulations to individual access to test reports maintained by laboratories subject to or exempt from CLIA" [13]. Thus, patients will have access to non-CLIA-compliant data, as well as any data that states had regulated under their own CLIA-type laws, even though clinicians may not be allowed to use such results to treat patients.

This rule poses particular issues with regard to multiplex data, such as genomic data, because this regulation requires that patients be able to obtain all the data generated on request so long as the test is complete [14]. Thus, it appears to mean that patients have access to *all* genomic data from a broad-based genome test if any part of the data is used for clinical care or if a third party payer is billed electronically for its generation.

This rule, however, conflicts with significant efforts over many years to create evidence-based processes to decide which data are useful clinically and to develop decision support strategies to ensure that clinicians and patients know how to use the results for prevention and treatment [15,16]. Admittedly, the concepts of clinical utility and actionability and who decides them are hotly debated [17], but most of the discussion assumes that at least some genomic data should not be returned because their meaning or what can be done with the data to improve health outcomes are not known.

What the impact will be for patients of having access to broad genomic data is unclear. The HITECH regulations state specifically that laboratories are not required to provide any interpretation of the data. The OCR leaves the responsibility for interpretation to clinicians, the vast majority of whom lack these skills. Until better support systems become available in the healthcare system, patients in the future may need to turn to commercial entities that are currently developing interpretation protocols.

Direct-to-consumer genetic testing raises a different set of concerns. People have been able to get a variety of tests outside the healthcare system, such as blood pressure measurements as well as blood glucose and cholesterol. Direct-to-consumer imaging became popular in the early 2000s, only to fall into disfavor [18]. In the last decade and a half, a number of companies have offered direct-to-consumer genetic testing, for a whole array of purposes, including determining paternity and ancestry, but some focused on providing genetic-risk information to purchasers, a move that has been quite controversial [19,20,21]. Here as well, the FDA has asserted authority to regulate, and on November 22, 2013, issued a warning letter to 23andMe instructing it to stop its Personal Genome Service until it obtained marketing approval from the FDA, which required, inter alia, demonstrating analytical and clinical validity [22]. New York, California, and other states had previously regulated these entities as well [23]. If these tests become available again, questions still remain about whether physicians are willing to use their data in their practices [24,25] and if they can do so in ways that improve healthcare outcomes [26].

Coverage and reimbursement

The United States has a complex "system" of healthcare coverage, with coverage by governments under Medicare and Medicaid and a wide variety of private payers, including health insurers and self-insuring companies. Medicare and most Medicaid programs cover some genetic tests, the latter varying widely among states. Other than state-run health newborn screening programs, the only genetic tests that are required to be covered by private insurers under the Patient Protection and Affordable Care Act are those for BRCA1 and BRCA2 for high-risk women [27]. This requirement exempts already existing or so-called grandfathered plans. Otherwise, private health insurers differ widely in their coverage of genetic tests, often requiring precertification and specific clinical features that indicate an elevated risk that the person carries a pathogenic variant [28].

Issues of coverage and reimbursement, obviously, are quite different in countries where the government is the only or the primary payer for health care and so makes decisions about coverage centrally.

Liability

Medical practice in the United States is shaped in no small part by the fear of liability, which is no surprise since it is one of the most litigious societies in the world. Anyone can go to the courthouse, although they are unlikely to get help from a lawyer unless they have a reasonably strong claim. In this section, we will discuss the realistic claims that patients can bring against clinicians and healthcare institutions.

None of the federal laws discussed earlier—CLIA, HIPAA, and FDA—create a so-called private right of action, which would allow patients to seek to enforce the law or to obtain damages if the laws were not followed. Thus, patients are largely left to pursue claims based on common law theories. The most important of these is the law of negligence, in which claimants must prove that the defendant(s) breached a duty of care proximately causing compensable injuries. One major issue, then, is the scope of the duty of care, which in the healthcare context is generally defined by what reasonably prudent, similarly situated practitioners would do. This is how federal laws may become pertinent even though they are not directly enforceable by patients because their goal is to shape the standard of care by setting the standards for, for example, how laboratories should run or what information should be released. Evolving practice patterns and professional guidance also affect the scope of the duty. As noted previously, much effort is being devoted to determining which results ought to be returned and how best to do so [29], which will shape practice and hence what providers and institutions will be expected to do. If failure to meet these obligations causes a patient to suffer physical harm, providers may well be liable for damages.

The use of multiplex technologies like genomic testing has the potential to reveal findings that are not pertinent to reason testing was undertaken, the so-called incidental or ancillary findings, raising questions about whether clinicians have a legal duty to return them. A recent review found no reported case in which patients have argued that they were harmed by not receiving incidental genomic results [30]. The few cases that have been reported in the context of medical imaging point out the importance of having a physician—patient relationship and defining the standard of care for the particular physician in order to impose liability. An additional issue in the context is whether there is a duty to hunt [31] for ancillary findings by examining all the genomic data, whether the clinician can examine only those parts of genome specific to the patient's clinical issues, or whether some subset of additional genes should be analyzed, as was recently recommended by the American College of Medical Genetics and Genomics [32,33]. Even limited inquiries will likely reveal pleiotropic or ancillary findings [34]. How far these duties to identify and return these additional findings will extend will depend upon how practice patterns evolve unless legislative bodies intervene to define the scope of liability.

One particularly challenging issue may arise when patients have access to more genomic test information than clinicians are prepared to interpret and rely upon, particularly information that is not included within practice guidelines for use. The standard of care should be an adequate defense for uses outside standard practice, but physician reluctance to do what the patient wants is the type of situation that leads to unhappiness, a major determinant of litigation. Referral will not always be an option due to the paucity of geneticists or other knowledgeable clinicians.

Obligations to family members are more complex. For the most part, clinicians do not have a physician–patient relationship with relatives and so arguably have no duty. Two cases in the United States in the 1990s, *Safer v. Pack* [35], and *Pate v. Threlkel* [36], raised questions about whether physicians have a duty to warn relatives. The more far-reaching case, *Safer*, which held that a physician could be required to warn his patient's young child specifically about her genetic risk, was overturned by that state's legislature, and a general consensus emerged that, with few potential exceptions, physicians were ethically required only to warn their patients about the importance of sharing genetic-risk information with relatives and to offer to assist them in doing so. No lawsuits asserting failure to warn relatives of genetic risk have been filed since that time, but this may change with the recent changes in HIPAA, described earlier, which permit physicians to warn the physicians of at-risk family members. Questions about whether the patient's physicians owe a legally enforceable duty to their patients' family members would still remain, but the regulatory change evinces a policy of informing relatives, which could lead to lawsuits if relatives were not warned and became ill.

COMPARATIVE ISSUES IN QUEBEC

When it comes to how legal traditions address the above-framed issues, the Canadian province of Quebec provides an interesting contrast to the United States. Being a civil law jurisdiction, Quebec puts much more emphasis on the role of laws, regulations, and codes in providing guidance. This, however, does not mean that these jurisdictions take diametrically disparate stances on the common issues facing the orderly implementation of genomic testing in the clinic.

What clinicians can do?

Quebec, being one of 10 provinces in Canada, finds itself governed by both provincial and federal laws and regulations. Currently, genomic testing is indirectly regulated by various sources, and a sharing of power occurs between the provincial and federal jurisdictions on regulatory matters. Indeed, Canada's Constitution endows both the federal and provincial legislatures with competence to regulate diagnostic tests. The federal government, through Health Canada, is empowered to regulate medical devices under its prerogative over criminal matters [37]. As for the provinces, their jurisdiction over the management of hospitals [38] provides them with the authority to regulate tests locally [39]. The regulatory process at the federal level can be stringent but allows for the marketing of the product across the country. In contrast to the United States, Canada recognizes four classes of medical devices (regulated by federal authorities) based on the level of control required to ensure safety and efficacy [40]. A similar classification can be found in the European Union [41].

According to Canada's *Medical Devices Regulations*, genetic tests are considered *in vitro* diagnostic tests and are classified as class III (Schedule 1, rule 4). "Genetic testing" is defined as "the analysis of DNA, RNA or chromosomes for purposes such as the prediction of disease or vertical transmission risks, or monitoring, diagnosis or prognosis." (see USA section, pp. 121) Genetic tests that are not sold as test kits but developed only for and in one laboratory will fall under the general category of LDTs, which are regulated provincially. LDTs can only be performed locally and "cannot be commercially marketed to consumers or sold to other laboratories" [39].

Section 31 of the *Medical Act* (chapter M-9) provides that the prescription of diagnostic tests generally must be done by a physician. The obligations to be carried out by these clinicians are outlined in the *Code of Ethics of Physicians* (chapter M-9, r. 17), which states that a physician's duty is "to protect and promote the health and well-being of the persons he attends to" (section 3), while practicing in a "manner which respects the life, dignity and liberty of the individual" (section 4), and which is "in accordance with scientific principles" (section 6). A physician must be sensible "in his use of the resources dedicated to health care" (section 12) and must "make his diagnosis with the greatest care, using the most appropriate scientific methods and, if necessary, consulting knowledgeable sources" (section 46). A physician must use diagnostic technologies that are proven and medically necessary (sections 48–50). However, he must respect the patient's diagnostic choices, all the while ensuring that sufficient information is provided to the patient concerning risks or disadvantages that could result (sections 48–50). The next section will address the nature of the information that will be disclosed to patients and the level of protection surrounding them.

What patients can get?

Section 38 of the *Civil Code of Quebec* states generally that "[. . .] any person may, free of charge, examine and cause the rectification of a file kept on him by another person." Additionally, the information contained in such a file should be made accessible in an intelligible transcript. According to the *Act Respecting Health Services and Social Services* (chapter S-4.2), every patient 14 years of age and over has a right to access his/her records held in a health institution (section 17). However, such a right is not absolute and patients could be denied access temporarily if the communication of the record "would likely be seriously prejudicial to the user's health" (section 17), among other limitations. It is safe to say that results from genetic tests found in the medical record will be made available to patients, even if not immediately.

What about results from whole-genome (WG) or whole-exome sequencing (WES)? Given the complexity of such information, can they be accessed by patients? If made available in the medical records, all indications point to the right of access of patients to such information based on the above-mentioned provisions of the *Act*. As a practical matter, no Canadian normative document exists on the type of information that should be made available in the medical record of a patient following WG or WES. Recently though, a statement endorsed by the Board of Directors of the Canadian College of Medical Geneticists (CCMG) has suggested that competent adult patients should be given an option to receive clinically significant and actionable results following clinical WGS/WES [42], which would mean that such information would be made available in the patient's medical records. "Medically actionable" and "clinically significant" have yet to be specifically defined, but in this document, a clinically significant finding is understood as being "of potential health or reproductive importance" and medically actionable results are broadly characterized as revealing "preventable or treatable conditions." Overreporting could potentially harm patients by creating undue anxiety and could be used as a justification to temporarily deny them right of access to their medical records, according to the *Act*, although this has yet to be tested in administrative courts. In any case, the *Act Respecting Health Services and Social Services* specifies that a health institution that provides a patient with information contained in his/her medical record, shall, at the request of the patient, "provide him with the assistance of a qualified professional" (section 25), something that online direct-to-consumer companies usually fail to do.

Like the United States, patients in Quebec and Canada have access to online services provided by direct to consumer (DTC) companies. Very recently, 23andMe has begun operating in Canada, even though Personal Genome Service was blocked by the FDA in the United States for lack of marketing approval [43]. According to 23andMe, after extensive discussions with the federal health authority in Canada (Health Canada), the latter has determined that given the nontherapeutic nature of their product, it falls outside the purview of the *Food and Drugs Act* [44], as well as the *Medical Devices Regulations*, and so needs no federal premarket approval.

This, however, will not protect them from eventual regulation in the provinces, especially when it comes to the expertise required for the interpretation of the results. Currently, the Quebec *Medical Act* provides that only physicians are allowed to diagnose illnesses (section 31(1)). While some DTC companies assert that their services are not for diagnostic purposes, Quebec courts have broadly defined the term "diagnosis" to include opinions and observations by a person on the health status of another individual [45]. Provincial medical colleges could eventually pressure provincial legislatures, such as Quebec's, to stop direct-to-consumer genetic testing companies from bypassing the involvement of a licensed physician during the interpretation process.

Costs and reimbursement

In Quebec, the universal Health Insurance Plan covers healthcare services that are medically necessary and that have been rendered by a licensed general practitioner or medical specialist [46]. The costs of any genetic test meeting this definition will be covered. Any healthcare service that is not medically necessary is not covered by the Plan. Patients are expected to pay for such services, which include most laboratory services, unless provided in a hospital [47].

Reimbursements can be made when the insured service is not available in the province of Quebec (Health Insurance Act, section 10) or when the individual did not present his/her health insurance card when receiving the service (section 13.1). In both cases, only covered medical services will be reimbursed provided the patient submits his/her claim within 1 year of the date on which he/she received the service (section 14.2).

Liability

Physicians in Quebec have four duties toward their patients: (i) duty to inform, (ii) duty to treat, (iii) duty to follow-up, and (iv) duty to uphold professional secrecy [47]. This section will focus on the duty to inform as well as the duty to uphold professional secrecy.

The essentials of the therapeutic duty to inform in Canada can be characterized as the provision of sufficient information (material risks, as well as, in the common law provinces, special or unusual risks) to patients in order for them to make the best decision possible when consenting to treatment. In the landmark common law *Reibl v Hughes* case [48], Judge Laskin of the Supreme Court of Canada wrote: "What the doctor knows or should know that the particular patient deems relevant to a decision whether to undergo prescribed treatment goes equally to his duty of disclosure as do the material risks recognized as a matter of required medical knowledge" (paragraph 16). In *Hopp v Lepp* [49], the same Court further specified the scope of the duty to inform of physicians, to include answering "any specific questions posed by the patient as to the risks involved without being questioned, disclose to [their patients] the nature of the proposed operation, its gravity, any material risks and any special or unusual risks attendant upon the performance of the operation" (paragraph 29). This has since become the minimum standard with which physicians are expected to comply in the common law provinces.

In Quebec civil law however, courts have tended to reject the "reasonable patient" threshold proposed in *Reibl v Hughes* and upheld a test focused on what a reasonable physician would disclose in the circumstances [50]. In civil law jurisdictions more generally, the duty to inform has been advanced through civil codes. In Quebec, the duty to inform has been incorporated in both the *Civil Code of Quebec* and under professional norms, such as the *Code of Ethics of Physicians*. The latter enshrines the legal duty to provide the patient with explanations that are pertinent to their "understanding of the nature, purpose and possible consequences of the examination, investigation, treatment or research which [the physician] plans to carry out" (section 29). The physician—patient relationship, which is classified as a contractual relationship, is bound by the *Civil Code of Quebec*'s chapter on contract for services, which specifies that a contractor "is bound to provide the client, as far as circumstances permit, with any useful information concerning the nature of the task which he undertakes to perform" (section 2102).

In France, the duty to inform is enshrined in the French *Code de la santé publique*. This law states that every person has the right to be informed about his/her state of health (section L1111-2), which would also include information pertaining to the proposed treatment, investigation, the potential benefits, and the foreseeable risks. In both jurisdictions, if a patient undergoes genetic testing prescribed by the physician, the latter is obligated to provide the patient with the results as well as how it affects his/her overall health assessment (e.g., diagnosis). But, is the clinician obligated to inform patients of incidental findings? A comprehensive answer to this question is made difficult by the lack of normative guidance on the topic although some argue that providing unverified and nonvalidated information to the patient is counterproductive and will most likely fail the standard of prudence and diligence required by the courts [51]. That being said, if a finding is validated and reveals a clinically significant and actionable condition, does the physician have a duty to return it? A recent proposal endorsed by the Board of Directors of the CCMG suggests that physicians should offer patients the option of receiving such findings [42]. Contrary to the position taken by the American College of Medical Genetics and Genomics, however, this proposal did not establish any duty to hunt for such findings.

Similar to common law jurisdictions, a breach of the duty to inform in civil law could result in the liability of clinicians for damages. Actions in such cases will not necessarily be those of "negligence," but of medical malpractice under the general rules of civil liability or "responsabilité civile." These actions will require the presence of (i) fault, (ii) injury, and (iii) a causal link. Plaintiffs must prove the existence of each of these components: "[e]very person has a duty to honor his contractual undertakings. Where he fails in this duty, he is liable for any bodily, moral or material injury he causes to the other contracting party and is liable to reparation for the injury" [52]. This same standard applies in France, where section 1382 of the French *Code civil* states, "Any act whatever of man [sic], which causes damage to another, obliges the one by whose fault it occurred, to compensate it." Section 1383 further explains that a person is "liable for the damage he causes not only by his intentional act, but also by his negligent conduct or by his imprudence."

As for the duty to uphold professional secrecy, it is based on the right of every individual to nondisclosure of their confidential information. The Quebec *Charter of Human Rights and Freedoms* (chapter C-12) (a quasi-constitutional normative document) states that: "no person bound to professional secrecy by law may, even in judicial proceedings, disclose confidential information revealed to him by reason of his position or profession" (section 9). This duty, however, is not absolute and can be breached by the physician if he/she is authorized to do so by the patient or by an express provision of the law (section 9).

What about family members? Similar to the United States, clinicians in Quebec do not have a physician–patient relationship with the patient's relatives and, consequently, have no duty toward them. However, the law allows physicians to breach their professional secrecy when "there are compelling and just grounds related to the health or safety of the patient or of others" [53] or if someone's life is in peril and requires immediate physical assistance (duty to rescue) [54]. The "compelling and just grounds" authorization is discretionary and not an obligation. As for the duty to rescue, most authors argue that it does not apply to genetic conditions [42]. In any case, until clear legislative guidance is made available, one cannot conclude that a physician has a duty toward the relatives of patients. A recent Quebec case, *Watters v White* [55], tackled this issue more specifically and arrived at the conclusion that physicians only owe a duty to inform to their patients. The facts of the case, however, dated back to the 1970s and may not be pertinent to today's standards. The recent proposal endorsed by the CCMG suggests that when "results reveal clinically significant and actionable conditions for identifiable family members, physicians should, on a case-by-case basis, encourage patients to communicate such results." Moreover, physicians are encouraged to make themselves available to discuss such matters with the family members.

CONCLUSIONS FOCUSING ON WHAT THIS WILL MEAN FOR NEXT GENERATION SEQUENCING

This brief comparison reveals the complexity of the legal issues raised by implementing new genomic technologies in the clinical setting, both within any particular country and among countries. The legislative and regulatory process is complicated, with jurisdiction often being spread among different levels of government. In many instances, legislators and regulators have failed to address issues raised by new technologies, especially in a comprehensive manner. When they have spoken, their actions are not always consistent. Direct-to-consumer genetic testing, for example, currently is treated differently in the United States and Canada. Privacy rules also differ widely around the world, as do which technologies are covered by healthcare systems. To make matters even more challenging, the relevant laws and regulations continue to change over time. Another source of diversity is in the scope of liability, a factor known to drive clinician behavior, which is defined differently in civil law and common law jurisdictions, the judge-made law in the latter injecting an additional element of uncertainty. Thus, those who are working in this field are advised to obtain expert legal advice about the legal landscape in the jurisdiction(s) in which they operate, recognizing that the law simply defines the outer limits of what is permissible and mandatory and that many of the most challenging issues will turn on ethical norms well within these legal boundaries.

This chapter was supported in part by 1R21HG00612-01 and by Genome Canada through the *Innovative chemogenomic tools to improve outcome in acute myeloid leukemia* project.

REFERENCES

[1] Katz G, Schweitzer SO. Implications of genetic testing for health policy. Yale J Health Policy Law Ethics 2013;10(1):90−134.

[2] 42 U.S.C. § 263a (2014).

[3] Evans BJ. The first amendment right to speak about the human genome. University of Pennsylvania. J Constitut Law 2014;16.

[4] 21 C.F.R. Part 812 (2014).

[5] Evans BJ, Dorschner MO, Burke W, Jarvik GP. Regulatory changes raise troubling questions for genomic testing. Genet Med 2014;16(11):799−803.

[6] N.Y. Comp. Codes R. & Regs., tit.10, § 58-1.7 (2014).

[7] Cal. Bus. & Prof. Code § 1288.5 (2014).

[8] ASHG statement. Professional disclosure of familial genetic information. The American Society of Human Genetics Social Issues Subcommittee on Familial Disclosure. Am J Hum Genet 1998; 62(2):474−83.

[9] 78 Fed. Reg. 5566 at 5668 (2013).

[10] Dupras C, Ravitsky V. Disclosing genetic information to family members: the role of empirical ethics. eLS. Chichester: John Wiley & Sons; 2013.

[11] 45 C.F.R. § 164.524 (2014).

[12] 45 CFR §164.501 (2014).

[13] 79 Fed. Reg. 7290 at 7296 (2014).

[14] 45 C.F.R. § 493.1291(l) (2014).

[15] Pulley JM, Denny JC, Peterson JF, et al. Operational implementation of prospective genotyping for personalized medicine: the design of the Vanderbilt PREDICT project. Clin Pharmacol Ther 2012;92(1):87−95.

[16] PharmGKB. CPIC: Clinical Pharmacogenetics Implementation Consortium. <http://www.pharmgkb.org/page/cpic> [accessed 9.5.2015].

[17] Eckstein L, Garrett JR, Berkman BE. A framework for analyzing the ethics of disclosing genetic research findings. J Law Med Ethics 2014;42(2):190−207.

[18] Goldrich MS. Direct-to-consumer diagnostic imaging tests, Report of The Council on Ethical and Judicial Affairs; 2005.

[19] Roberts JS, Ostergren J. Direct-to-Consumer genetic testing and personal genomics services: a review of recent empirical studies. Curr Genet Med Rep 2013;1(3):182—200.

[20] Caulfield T, McGuire AL. Direct-to-consumer genetic testing: perceptions, problems, and policy responses. Annu Rev Med 2012;63:23—33.

[21] Borry P, Cornel MC, Howard HC. Where are you going, where have you been: a recent history of the direct-to-consumer genetic testing market. J Community Genet 2010;1(3):101—6.

[22] Food and Drug Administration. Warning Letter to 23andMe, Inc.; 11/22/2013.

[23] Dick HC. Risk and responsibility: state regulation and enforcement of the direct-to-consumer genetic testing industry. J Health Law Policy 2012;6(1):167.

[24] Mainous III AG, Johnson SP, Chirina S, Baker R. Academic family physicians' perception of genetic testing and integration into practice: a CERA study. Fam Med 2013;45(4):257—62.

[25] Goldsmith L, Jackson L, O'Connor A, Skirton H. Direct-to-consumer genomic testing from the perspective of the health professional: a systematic review of the literature. J Community Genet 2013;4(2):169—80.

[26] Bloss CS, Schork NJ, Topol EJ. Direct-to-consumer pharmacogenomic testing is associated with increased physician utilisation. J Med Genet 2014;51(2):83—9.

[27] Public Health Service Act § 2713 (2010).

[28] Michael DG, Needham DF, Teed N, Brown T. Genetic testing insurance coverage trends: a review of publicly available policies from the largest US payers. Pers Med 2013;10(3):235—43.

[29] ACMG policy statement: updated recommendations regarding analysis and reporting of secondary findings in clinical genome-scale sequencing. Genet Med 2014.

[30] Clayton EW, Haga S, Kuszler P, Bane E, Shutske K, Burke W. Managing incidental genomic findings: legal obligations of clinicians. Genet Med 2013;15(8):624—9.

[31] Holm IA. Clinical management of pediatric genomic testing. Curr Genet Med Rep 2014;1—4.

[32] Green RC, Berg JS, Grody WW, Kalia SS, Korf BR, Martin CL, et al. ACMG recommendations for reporting of incidental findings in clinical exome and genome sequencing. Genet Med 2013; 15(7):565—74.

[33] Incidental findings in clinical genomics: a clarification. Genet Med 2013;15(8):664—66.

[34] Kocarnik JM, Fullerton SM. Returning pleiotropic results from genetic testing to patients and research participants. JAMA 2014;311(8):795—6.

[35] 677 A.2d 1188 (N.J. App. Div. 1996).

[36] 661 So.2d 278 (Fla. 1995).

[37] Section 91(27) *Constitution Act*.

[38] Section 92 (7) *Constitution Act*.

[39] Joly Y, Koutrikas G, Tassé AM, Issa A, Carleton B, Hayden M, et al. Regulatory approval for new pharmacogenomics tests: a comparative overview. Food Drug Law J 2011;66(1):1—24.

[40] *Medical Devices Regulations*, SOR/98-282, section 6.

[41] 98/79/EC; 93/42/EEC; 90/385/EEC.

[42] Zawati MH, Parry D, Thorogood A, Nguyen MT, Boycott K, Rosenblatt D, et al. Reporting results from whole-genome and whole-exome sequencing in clinical practice: a proposal for Canada? J Med Genet 2014;51(1):68—70.

[43] Picard A. Controversial genetic self-testing kits coming to Canada. Globe and Mail online: <http://www.theglobeandmail.com/life/health-and-fitness/health/genetic-self-testing-kits-to-come-to-canada/article20885678/> [accessed 9.5.2015].

[44] R.S.C., 1985, c. F-27.

[45] Zawati MH. Les conseillers en génétique et les professions médicales et infirmières au Québec: des frontières brouillées ? MJLH 2012;137—87.

[46] *Health Insurance Act*, chapter A-29, section 3a.

[47] Régie de l'assurance maladie du Québec (RAMQ) (Quebec Health Insurance) website: <http://www.ramq.gouv.qc.ca/en/citizens/health-insurance/healthcare/Pages/medical-services.aspx> [accessed 9.5.2015].

[48] Reibl v Hughes, [1980] 2 SCR 880, 114 DLR (3d) 1.

[49] Hopp v Lepp ([1980] 2 SCR 192, 112 DLR (3d) 67.

[50] Pelletier c Roberge, 41 QAC 161, [1991] RRA 726 (Qc CA).

[51] Philips-Nootens S, Lesage-Jarjoura P, Kouri R. Éléments de responsabilité civile médicale: le droit dans le quotidien de la médecine. 3rd ed. Cowansville: Yvon Blais; 2007.

[52] *Civil Code of Quebec*, section 1458.

[53] *Code of Ethics of Physicians*, section 20(5).

[54] *Charter of Human Rights and Freedoms*, section 2.

[55] *Watters v White*, 2012 QCCA 257.

CHAPTER 7

Genomics, Patents, and Human Rights

Michiel Korthals
Wageningen University, Wageningen, The Netherlands

Contents

INTRODUCTION: GENOMICS TECHNOLOGIES AND LEGAL PROTECTION

Genomics has changed the domain of the life sciences enormously, and it can be expected that more radical changes will follow. New genomics diagnostics, drugs, treatments, and novel foods are already on the market. The life sciences have changed enormously: new disciplines, such as genomic and metabolomic technologies, have revolutionized the descriptive and normative power wielded by these disciplines. The technological developments accompanied by new scientific approaches and positions make the daily practices in the laboratories of the life sciences radically different from life science practices before these developments. New organizations of scientific work emerge and this has a deep social and normative impact. In these new life science approaches and practices, new norms and values are incorporated which are significantly different from the earlier forms of life science practices; however, they are also accompanied by new social and normative behavior patterns, rules, and institutions. Both internally as well as externally these new sciences have acquired new forms of descriptive and normative impact. These impacts affect human rights, both in a positive and in a negative way, but they also affect the ownership issue. In this chapter we will deal with the impact of genomics on intellectual property issues and in particular ask in how far the human rights to health, food, and participation in science are affected. Moreover, we will ask, can one claim that to own a certain product or services of genomics is a human right? We will first discuss the role of human rights focused on the life sciences, and then discuss the way life sciences are organized, their societal impact, and the current patenting regime. Although currently ownership issues of the life sciences are regulated via the worldwide agreed-upon Intellectual Property Rights regime, it is doubtful as to how far this regime can fruitfully organize life science innovations, both from the view of the progressive developments of the life sciences as well as from a human rights' perspective. The function of patents and other types of ownership will therefore be extensively discussed. Finally, we will finish with a short discussion of several alternative or complimentary proposals to the current patenting regime that are more firmly based on human rights.

HUMAN RIGHTS TO HEALTH, TO FOOD, AND TO SCIENTIFIC PARTICIPATION

Several types of human rights are relevant for the new life sciences, starting with the human right to health, but also the right to food, and the right to share in scientific advancement and its benefits. First, the human right to health, which is interpreted quite broadly according to The International Covenant on Social, Economic, and Cultural Rights (1976 in force) which states in *Article 12*:

1. The States Parties to the present Covenant recognize the right of everyone to the enjoyment of the highest attainable standard of physical and mental health.
2. The steps to be taken by the States Parties to the present Covenant to achieve the full realization of this right shall include those necessary for: ... (b) The improvement of all aspects of environmental and industrial hygiene; (c) The prevention, treatment and control of epidemic, endemic, occupational and other diseases; (d) The creation of conditions which would assure to all medical service and medical attention in the event of sickness.

The right to health is not interpreted as freedom from disease, but as the right to the highest attainable standard of physical and mental health (cf. UN Committee on Economic, Social, and Cultural Rights 2000).

Next, the right to food, which received canonical form in a decision of the United Nations Social and Economic Council of the Human Rights Commission in 1996. According to this council the right to adequate food covers several issues:

1. Article 11. General comment on its implementation 1. The States Parties to the present Covenant recognize the right of everyone to an adequate standard of living for himself and his family, including adequate food, clothing and housing, and to the continuous improvement of living conditions. The States Parties will take appropriate steps to ensure the realization of this right, recognizing to this effect the essential importance of international co-operation based on free consent. General comment on its implementation

2. The States Parties to the present Covenant, recognizing the fundamental right of everyone to be free from hunger, shall take, individually and through international co-operation, the measures, including specific programs, which are needed:

a. To improve methods of production, conservation and distribution of food by making full use of technical and scientific knowledge, by disseminating knowledge of the principles of nutrition and by developing or reforming agrarian systems in such a way as to achieve the most efficient development and utilization of natural resources;

b. Taking into account the problems of both food-importing and food-exporting countries, to ensure an equitable distribution of world food supplies in relation to need.

The committee officially commented on the right to adequate food by stating that the right to food is an inclusive right and that it is not a right to be fed but primarily the right to feed oneself in dignity. It is not simply a right to a minimum ration of calories, proteins, and other specific nutrients. It is a right to all nutritional elements that a person needs to live a healthy and active life, and to the means to access them. In other comments it explicitly stated that the cultural acceptability of food is included in the right and that it is not the same as a right to be fed and the right to safe food (articles 8 and 11). The right to food is also undersigned by the Food and Agriculture Organization of the United Nations (FAO):

> We, the Heads of State and Government, or our representatives, gathered at the World Food Summit at the invitation of the Food and Agriculture Organization of the United Nations, reaffirm the right of everyone to have access to safe and nutritious food, consistent with the right to adequate food and the fundamental right of everyone to be free from hunger. (Rome Declaration on World Food Security, 1996, FAO)

The third right involved, when speaking about the life sciences, is the human right to participate in scientific advancements. This right is not only a right to enjoy the fruits of sciences, but also to participate in the scientific endeavor because Article 27 of the Universal Declaration of Human Rights states "[e]veryone has the right freely to participate in the cultural life of the community, to enjoy the arts and to share in scientific advancement and its benefits" (1948, hereinafter UDHR; see also Ref. [1]).

It is doubtful that there is a human right to patent an innovation. Both UDHR article 27 and the International Covenant on Economic, Social, and Cultural Rights (1976, hereinafter ICESCR) article 15.1 do not mention scientific innovation unambiguously by name: "The States Parties to the present Covenant recognize the right of everyone: (a) To take part in cultural life; (b) To enjoy the benefits of scientific progress and its applications; (c) To benefit from the protection of the moral and material interests resulting from any scientific, literary or artistic production of which he is the author."

HOW TO INTERPRET HUMAN RIGHTS?

Some philosophers interpret these rights including basic rights, liberties, decent educational and employment opportunities, and wealth and income as resources or social primary goods. These goods are according to they what every rational person can want for everyone. They circumscribe standard needs and endowments that every rational person would assume morally important for everyone. Rawls argues that people have two separate moral powers, one is people's capacity for a sense of justice and the other is a belief in a good life. The social primary goods provide the "all-purpose means" [2] for the development and exercise of these powers; they are resources. Others have criticized this interpretation because these standard needs and endowments would not take seriously the pluralism of human needs. However, Thomas Pogge, one of Rawls pupils, stated that a resourcist like Rawls can still "take full account of the full range of diverse human needs and endowments" [3]. However, in particular from the side of the capability approach it is stressed that ethically seen capabilities are more important than resources, because they cover what set of opportunities are effectively available for a person, both internally (what the person is able to do) as well as externally. Sen therefore also calls capabilities freedoms, conceived as real opportunities [4]. In interpreting Human Rights in this way Sen expands the meaning of public goods.

THE STRUCTURE OF GENOMICS

Genomics is a branch of the life sciences, oriented to the structure of living organisms, in particular the genetic structure and the interaction between genes and (biological and nonbiological) context. The evolution of living organisms according to different degrees and paces started billions of years ago and it still continuing. Living organisms are constantly adapting to changing environments. The genetic code of an organism is shaped accordingly and this dynamic process has for a decade or more been the subject of genomics.

Genomics research, in particular for drugs and food, is so complex and needs so many data, that these research problems are not considered to be feasible for single laboratories to solve. Curing diseases with drugs and producing healthy food covers so many variables and unknowns that only large-scale research laboratories in cooperation with each other can deal with them. These research problems have first to be divided into smaller ones, which can then as modules be fit together in the final research product [5]. Therefore, large genomics consortiums such as NuGO combine available skills, new expertise, animal models, scanning and diagnostic machinery, and biological material, such as proteins [6]. Moreover, neoliberal credo of the privatization of knowledge and the rise of public—private partnerships support this trend toward networking in the life sciences.

The overall effect of the modular design is that the research product is the summary of all the diverse and fragmented contributions (modules); the people working at the different modules do see the others often, and their interaction is about the overall effect, not about the singular contribution of each. Because of this fragmentation, it can happen that particular researchers claim to have constructed a certain genomics pathway along which drugs can be promising, but that later others will claim something different. Promises and hypes seem to be endemic. A second issue is also important: the research object of medical and food genomics, although constructed as one contributing to a societal defined health problem, is framed according to the needs and complexities of the network. Penders and Korthals [7] conclude that:

> Current large-scale research, including large-scale nutrigenomic research is confronted with a research situation in which problem modules take precedence over overall problems and the notions of health and food are framed and reframed based upon the context in which they are used. Health and food, although esoteric in the laboratory context, are (re)framed based upon the context in which it is present. This context is, on the one hand, laboratory-specific, thus creating a diversification, and on the other hand a number of context-characteristics can be recognized in many of the laboratories devoted to nutrigenomic inquiry.

The network character of genomics research makes it very difficult to organize the modules from a juridical point of view that assumes that property, inventions, and owners can be very strictly identified. Nevertheless, the privatization of life sciences also means that private investors want a return on their investments, and that propels a sometimes very strict property regime with respect to DNA codes and other nontangible objects. We must ask how far those rules might be outdated and block progressive innovations, and what negative costs for public welfare and business opportunities are bound to such anachronistic legislation.

PROPERTY RIGHTS REGARDING INTANGIBLE OBJECTS

Patenting and organizing property rights have been regulated at first for tangible objects, such as machines. In Western philosophy, property rights of tangible objects are firmly legitimized by Locke's theory of labor: according to Locke appropriating an object through your own labor is the main criterion of the claim that you own the object. If an inventor only changes a small part of a genome (the other parts are not appropriated by the inventor) and therefore also according to Locke a patent on a product should never exceed to things an inventor has not changed. Others, like Hegel, interpret property as a kind of expression of personality. But in the twentieth century a more utilitarian interpretation dominates [8].

The idea is that when private inventors are not properly remunerated for their efforts, and made dependent in the free market, they always will lose and will not be stimulated to invent again. Inventors were perceived as potential victims of market failures that should be protected by law, and patenting gave them for a while the possibility to recoup their investments and to charge a higher price than the cost price. The Paris Convention of 1883 largely set the "rules of the game" for patenting. Although the late nineteenth century was an era where differences between bigger and smaller companies were less pronounced, the twentieth century saw a trend for bigger and bigger companies, which in particular happened in the field of medical and agricultural biotechnology. These events reversed a development of thousands of years. From the beginning of agricultural and medical research, new herbs, drugs, seeds, and foods have been exchanged and traded without any legal regulation of intellectual property. Huge improvements in seed breeding resulting in increased harvests (like the ones resulting from the so-called Green Revolution in the 1960s) have been achieved just by "brown bagging," sharing, experimenting, and sometimes stealing and imitating.

However, at the end of the 1970s, biotechnology was increasingly perceived by governments as an exciting field of technological innovation that would lead to renewed economic growth and that would restore international competitiveness for Western countries. The huge economic potential of this new field of technology would however, only be unlocked, so it was thought, if biotechnological inventions were to receive proper legal protection. In the landmark case of *Diamond v. Chakrabarty* concerning the patentability of a genetically modified oil-consuming bacterium, the US Supreme Court ruled in 1980 that "anything new under the sun that is made by man," whether living or nonliving, can be patented. In subsequent years US jurisprudence explicitly extended patentability to multicellular organisms like plants (1985), oysters (1987), and mammals (1988). Other Western countries ultimately followed the American example, albeit with some delays and hesitations. In 1988, the patent offices of the United States, the European Union, and Japan proclaimed the new policy line that DNA sequences and genes would also be eligible for product patents. Their justification was that sequences and genes, when isolated and purified, would be essentially different from their natural counterparts and therefore qualify as inventions rather than discoveries.

The globalization of intellectual property rights received its juridical form with the establishment of the Trade-related Aspects of Intellectual Property Rights Agreement (TRIPS) in 1994 as part of an overall World Trade Organization (WTO) package, which introduced a system for the global enforcement of intellectual property rights with sanction possibilities for noncompliance. The TRIPS agreement sets worldwide minimum standards for the protection of intellectual property rights (including patents, copyright, and breeder's rights). It mandates that, with few exceptions, "patents shall be available for any inventions, whether products or processes, in all fields of technology" (article 27.1). In the TRIPS Agreement (1994), article 27(2), it is stated that: "Members may exclude from patentability inventions, the prevention within their territory of the commercial exploitation of which is necessary to protect *ordre public* or morality, including to protect human, animal or plant life or health or to avoid serious prejudice to the environment, provided that such exclusion is not made merely because the exploitation is prohibited by their law." Also, the agreement affirms compulsory licenses, that is state-organized obligations to produce and sell products with a lower price than the private company originally proposed (article 31). In 2001, following the DOHA declaration, it as stated that "TRIPS Agreement can and should be interpreted and implemented in a manner supportive of WTO Members' right to protect public health and, in particular, to promote access to medicines for all." The drafters of the TRIPS agreement, as well as the signatories of the Doha Declaration (still not worldwide accepted), acknowledge that intellectual property rights can clash with higher societal goals, most notoriously public health needs. Signatories agreed that in case of conflict, urgent public health interests supersede private interests [9].

Property rights and free trade got a lot of support, in particular by large international organizations such as WTO and large countries including the United States. The proponents argue that, for instance, food security should not be conflated with "food self-sufficiency" and that free trade was the vehicle to effectively distribute food products [10]. It is interesting to see that in the FAO a different position is sometimes defended, in which food sovereignty (and limits to IPR) plays an important role. However, "the FAO is under heavy pressure from powerful member states to stay out of the WTO's space and to avoid excessive critiquing of the neoliberal agenda" [11].

OBJECTIONS TO PATENTING FROM A HUMAN RIGHTS' PERSPECTIVE

Biotech companies demand respect for intellectual property; they argue that extensive research and development (R&D) enterprises are made profitable as rights holders can market their products exclusively, securing the existence of new commodities and due to the temporary nature of IP, also the provision of future public goods, as the invention becomes part of the public domain. Nevertheless, criticisms abound and many other stakeholders raise ethical objections against this type of legal protection. Particularly in the life sciences, IP rights regulate objects such as food and medicines that are keys to securing human rights, especially the right to adequate food and the right to health. IP rights serve private and public interests. Private interests consist of being able to enjoy the fruits of one's labor and to control products that stem from one's creative inventiveness. The public interests entail the provision of current and future public goods. Many objections regard the provision of current and future public goods.

We will examine five objections, which are specific to the field of life sciences and human rights: *public health and food security, expansive patents, traditional knowledge and environmental sustainability, social justice,* and *loss for the public domain.*

Firstly, the current IPR regime provides, according to many, insufficient incentive for providing innovations that will alleviate the problems that predominantly affect the poor and to make those innovations widely accessible. Objects, such as drugs and seeds, predominantly needed in resource-scarce markets will not be developed, as R&D expenses cannot be recovered. European states have a long-standing tradition of securing for their citizens the minimum requirements for an adequate living standard. The success in eradicating extreme poverty in Western Europe has not only led to viewing this harm to human welfare as something that is unacceptable, but also as preventable. As the sums needed to alleviate this welfare burden globally are relatively low compared to other expenditures made in developed countries, not helping is seen as an unacceptable moral menace by a growing percentage of the citizens of developed and developing countries. With the excessive income inequalities all over the world it is evident that if the economically worse-off people, such as farmers, are not allowed to make use of the technological innovations of developed countries, they will end up even poorer [12, 13].

Moreover, the regime implies an unacceptable form of exclusion of the many poor farmers that can't afford to pay for the patents and many fear that the *autonomy* and *independence* of farmers will be increasingly undermined by more stringent IP restrictions on saving seed. The famous report on the International Assessment of Agricultural Knowledge, Science, and Technology for Development (IAASTD) expresses "concern about present IPR instruments eventually inhibiting seed-savings and exchanges" ([13], p. 42), thereby assuming the capabilities of farmer communities to develop locally adapted varieties and to maintain gene pools through *in situ* conservation—essential to local practices that enhance food security and sustainability [14].

Secondly, as costs of bringing out a saleable product in the life sciences have constantly risen, a stricter market orientation has become even more mandatory, with the effect that industry mergers were seen as necessary in providing goods in the food and health sectors. Patents have actually been key drivers behind the increasing economic concentration (and vertical integration) of the global agrifood industry [15]. Often goods that were formerly free, like seeds, have now to be paid for by end-users due to high product development expenses. The current system has the unintended consequence that increasingly larger-scale players dominate the markets, with a foreseeable adverse effect on the rate and quality of inventions and the survival of small and medium-size enterprises. Here particularly, newcomers from the developing world face a difficult start. One urgent problem is that of patents that exceed the intentional uses of a patented product. The World Intellectual Property Organization describes this worry when noting that "there is concern that some gene patents are, for example, drafted too broadly, with the effect of overcompensating the patentee by covering all future applications."

Thirdly, there are, up to this moment, unresolved issues such as how indigenous knowledge should be treated and in what way biodiversity should be maintained, as well as questions concerning the regulation of biosafety dossiers. It has to be assessed to what extent those issues should be addressed by IP regimes themselves or how far the existence of these regimes has created a situation that demands those issues to be dealt with. The first two subject matters are often brought under one umbrella as interests of the developing world. Biodiversity is often seen as something that is vital, but there are insufficient empirical studies that provide clear evidence that industry needs biodiversity as much as is commonly stated. Success in conserving (or even enhancing) biodiversity depends very much on the outcomes of such studies, as such evidence is a huge leverage for bargaining deals for its protection and therefore the establishment of proper incentives. Something similar holds for traditional knowledge. Stating its importance as cultural heritage of mankind might not be enough to find sufficient infrastructural support—studies showing how industry has benefited from traditional knowledge will help to gather a much wider involvement in initiatives that seek to conserve and recognize indigenous scientific practices and knowledge.

Fourthly, one of the negative effects is that the system of protecting intellectual property has become extremely expensive in its demand of researchers' time and resources. Add to this that a wide contingent of legal experts has to be financed by reallocating funds originally destined for R&D, which only rich stakeholders can afford. This trend is contrary to the plea for a democratization of science, a demand for openness and inclusion, both in active participation and decision-making, elements that in the human rights discourse are encompassed in the right to share in the advancement of science (UDHR, article 27.1). So, the right to adequate food and the right to health (UDHR, article 25.1) are not the only two human rights that collide with liberties granted by the use of exclusive rights secured by IP regimes. The way the current IPR regime works restricts the freedom to operate for innovators and many comment that high-level science is treated as a luxury reserved for the developed world alone. Social justice demands more than a fairer distribution of objects of innovation and the availability of biotechnological solutions for the problems that the poorest people in the world are predominantly confronted with. Being able to participate at all levels of the innovation process and having a say on research agendas is something completely out of reach for most of the world's population. For some human rights advocates and scientists, wider participation cannot be sacrificed even for higher efficiency in making technologies available to more people, this coming in line with what is known as the right to share and participate in the advancement and benefits of science in the human rights discourse.

The fifth objection concerns the bug–drug herbicide spiral that may be summed up in the slogan: Nature fights back. This spiral debunks the so-called Patent Bargain which implies that in applying for a patent the inventor receives a *temporary* monopoly on exploitation in return for disclosing his invention and that *after* the expiration of the patent term his invention will fall into the public domain. However, the (fast) emergence of resistant bugs or weeds as a "natural" response to the widespread use of drugs and GM crops call the rationale of the patent system into question [16]. Antibiotic resistance makes the patent bargain an illusion and even favors patented pharmaceutical innovation [17]. The obsolescence of genomics drugs and transgenic crops at the end of the patent term due to the evolution of weed resistance also vitiates a key justification for the protection of intellectual property. It now transpires that once it becomes available for free public use, the invention may have almost entirely lost not only its economic value, but also its technical efficacy. Even more, a patent may induce its owner to socially *waste* a finite resource, to wit, the depletable effectiveness of means for crop protection.

ALTERNATIVES FOR THE CURRENT IPR REGIME: OPEN SOURCE, OPEN ACCESS, FOOD IMPACT FUND

There are at least three alternatives (or complements) to the current IPR regimes covering production, participation, and consumption of knowledge that take human rights as capabilities more seriously. "Open Source" or "Open Innovation" does not mean for free, but free in the sense of being transparent and unrestricted—business models are compatible with open innovation. The question of whether more openness leads in itself to fairer distribution is something that remains unresolved until further research. Many emphasize the role of commons (resources that are not owned privately but held in common) for the future production of knowledge and their potential to rebalance uneven power relations [18–21].

The second, "Open Access" [22], covers two currents in the access to knowledge movement, one that aims to build an information society where knowledge is openly available without restriction at all and the second that seeks a general expansion of the public domain. These currents are in particular relevant for initiatives for protecting traditional knowledge through exclusive rights. The attempt to protect traditional knowledge by exclusive rights is at odds with approaches based on sharing rather than appropriating knowledge. The compatibility of predominantly Western conceptions of intangible property with customary laws, and the extent to which they adequately consider the static and dynamic nature of traditional knowledge, is rather unclear.

The third approach aims at an "Impact Fund" [23]. Linking profits to positive impact on alleviating an urgent problem is of particular interest for targeted products not covered sufficiently by market incentives, such as medicines for the so-called neglected diseases or improvements in seed varieties especially targeted for the needs of the poor. The idea behind the fund is to offer a reward to companies that aim at maximizing quality-adjusted life years (QALY) of people suffering a disease or disorder. While retaining its IP rights, the company has to commit itself to sell the medicine at cost-price in order to be rewarded monies proportional to the impact in increasing QALY its medicine has. The main criticism of the impact fund idea questioned the prerequisite of patents for fund rewards and its maintaining of current power relations [24].

SUMMARY

The network character of genomics research makes it very difficult to organize the modules from a juridical point of view that assumes that property, inventions, and owners can be very strictly identified. Nevertheless, the privatization of life sciences also means that private investors want to have a return on their investments, and that propels a sometimes very strict property regime with respect to DNA codes and other nontangible objects. Those rules might be outdated and block progressive innovations, and such anachronistic legislation can produce negative costs for business opportunities.

Moreover, there is reason to be cautious about claimed and expected public domain benefits of patenting genomics products. We showed that human rights are not always stimulated by the current system. It is recommendable that debates start about the socially desirable balance between types of ownership and patenting that innovative enterprises require and the inclusive public goods protection these types are said to serve, such as human rights. Several ways are dealt with, including the current IPR system and new ideas such as Open Source and the Access to Knowledge movement, and we argued that a discussion is needed about ownership and patenting of genomics process and products and their optimal integration with the public good of human welfare, fair trade, and fair invention. The patenting of biological inventions thus raises two broad ethical issues, such as in what way to develop an intellectual property rights system that is inclusive, not only listening to the voices of patent holders but also to those stakeholders that are affected by the patent system? And secondly, how can inventions be stimulated that are specifically designed to alleviate urgent problems and reach global targets, such as the Millennium Development Goals and caps in climate change gas emissions?

REFERENCES

[1] De Schutter O. The right of everyone to enjoy the benefits from scientific progress and the right to food: from conflict to complementarity. Hum Rights Q 2011;33:304—50.
[2] Rawls J. Political liberalism, 76. New York: Columbia Press; 1993.
[3] Pogge T. A critique of the capability approach. In: Brighouse H, Robeyns I, editors. Measuring justice, primary goods and capabilities, 31. Cambridge: University Press; 2010.
[4] Sen A. The idea of justice. London: Allen Lane; 2009.
[5] Afman LA, Müller M. Human nutrigenomics of gene regulation by dietary fatty acids. Prog Lipid Res 2012;51(1):63—70.
[6] Penders B, Horstman K, Vos R. Large-scale research and the goal of health: an analysis of doable problem construction in "new" nutrition science. Interdiscip Sci Rev 2009;34(4):332—49.
[7] Penders B, Korthals M. Harvesting normative potential for nutrigenomic research. In: Ferguson L, editor. Nutrigenetics and nutrigenomics in functional foods and personalised nutrition. Boca Raton, London & New York: CRC Press; 2013. p. 361—74.
[8] Rimmer M, McLennan A. Intellectual property and emerging technologies. New Biology. Cheltenham: Elgar; 2010.
[9] Timmermann C, van den Belt H. Global justice considerations for a proposed "climate impact fund". Public Reason 2012;4(1_2):182—96.

[10] Esposo Guerrero BJ. Politics, Globalization, and Food Crisis Discourse. Economics Discussion Papers, No 2010-22, Kiel Institute for the World Economy. < http://www.economics-ejournal. org/economics/discussionpapers/2010-22 > ; 2010.

[11] McKeon N. The United Nations and civil society. Legitimating global governance whose voice? London & New York: Zed Books; 2009.

[12] Korthals M. Global justice and genomics: toward global agro-genomics agency. Genomics Soc Policy 2010;6:13–25.

[13] IAASTD, Synthesis Report of the International Assessment of Agricultural Science and Technology for Development, Washington, <http://www.agassessment.org>; 2008.

[14] De Schutter O., Seed policies and the right to food: enhancing agrobiodiversity and encouraging innovation (Report presented to the UN General Assembly, 64th session, UN doc. A/64/170); 2009.

[15] Glenna LL, Cahoy DR. Agribusiness concentration, intellectual property, and the prospects of rural economic benefits from the emerging biofuel economy. South Rural Sociol 2009;24(2):111–29.

[16] Mortensen DA, et al. Navigating a Critical Juncture for Sustainable Weed Management. BioScience 2012;62:75–84.

[17] Outterson K. The legal ecology of resistance: the role of antibiotic resistance in pharmaceutical innovation. Cardozo Law Rev 2009;31(3):613–78.

[18] Commission on Intellectual Property Rights. Innovation and Public Health (CIPIH). Public health, innovation and intellectual property rights. Geneva: World Health Organization; 2006.

[19] Jefferson R. Science as social enterprise. Innovations (Fall edition) 2006:13–44.

[20] DNDWG, Drugs for Neglected Diseases Working Group. Fatal imbalance, p. 10. http://www.doctors-withoutborders.org/publications/reports/2001/fatal_imbalance_short.pdf.; 2001 [accessed 19.05.11].

[21] Wallace H. Bioscience for life. London: Genewatch; 2010.

[22] Krikorian G, Kapczynski A. Access to knowledge in the age of intellectual property. Cambridge. MA: MIT Press, Zone Books; 2010.

[23] Pogge T. World poverty and human rights. 2nd ed. Cambridge and Malden, MA: Polity Press; 2008.

[24] Timmermann C. Limiting and facilitating access to innovations in medicine and agriculture: a brief exposition of the ethical arguments. Life Sci Soc Policy 2014;10(8).

CHAPTER 8

Teaching Genetics and Genomics for Social and Lay Professionals

Sandie Gay, Michelle Bishop and Stuart Sutherland
NHS National Genetics and Genomics Education Centre, Birmingham, UK

Contents

INTRODUCTION

Society at large is increasingly gaining knowledge and appreciation of the scope and power of genetics and genomics in many aspects of life. However, we are still in the very early days of what could turn out to be a revolution in genomic research and healthcare applications. Rapid advances in genetic technologies and ever faster and cheaper genome sequencing have generated many questions and high expectations. The impact of this revolution is likely to be felt across many professional

Genomics and Society
DOI: http://dx.doi.org/10.1016/B978-0-12-420195-8.00008-2

groups, from social science lecturers to lawyers to social workers, which are not embedded in the world of genetics. Such professionals may find themselves having to educate their colleagues, clients, students, and others about various aspects of this often complex, sometimes controversial and occasionally misunderstood subject.

BASIC TEACHING SKILLS

Social and lay professionals who find themselves with the task of teaching in this area need basic teaching skills and possibly some understanding of the fundamental concepts of genetics and genomics. The current chapter aims to provide:

- a range of practical advice and tips drawn from good, evidence-based teaching and learning practice that can inform educational activities about genetics and genomics (Section 1);
- an overview of the key genetic and genomic concepts that are likely to be required to be covered in such educational activities (Section 2).

Readers with less experience in education and training are encouraged to work through the chapter in order, although both sections can be read independently of each other. Readers with more of a background in education and training may wish to focus mainly on Section 2, although they are still likely to find useful, practical advice about educational activity in Section 1.

SECTION 1—PRACTICAL ADVICE FOR TEACHING AND LEARNING ABOUT GENETICS AND GENOMICS

A genetics teaching scenario

Teaching event: A training day at a professional body for lawyers

Topic: Contemporary issues in health care

Audience: Legal and paralegal professionals

Teaching session: One hour to facilitate an understanding and debate on the implications of direct-to-consumer (DTC) genetic testing for families using the newspaper headline seen in Figure 8.1 as a starting point

Teacher: Kromilla Osome, family lawyer specializing in health-related cases

We begin with a scenario as an approach to teaching you, the interested reader, about aspects of good practice in teaching. As we outline later, a scenario puts the knowledge and skills to be learned into a real context and allows learners to more readily visualize the concepts being covered and understand their application. As we work through the fundamental aspects of teaching and learning shortly, we will attempt to relate teaching and learning theory and evidence-based practices to this initial scenario so you can see how they relate to a 'real' context. However, once you have an overview of the

Warning over £125 test that claims to show risk of cancer

» Personalised genetic reports go on sale in Britain today   » Department of Health warns public: 'No test is 100% reliable'   » Google-backed company has already faced crackdown in US

Figure 8.1 Contemporary news stories like the one above may trigger or help to frame educational activities.

underpinning principles relating to teaching and learning, it is essential that you take a proportionate and appropriate approach when preparing for any teaching event. In order not to overload your learners, a good rule of thumb is to plan your session with a maximum of three major points you want your learners to take away.

The value of scenarios in genetics education

Stories are like flight simulators for the brain.

Chip and Dan Heath, Made to Stick (2007) p. 213

When a clinician begins a discussion with a patient about a genetic condition and when an educator initiates a debate about the ethics of genomic testing, the conversations that ensue are rarely simple and uncomplicated. For example, they are likely to require some communication about a number of key scientific concepts, about the processes involved in the testing and the use of genomic information, about values and priorities, and about the implications for other people. As we will see when we discuss the case of Angelina Jolie in Section 2, cases involving genetics and genomics, by their very nature, are multilayered and complex. This is one good reason why it is very valuable to use scenarios in education and training about genetics and genomics.

Scenarios add context and detail that make content relevant and relatable. When done well, scenarios help learners to visualize and put themselves into a story in a way that can help to forge connections with important knowledge. Scenarios encourage learners to apply their understanding to the real, multilayered, 'complex' challenges of genetics and genomics [1].

So what might this look like in the teaching of genetics? Here are a few suggestions.

Simple scenarios for application of the basics

Short patient scenarios or family histories can form the basis of exercises that can be used to encourage application and grasp of some of the fundamentals of genetics. Given the narrative detail provided, learners can be invited to apply their understanding of inheritance, of risk and probability, and of how to create a formal family history pedigree, for example. This initial, practical application of these ideas and habits can help to foster their retention more than if they were simply or only 'transmitted' or described in the abstract.

Challenging scenarios for decision-making

Scenarios that invite people to not only apply some genetic knowledge but also to conduct some analysis of the scenario and make decisions or recommendations about the scenario can really push people to learn a lot. Such scenarios can be developed with a little more effort than that required to construct basic patient narratives or family histories. Ideally, they should push the learner to grapple with some grey or contentious areas and either recommend a decision or choose from a set of plausible decisions.

A useful tactic when creating challenging, decision-focused scenarios is to introduce options or actions which are plausible but ultimately wrong or inappropriate. Scenarios should not necessarily show or push learners to what good practice looks like; they can draw learners into valuable learning by showing or spelling out the consequences of making wrong or even catastrophic decisions—the kind of outcomes that your learners will be motivated to avoid.

A couple of hours spent writing challenging, decision-making activities that help people learn through experience can be rewarded several times over. As soon as your scenario is sufficiently detailed, it is likely that it can be interrogated from a number of different angles and reused or revisited as you and your learners move on to other topics.

The characteristics of a good scenario

What does a good challenging scenario look like? Ideally, it should:

- be sufficiently concrete; for example, using realistic names, dates, ages, and locations;
- contain some telling human detail; for example, about behaviors, feelings, and conversations;
- offer plausible options for the learner to select from and provide well-crafted, detailed feedback about the consequences of each option;
- draw the learner into a situation which requires some weighing up of alternatives or analysis of grey areas where they have to draw on a range of knowledge or skills.

A good decision-making scenario might be extended further by being turned into a staged or sequential scenario in which, much like in the real world, not everything happens or is discovered at once. In sequence, learners can be provided with parts of the story and invited to make decisions and anticipate consequences. The consequences of those decisions (right or wrong) can then be explored further as more detail is provided and new decisions need to be taken. Wrapping this series of analyses and decisions around a single, developed narrative provides learners with a strong set of hooks into the knowledge we wish them to acquire and retain.

Starting a teaching session with a scenario

It may seem counter-intuitive to suggest that you should start a teaching session (whether face-to-face or online) with a challenging scenario before you have covered the information that needs to be known to successfully deal with the scenario. However, this approach has many benefits and several studies suggest that when we challenge learners first and then provide instruction, we can improve their ability to apply what they have learned to other situations [2].

Rarely will you encounter a group of adult learners who all have exactly the same level of knowledge about genetics and genomics. To cover everything before you let your learners loose on scenarios might both overwhelm some of your learners with new information at the same time as patronizing or boring others with material they already know. So why not throw them in at the deep end with a challenging scenario?

Starting with a scenario can achieve a number of things:

- It can help learners to determine the level of their pre-existing knowledge.
- From the mistakes they make, it can show learners where they need to focus.
- Carefully crafted feedback supplied in response to mistakes can foster understanding of the basics.
- A concrete, detailed narrative can provide learners with a memorable set of hooks on which to hang the more abstract information that may follow in the teaching session.

As the instructional designer Cathy Moore writes: "If you want to make sure everyone has the same basic knowledge before continuing, design activities that let people either prove they know the basics or discover the basics through feedback" [3].

Preparing for teaching

For any teaching event, including the one described in our scenario earlier, there are many factors that need to be considered in order for the event to be effective. As we work through these factors, it is essential for you, the reader and teacher, to bear in mind that the factors covered here are not exhaustive and often context dependent. Also they are not sequential, so, although we begin thinking about preparing to teach in this particular scenario with knowing your learners and their needs, in other situations it may be the aim or purpose of the session that drives the preparation. And, of course, a number of factors may already be defined in the teaching assignment you undertake; for example, for Kromilla Osome in our scenario, the overarching topic, the purpose of the session, the audience, and the duration have been specified.

Knowing your adult learners

We are assuming that, as in our scenario, your learners will mainly be adults or young adults (16–18 years of age) so it is important for teachers and educators to have some awareness of how adults learn, and of the purpose of their learning, which will influence their motivation and aspirations for their learning.

There has been much research in many aspects of the human psychology that lend themselves to learning. Among others, a well-accepted theory of how adults learn is that propounded by psychologist Lev Vygotsky (1896–1934), namely social constructivism. Briefly, Vygotsky found that people learn more effectively in groups where they can bounce already known knowledge off each other to construct new learning built on discussion and collaboration. Vygotsky also described the Zone of Proximal Development suggesting that learners are able to achieve more with support from a teacher close to hand, than if left to learn by themselves [4].

We now, also, accept that adult learners have fundamentally different characteristics to younger learners. Malcolm Knowles (1913–1997), learning theorist and practitioner, in his book "*The Modern Practice of Adult Education*" [5] initially distilled the differences into four characteristics and a fifth one was added later. These should be acknowledged and accommodated when preparing a teaching session:

- Adults need to know why they are learning, what use it will be to them.
- Adults have had previous life experiences and undertaken learning.
- Adults generally have a reason for learning so are said to be self-motivated.
- Adults like to direct their own learning so what they learn makes sense to them and gives satisfaction.
- Adults are usually ready to learn as they have chosen to do so.

Although introduced by Wolfgang Kapp in 1833, Knowles adopted the term 'andragogy' to embody the characteristics of adult learners to mean teaching and learning that is 'learner-led;' that is, it is the needs of the learner, as described earlier, that are the primary consideration when developing teaching sessions for adult learners. Andragogy has overtaken the concept of 'pedagogy,' the latter meaning teaching and learning that is essentially 'teacher-led;' that is, the teacher decides and is responsible for what is to be learned, and it is andragogy that underpins the preparation for adult learning sessions.

In our scenario, Kromilla, more often known as Krom, will need to assure her understanding of her learners by considering the profile of her audience and their needs—what do they want to get out of the session?

Learner profile

Depending on the circumstances and teaching context, you can use various sources of data to build up a picture of your learners and their learning needs including specific data (age, gender, qualifications, job, disabilities, and requirements for reasonable adjustments) such as that given on application forms and specific needs (motivations, interests, and aspirations) from discussions with individuals. Additionally, social and population data can highlight learners' backgrounds and environments which may have influenced their attitudes, behaviors, and opinions.

In our particular scenario, Krom is in a good position to know the profile of her learners as they form a well-defined group of people; that is, legal and paralegal professionals. By knowing her audience, the amount, type, and level of learning undertaken so far, Krom will be able to estimate:

i. *The audience's level of general and specific scientific knowledge essential for the intended session.*

Krom could assume some knowledge and understanding of general scientific terms such as 'genetic,' 'DNA,' 'genes,' as well as understanding of some concepts such as how some of our genetic information is duplicated among members of our family members. Additionally, for the debate to make sufficient sense for participants to engage, they will need to have a degree of scientific knowledge related to the topic of the debate; for example, some understanding of what the direct-to-consumer (DTC) tests will be analyzing and what the results will be telling the recipients of the report. Krom will also need to help her audience have some idea about why there is a debate at all; that is, the questions to consider about DTC tests: the validity, reliability, and accuracy of the tests, implications, and ethical issues arising from undertaking tests for the individual and their family, concerns around storage and access to their personal data, confidentiality and privacy issues, access to counseling that might be required once results are returned, societal implications of individuals having knowledge about themselves previously not available, possible questions about financial implications such as life insurance policies, possible treatment options if any, etc. We explore ethical, legal, and social implications of genetic and genomic testing in more detail in Section 2.

Taking into account the duration of her session, Krom will need to choose carefully the relevant knowledge and understanding needed and the depth and breadth of her coverage. Refer to Section 2 for further details on gauging the depth of scientific and clinical aspects in your teaching.

ii. *Metaphors, similes, and analogies that she can use to facilitate the audience's understanding.*

The scientific concepts and the language of genetics can often appear dense and complex for learners, just as they can for patients. To unlock the important knowledge contained in the language of genetics and genomics, good genetic counselors and clinicians have a store of metaphors that they call upon to help them support their patients. Good genetics education can really benefit from a similar approach.

The power of metaphor lies in the fact that it can draw its audience into the sense-making involved. For example, by suggesting in a metaphor that "A is B," or in a simile that "A is like B," we are inviting the listener or reader to bring their knowledge of B to bear in order to make sense of A. The astute clinician who knows their patient well, and the astute educator who knows their learners well, can really help patient and learner to grasp important knowledge when they craft metaphorical descriptions that connect with their patients' or learners' worlds.

Knowing her audience's profile will help Krom to identify which metaphors, similes, and analogies will help her audience to grasp the essential concepts. For example, in our scenario, Krom could use the metaphor of the blueprint to help her learners understand replication and possible errors which may lead to disease.

Learning styles

Literature abounds about individual people's learning styles. For example, Honey and Mumford say that most learners fall into one of four categories of learning style [6]:

Activists—these learners like to try things out and understand more by doing things.

Reflectors—these learners need to have things explained and time to reflect before trying to do it for themselves.

Theorists—these learners want to know all the pros and cons before they want to have a go.

Pragmatists—these learners have a watch and learn approach.

Honey and Mumford learning styles (Figure 8.2) are based on David Kolb's experiential learning cycle [7] (variously phrased as theory, inventory, or cycle). One can be superimposed on the other.

VARK is another popular model of learning preferences [8]. Individuals are often said to be visual (V) learners preferring to have pictures and graphs to aid their understanding, others are predominantly aural (A) and like verbal explanations and discussions. Readers (R) gain most by doing just that and kinesthetic (K) learners feel that they need to get a 'hands-on' experience in order to really build their knowledge and understanding about a subject. There are obvious similarities between the learning preferences outlined by the VARK model and Honey and Mumford's model.

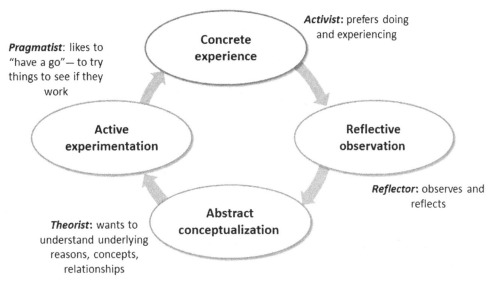

Figure 8.2 Honey and Mumford: typology of learners. *Adapted from The Experiential Learning Cycle (http://www.learningandteaching.info/learning/experience.htm), last accessed 01.12.14.*

More can be learned about learning styles from literature found in the library or the Internet; however, although, adults will have a dominant preference for learning, mostly, these will overlap with the styles that best fits the learning they wish to engage in.

Often, your own learning style will be reflected in your teaching approach but in general, if you can incorporate variety in your approach to teaching and learning, it will help to accommodate a mix of learning styles and at the same time, allow learners the flexibility in the structure that will help them to help themselves.

Session aim and learning outcomes

Defining the aim can help drive the preparation of a teaching session. If you don't know where you want to get to, you won't know how to work out, how to get there, or if you have got there.

For Krom, the aim would be to 'educate' her learners about DTC tests and provide them with sufficient understanding about their purpose and potential issues for learners to engage and participate in the debate.

Learning outcomes break down the aim into manageable and measurable steps; that is, a learner must be able to know/understand/do these steps in order to achieve the aim (for their learning). Developing measurable learning outcomes allows you and your learners to know if they have been achieved. Teachers can use the SMART framework shown in Figure 8.3 to help develop the learning outcomes for their teaching and learning sessions.

Specific — exactly what is to be learned, this often involves providing a way to measure the learning, for example three types of genetic tests
Measurable — allows the learner and teacher to have a concrete measure of what is learned
Achievable — by the learners taking into account the prevailing circumstances including ability and available time
Relevant — to the learner, their purpose for learning (see Knowle's characteristics of adult learners)
Time bound — by when will the learning outcomes be achieved

Figure 8.3 SMART framework.

What happens after your teaching session? Preventing forgetting through spaced practice

An ounce of practice is worth more than tons of preaching.

Mahatma Gandhi

While the ideas given earlier can help to shape a positive teaching session, and grappling with scenarios can encourage strong cognitive connections with important knowledge, these connections are liable to be weakened if the material is not revisited and its application is not repeatedly practised. A common weakness of so much training and education is to run the class or the 'sheep-dip' training event or the online session as a single, one-off event without any planned, later activity to promote reinforcement through spaced practice. This flies in the face of research into memory and forgetting which suggests that most of what we supposedly teach and learn is lost within minutes and hours.

In 1885, Hermann Ebbinghaus, a German psychologist who pioneered studies of memory and learning, developed his thesis of the exponential nature of forgetting and the speed with which information is lost when there is no attempt to retain it. Ebbinghaus is known for his 'forgetting curve' which suggests that people tend to continually halve their memory of newly learned knowledge in a matter of days or weeks unless they actively review the learned material [9]. Ebbinghaus also studied and suggested the value of practicing at spaced intervals over time to promote retention (Figure 8.4).

If so many of our 'traditional' means of teaching and training are as unproductive as the forgetting curve suggests, how might we design some spaced practice into our genetics education and training to promote retention? Here are a few suggestions:

- If you have more than one teaching or training session, beginning the session by requiring learners to apply what was covered in the previous session to a new scenario or context is a very worthwhile activity.

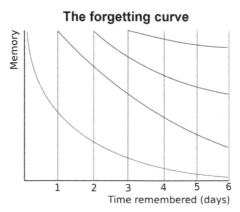

Figure 8.4 A typical representation of the forgetting curve. *(Image by Icez, from the Wikimedia Commons.)*

- Use e-mail. A group e-mail to all of your learners, pre-planned to be sent at a specific point in time after your teaching session, can push your learner to a reinforcement activity, which could be as simple as text in the e-mail, a link to an article, graphic, or video or a new scenario or piece of e-learning. Any activity which prompts your learners to further thinking about, and application of, important knowledge will help to fix the learning in the long-term memory.
- Use online discussion or social media. Set up a discussion board or an online group which you invite, or better still require, your learners to visit to grapple with a new question or scenario that you post for them. If you do not have access to any such tools and your learners are familiar with Twitter, you could create a hashtag for your course or session and then ask learners to respond to a question or activity, quoting the hashtag in each of their posts. When they click on the tag in any post, your learners can then see all of the posts containing the tag grouped together in one place. The social media hashtag is a simple tool for keeping a group of learners together for some spaced practice, revisiting and rethinking a topic over time.

The tone of voice and the approach adopted by BMJ Learning in its promotional e-mails about its online courses for medics offer an excellent, generalizable model for how both scenarios and spaced practice can be used to promote learning. Several days after your teaching session, why not contact your learners with a piece of communication—with less clinical detail but with the same problem-focused approach—like the following opening paragraph of a BMJ Learning e-mail (24/11/2014).

"Dear Colleague,

A 69 year old man attends your clinic for a CVD risk check. He has a history of ischaemic heart disease and stenting four years ago. You are surprised to note that he has never been on a statin. His blood tests demonstrate chronic kidney disease stage 3a with an estimated GFR of 52 ml/min/1.73m^2. Which statin should you start him on and at what dose?"

SECTION 2—KEY GENETIC AND GENOMIC CONCEPTS TO COVER IN A TEACHING SESSION

Later, we explore some of the core genetic and genomic concepts that are likely to inform or be covered in a teaching session. In doing so, we will make the assumption that you already have a decent understanding of the fundamentals of genetics and genomics. (If you do not, visit www.geneticseducation.nhs.uk and www.genomicseducation.hee.nhs.uk.)

As we outlined earlier, using scenarios can be a powerful tool to provide structure and context to a teaching session about genetics. Therefore, to illustrate how key concepts in genetics and genomics might be explained, we will use a recent high-profile case of genetic testing—that of Angelina Jolie.

In May 2013 Angelina Jolie wrote about her experience of finding out that she had a gene alteration in the *BRCA1* gene and her decision to undergo a preventative double mastectomy to reduce her chance of developing breast cancer [10]. Using this example, we outline the key concepts that could be covered in a teaching session, and how to relate these topics to Angelina Jolie's story. In each situation, we consider the audience and how this may change the focus of the session.

Scientific and clinical aspects

When teaching about genetics and genomics, scientific concepts will often be a core component of your teaching session. However, as we suggested Section 1, the depth and breadth that you need to cover will be determined by your audience (and their presumed prior knowledge on the subject), as well as the key messages you want to address. While it can sometimes be tempting to overload your session with scientific information, unless you have multiple sessions with your audience, your key messages may get lost in the science. Where you do have multiple sessions, as we outlined earlier, spaced practice and review over time aids retention, rather than overloading an individual session with detail that your learners are likely to forget. In our experience, a good rule of thumb in determining how much science to include is to ask yourself: What genetic knowledge does my audience need to know in order to understand the key messages?

Let us consider the Angelina Jolie case. If your key messages relate to the personal decision-making process for women in similar situations, simple information about genes (and the fact that we all have two copies of the *BRCA1* gene) and the basic effect of different genetic alterations may be sufficient information for your audience to engage with the subject matter. Specific clinical information about breast and ovarian cancer may also be required to provide context around the different treatment and management options. If, however, you wanted to use this scenario to introduce cancer genetics more broadly, you may consider introducing topics such as the role of tumor suppressor genes and the differences between germline mutations (those genetic alterations that we have usually inherited from our parents) and DNA alterations that we accumulate during our lifetime (somatic mutations).

Inheritance and family history

Genetic information is shared information. Therefore, teaching sessions that use patient scenarios will inevitably involve the discussion of the inheritance of particular genetic conditions. We inherit our genetic information from our parents and pass one of each copy of our chromosomes to our children. Therefore, when someone finds out they have a genetic alteration in one of their genes, this often has implications for other family members.

In the case of Angelina Jolie's story, a teaching session may cover the implications for her biological children, including working out the probability that they have also inherited the same genetic alteration. For the audience to understand these probabilities, you may first need to cover the different modes of inheritance of single gene conditions and how to calculate the recurrence risks for each mode of inheritance. In our experience, using graphics or animations to explain the passing down of genetic information from a parent to a child can be helpful teaching tools (Figure 8.5). (For example, have a look at

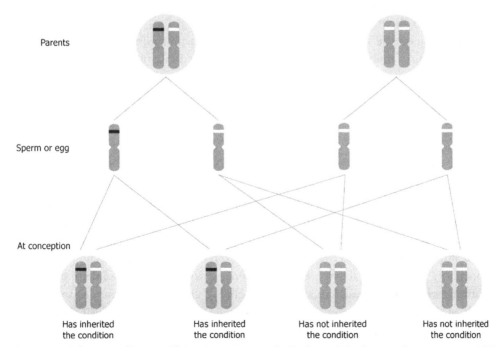

Figure 8.5 Schematic diagram illustrating autosomal dominant inheritance. *(Image source: NHS National Genetics and Genomics Education Centre)*

Table 8.1 Common misconceptions

Common misconceptions

Chance has no memory	With an autosomal recessive condition there is a one in four chance the child of two carrier parents will be affected. Some people misinterpret this as one in four of the children will be affected. Chance has no memory so this probability is for each pregnancy.
Independent segregation of genetic information	Just because you look like your parent doesn't mean that you will have the same genetic condition.
Inheritance of conditions related to sex specific organs (e.g., breast and ovarian cancer)	Misconception that the genetic alterations associated with breast and ovarian cancer can only be inherited through the maternal line.

animations from the National Health Service (NHS) Genetics and Genomics Education Centre (NGGEC) on YouTube at http://bit.ly/1gFS7Zt.)

Talking about inheritance can serve a dual purpose. Your audience will gain an understanding of the inheritance of genetic conditions and it provides you with an opportunity to correct any misconceptions around inheritance. Some common misconceptions about inheritance are outlined in Table 8.1.

Depending on the scope of the teaching session, further information could also be presented about family history in general. Practicing drawing a family history can be used as a group activity to engage the learners in the teaching session. The exercise then allows the teacher to demonstrate how a genetic family history can be used in clinical practice for genetic risk assessment and for making clinical decisions regarding management.

Genetic and genomic testing

Depending upon your audience and the aim of your session, discussion around genetic and genomic testing could range from the scientific methods employed through to the availability and cost of each test. It may also be applicable to present and discuss the range of clinical information that these tests provide. If your audience are healthcare professionals, reading and interpreting test results may provide an opportunity for a group activity. For an audience made up of lay people, a discussion regarding what type of tests are available, eligibility criteria for testing, as well as the length of time taken for results may be appropriate. For instance, when talking about the Angelina Jolie case, it would be important when talking to a UK audience to stress the referral criteria for *BRCA1* in the NHS whereas in the United States the focus may be on eligibility criteria defined by insurance companies.

Ethical, legal, and social implications

Many of the genetic topics that may be taught will often lead to a discussion regarding ethical, legal, and social implications. While there isn't the scope in this chapter to discuss all of these in detail, we will highlight a few specific topics that could be discussed using the Angelina Jolie case study. Often, introducing an ethical dilemma as part of your teaching session will provide an opportunity for the audience to discuss the situation within the group. As a starting point you could pose a series of statements and ask your audience to rate how much they agree with each statement. Alternatively, as we suggested earlier, you might start your session with a scenario containing an ethical dilemma and invite your audience to make decisions about it. If you can, use an electronic voting system. This allows your audience to provide their opinion in an anonymous fashion. In our experience, this method opens up the floor for discussion, and in some cases, heated debate. During the discussion, encourage your audience to see the issue from different perspectives; this often allows people to appreciate the complexities around genetics and genomic testing.

In the case of the Angelina Jolie story, some of the issues that may be raised include:

- Access to genetic testing: Is there equality in who has access to testing? What referral criteria are in place (if any)? What financial cost is there to the patient?
- Implications for family members: Who else in the family may have inherited the genetic alteration? How will this information be shared with family members?
- Testing of children: Should children be tested for this genetic alteration? What about their right to make an autonomous choice?
- The right "not to know": What if someone in the family doesn't want to know if they have the genetic alteration? Will this affect their medical management in screening for breast and ovarian cancer?
- Respecting individual choice regarding management and treatment: Different people may have different choices about what they do with their genetic information. What if you, as a health professional, don't agree with the patient's choices?
- External implications: Will knowing you have a *BRCA1* gene alteration affect your current or future insurance policies?

For the majority of these situations, there are guidelines and policies in place to aid healthcare professionals to facilitate the testing process and answer patient questions. However, teaching sessions offer an environment where these issues can be debated and discussed, allowing people to gain a greater understanding of the issues involved and their own views on these topics.

With the advent of whole genome sequencing, there are additional considerations that need to be made and debated. Two of the main issues currently being discussed (at the time of writing) relate to the data obtained from the sequencing process. Previously, the healthcare system would 'look' at a specific region of the genome to ask a question about whether there was a genetic alteration present or not. Now, scientists are able to look at the whole genome, so you may find out about genetic alterations in a number of genes that were not part of your original diagnostic question. Should all this information be provided to the patient? What if it is not clear what the clinical effect is of these genetic alterations? What if there is no treatment or clinical management available?

Another issue relates to the storage and access of this data. The data generated by whole genome sequencing is vast. This information needs to be stored securely to protect individuals' health information. How is this going to be achieved and who will be responsible? In addition, thought needs to be given to how and when people can access this information. Should it be available for all health professionals involved in a person's care and even to researchers investigating similar conditions? Does the individual need to give consent each time, or should it be treated as part of their medical record?

Language

When teaching about genetics and genomics you are often modeling good practice in how to describe scientific concepts. Therefore, it is important to consider the words you use in your teaching session. Many of the words used to describe scientific concepts in genetics can be emotive. For example, you may hear people describing a gene that has a genetic alteration as a 'bad gene' or an 'abnormal gene.' Angelina Jolie described herself as having a 'faulty gene.' In research conducted by the NGGEC in the United Kingdom, people directly affected by a genetic condition stated that these terms were often unhelpful descriptors and in some cases hurtful [11]. To aid those people who discuss genetic concepts, the NGGEC compiled a list of alternative ways in which genetic terms can be explained (Table 8.2).

Table 8.2 Alternative ways to describe genetic terms

Genetic term	Recommended alternative
Genetic disease/genetic disorder	Genetic condition
Normal gene	Usual gene
Bad/abnormal/faulty gene	Altered gene
Mutation	Alteration
Risk	Chance/probability
Normal person	Unaffected by
Cystic fibrosis carrier/carrier of cystic fibrosis	Carrier for cystic fibrosis
Sufferer	Affected by
An achondroplast/a Down syndrome baby	A person with achondroplasia/a baby with Down syndrome

There are also words that are used in a genetic context that are used in different health situations. One word that often causes confusion is 'carrier.' This word has different meanings depending on the context in which it is being used. For example, the word can be used to describe the health status of someone with an infectious disease (such as an HIV carrier). This can lead to the misconception that when the word is used in a genetic context (e.g., a carrier for cystic fibrosis) people assume that this condition is also contagious or that transmission is more likely. In these situations, taking the time to explain the terminology in the specific context will negate any misconceptions.

SUMMARY

To run an education and training activity about genetics and genomic is to lead your learners to explore a series of complex, multilayered and potentially controversial issues and questions. Some key takeaway points from this chapter for professionals leading such an activity are:

- Go into your educational activity with a maximum of three major points you want your learners to take away.
- Do your utmost to know who your learners are and what prior knowledge and experience and preferences they will be bringing with them.
- Make time for sufficient and appropriate preparation when undertaking teaching about genetics and genomics.
- Consider using scenarios to allow your learners to grapple with the real, complexity of genetic and genomic cases.
- Consider how to prevent your learners from forgetting key ideas after your teaching session and think about how your learners can benefit from repetition, revisiting, and spaced practice.
- That education and training should seek to address misconceptions and misunderstandings surrounding genetics and genomics.
- The language that you use to describe key genetic concepts can be rich in connotation and implications, both positive and negative. Appropriate metaphors and analogies can be very powerful in successfully communicating key genetic and genomic concepts.

REFERENCES

[1] Bean C. The accidental instructional designer. Alexandria, Virginia: ASTD Press; 2014. p. 133–143
[2] Clark RC, Mayer RE. Scenario-based e-Learning: evidence-based guidelines for online workforce learning. San Francisco, California: Pfeiffer; 2012.
[3] Moore C. 5 quick ways to pull learners into a course. <http://blog.cathy-moore.com/2014/05/5-attention-grabbing-ways-to-start-a-course/> [accessed 14.12.14].

[4] Vygotsky LS. Thought and language. Cambridge [Mass]: M.I.T. Press; 1962.

[5] Knowles M. The modern practice of adult education. New York, New York: Association Press; 1980.

[6] Honey P, Mumford A. The learning styles questionnaire. Peter Honey Publications. <http://www.peterhoney.com/content/LearningStylesQuestionnaire.html> [accessed 15.12.14].

[7] Kolb D. Experiential learning: experience as the source of learning and development. Englewood Cliffs, New Jersey: Prentice-Hall; 1984.

[8] Marcy V. Adult Learning Styles: How the VARK©learning style inventory can be used to improve student learning. Perspective on Physician Assistant Education, Journal of the Association of Physician Assistant Programs 2001; 12(2).

[9] Ebbinghaus H. Translation of memory: A contribution to experimental psychology. (H.A.Ruger & C.E.Bussenius Trans.) New York: Teachers College, Columbia University, 1913 (Reprinted Bristol: Thoemmes Press, 1999).

[10] Jolie A. My medical choice. 2013. The New York Times. <http://www.nytimes.com/2013/05/14/opinion/my-medical-choice.html?_r=0> [accessed 15.12.14].

[11] Burke S, et al. The experiences and preferences of people receiving genetic information from healthcare professionals. NHS National Genetics Education and Development Centre. Birmingham, UK; 2007.

CHAPTER 9

Engaging and Empowering Public and Professionals in Genomics

Maggie Kirk, Rachel Iredale, Rhian Morgan and Emma Tonkin
Genomics Policy Unit, Faculty of Life Sciences and Education, University of South Wales, Pontypridd, South Wales, UK

Contents

Genomics and Society
DOI: http://dx.doi.org/10.1016/B978-0-12-420195-8.00009-4

INTRODUCTION

Recognizing that advances in genetics and genomics would radically alter how we understand health and disease, the Genomics Policy Unit (GPU) has been exploring the preparedness of the public and healthcare professionals for the "new genetics" since 1996. The aim of our research has been to show audiences who might not typically engage with genomics, what new opportunities are being offered to improve human health and the social and ethical issues that surround these. For health professionals such as nurses, this is particularly important because advances in genomics mean it is moving out of its specialist sphere to all areas of clinical practice. In this chapter, we shall be describing major strands of our work through two initiatives, one focusing on members of the public and one on health professionals, which have centered on engaging and empowering people in relation to genomics.

Engaging nurses in genomics

Engaging nurses to incorporate genomics within mainstream healthcare practice is a global challenge for nursing. Inconsistent education provision, a dearth of relevant resources, limited knowledge, and lack of confidence are some of the hurdles in its implementation [1,2]. The GPU has taken a systematic approach to address these issues at national and international levels and from 2004 to 2012 it formed the basis for the nursing program at the National Health Service (NHS) National Genetics and Genomics Education Centre. As well as developing education frameworks to guide curricula, in response to needs analyses, we have also developed resources to support teaching and learning.

Our work uses real-life stories from individuals and families to promote understanding of how genomics impacts on people's lives, helping health professionals understand its relevance to their practice. Led by the GPU, in partnership with Plymouth University, Genetic Alliance UK and the NHS National Genetics and Genomics Education Centre, the web-based education resource *Telling Stories, Understanding Real Life Genetics* was the first to apply thematic analysis to set patient/carer stories within an education framework. How those stories can be used to engage learners and empower educators is the focus of the second section of this chapter.

Public engagement

Studies into the public understanding of genomics have often reported that the public's understanding is poor. By using highly interactive research approaches, we have demonstrated that "ordinary people" can hold and discuss complex social and ethical views about genomics. One outcome of a study seeking to engage young people in exploring the relation between genes and cancer (with twenty 12- to 13-year-old schoolchildren) was the production of their own genetics rap (Let's Talk About Genes; www.youtube.com/genomicspolicyunit). This study forms the basis for the first section of this chapter.

SECTION 1—LET'S TALK ABOUT GENES AND I DON'T MEAN TROUSERS: CHILDREN'S VIEWS ON GENOMICS AND CANCER

Much of the research into the public understanding of genomics has been quantitative and short-term, using traditional social scientific methods of inquiry. Research with young people concerning genomics has been limited, concentrating mainly on risk or communication between parents and children [3,4]. Young people are the generation for which genomics and reproductive technologies will have the most impact in the twenty-first century. It has been demonstrated using innovative methods, such as Citizens' Juries [5] and Mock Trials [6], that in-depth research with young people is both possible and necessary, and that young people have sophisticated, multilayered understandings of the concepts and issues surrounding genomics which they can express in a variety of ways, through a variety of media.

Acquiring genomics literacy is one of the most important things a person can do to promote their own and their family's health [7]. A family history, reflecting genetic susceptibility as well as shared environmental and behavioral factors, is a significant risk factor for many diseases, including cancer. We know very little about children's attitudes to cancer genetics, particularly in relation to the risk of inheriting a predisposition to a common cancer like breast, ovarian, or bowel cancer. Previous research with teenagers demonstrates that most have well-formed attitudes to, and habits, regarding, diet and lifestyle [8]. If such attitudes and habits are poorly informed yet entrenched by the teenage years, this has important implications for public health and health promotion. Involving children in discussions about genetics, risk, and common disorders can facilitate better communication within families about health and disease, and is likely to help families cope better with illness in the future.

METHODOLOGY

"*Let's Talk About Genes and I Don't Mean Trousers*" is a research project funded through a Tenovus Cancer Care Innovation grant, involving a team from the University of South Wales, working with a group of twenty 12 and 13 year olds from Stanwell School in Penarth participating in 3 days of workshops. The children were exposed to cancer genetics learning using fun activities and games, as well as creative exercises. Considerable attention was paid to encouraging self-reflection and self-understanding among participants.

Aim

The aim of the project was to explore how feasible and how acceptable it is to discuss cancer and genetics with schoolchildren. In particular, the following questions were of interest:

- How aware are children of genetics?
- How aware are they of cancer?
- Which cancers are they most informed about?
- Are they able to explore the role that diet and lifestyle plays in health and disease?
- Are they able to explore the role that family history plays?
- Can children act as disseminators of information to their peers?

Assumptions

Research with children can be difficult [9] and there were a number of ethical issues which had to be overcome [10]. Consequently, were a number of assumptions that underpinned our methodological approach, namely that these children will:

- Engage with cancer and genetics over an extended period of time, given enough time, opportunity, and support
- Think creatively about family history and cancer
- Acquire a cancer genetic literacy that makes personal sense to them and is age appropriate
- Be the coproducers of rich qualitative data that reflects their thoughts in a way that can be shared with other children of a similar age.

Recruitment

We identified named staff within the school to assist in the recruitment of participants and as points of contact for the duration of the project. Recruitment was undertaken through the school over a period of 4 weeks. Letters promoting the project were sent to the home address of all parents of year 8 children (n = 350), inviting them to consider participation for their child. Each letter provided details of what would be involved, as well as parental and child consent forms. Parents and children were also provided with an opportunity to meet with project staff to discuss participation and the proposed activities in more detail prior to consenting; 65 parents registered with the project for their child to take part, 36 boys and 29 girls. Of these, 20 children were randomly selected to participate (10 boys and 10 girls). Three full-day workshops were organized and all 20 children attended on days one and two, and one boy was absent on day three.

RESULTS

Children's attitudes

Voting handsets were used to gather information from participants about their attitudes to, and understanding of genetics, health, and cancer, at both the beginning and end of the project. The data was collected electronically via Turning Point technology and the group engaged quite quickly with this activity. We also issued attitudinal surveys at the start of day one and at the end of day three; 20 children participated in the survey on day one and 19 on day three.

Views on health

Participants indicated a positive interest in the topic of health in both rounds of voting. At the beginning of the project, 70% of the group indicated that they would describe themselves as healthy with two participants indicating they did not see themselves as healthy at all. At the end of the project no one disagreed with this statement suggesting they had a more positive perception of their own health after engagement in project activity. Results from the first vote indicated that participants did not discuss their own health with their peers. Children may be reluctant to discuss their health with friends perhaps through fear of teasing or concern at what their peer group may think. By the end of the project, data were a little more positive in that there was a reduction in negative responses to this question. Voting results also suggested that project activity triggered participants to think about their own health, but also provided them with some reassurance and confidence when discussing health issues.

Views on inheritance

On day one, 75% of participants either agreed or strongly agreed that some diseases run in families and this increased to 100% at the end of the project; 95% of participants indicated they would want to know if there was a disease running in their family. Most participants responded favorably to it being useful to know about the health of their relatives in both rounds of voting. Most disagreed that if your parents are healthy, you do not need to worry about getting a disease that has affected other family members; 70% of the group at the start of the project said they would change their lifestyle if a disease was running in their family, with all but one person indicating they would make changes by the end of day three. This increase in the level of agreement suggests that participants responded positively to messages about lifestyle factors and health.

Views on genetics

Eighty-five percent of the group indicated they were interested in genetics with only one participant disagreeing at the start of the project. Results were very similar at the end of the project, again with one participant indicating they were not interested in genetics. Half of the group rated their knowledge of genetics as about average at the start of the project; 10% felt less informed and 20% did not know. A significant result at the end of the project is that all participants said they now rated their knowledge of genetics as more informed than the average person.

Results from the electronic voting show that the family is children's primary source for talking and learning about genetics at this age. Before the project, 45% of the participants had talked to someone in their family about genetics and only 25% indicated they had spoken to friends about genetics; 15% said that they had learned about genetics in school and only 5% indicated they had learned about genetics from the Internet. Considering the high usage of the Internet among this age group, it is surprising that more did not report accessing learning about genetics in this way, indicating that their online behavior excludes them from any exposure to the topic of genetics.

At the start of the project 90% indicated that they thought it would be useful to know more about genetics. This is unsurprising considering they had volunteered for the project in the first instance. However this rose to 100% by the end of the project indicating participants had a positive experience of engagement in genetics. When asked if genetics is easy to understand if explained properly, results are similar from both voting rounds with the majority of participants in agreement with this statement; 40% of participants at the start of the project indicated that they think about their genes in relation to their own health and by the end of the project none of the participants disagreed with this statement.

Views on cancer

Before day one of the project, 80% of participants said they had talked to someone in their family about cancer and 65% had talked to friends; 85% said that talking about cancer is upsetting. Participants were asked if they know someone who has cancer and results were similar for both rounds of voting with over 80% indicating they did know someone with cancer. No signs of distress were observed or reported from cancer-related discussions and activities undertaken by participants, suggesting that discussing this topic can be upsetting rather than being personally distressing for children.

Half of participants at the start of the project said that getting cancer is more to do with genes than lifestyle; however, after exposure to cancer genetics learning over 70% of participants said they did not think that genes were more important than lifestyle choices indicating recognition of the role of external factors on cancer and perhaps an appreciation of preventative behaviors. An overwhelming majority of participants agreed that sun beds, sun bathing, and sun burn increases their chances of

Figure 9.1 Q&A sessions.

developing cancer. Alcohol and binge drinking were also strongly indicated as risk factors likely to increase their chances. Initially, 95% of participants said that smoking increases their chances of developing cancer and this rose to 100% by the end of the project. There seemed to be some uncertainty within the group about the effect of eating genetically modified or organic foods and there appeared to be confusion for some participants as to what these were.

In the first vote 40% of the group thought that processed meats such as burgers, sausages, and ham increased their chances of developing cancer. By the end of the project this rose to almost 90%. Messages delivered about diet and cancer on group days were accepted by the majority of participants following engagement with the project. There were mixed reactions to the idea of mobile phones, telephone masts, electricity pylons, and X-rays increasing your chances of developing cancer, and responses were perhaps influenced by urban myths surrounding these environmental factors.

During a Q&A session with a cancer expert, the group asked about some of these factors and their learning is reflected in the change of responses at the second vote (Figure 9.1). At the start of the project, 90% of the group said they think that being overweight increases your chance of developing cancer and this rose to 100% of participants for the 16 who responded at the end of the project. Most (85%) said that family history was a factor for developing cancer and this rose to 100% again at the end of the project, indicating a raised awareness and appreciation of the role of family history in cancer.

Participant evaluation

Participants' evaluations of their learning show some positive results. Most of the children responded favorably when asked had the sessions increased their understanding of genetics and inheritance. Most children also indicated that the sessions had been useful for increasing their understanding of risk. Participants indicated they were also more confident to talk about both family health history and cancer with 100% of participants reporting that they know about behaviors that can reduce the risk of developing cancer.

Participants were asked to rank the top three ways they think are best for communicating to young people important messages about cancer prevention. Celebrities came top with 26% and Internet resources came second with 22%. Specialist workshops and special events both scored around 13% as the group's third favorite method of communicating messages to others.

When asked how much fun and engaging the activities were, the majority of the group gave positive responses. Only one participant indicated that resources used were "*rubbish*," however the majority of the group described resources as either "good" or "really good." The level of acceptance and engagement in project activity is reflected in the final voting result with all participants indicating they would recommend this project to other young people.

FAMILY HISTORY AND INHERITANCE: GROUP ACTIVITIES

Each day of the project involved individual and group work, as well as games and creative activities. Some of these are described in the following sections.

Group collage

In four groups, participants were asked to create a collage to represent what they understand about family history and inheritance (Figure 9.2). Groups used magazine cuttings, hand drawn images, and words to reflect their ideas. All four collages were then reproduced in a word document and imported into NVivo for analysis.

Figure 9.2 Making a group collage on inheritance.

The majority of references obtained from this activity demonstrate an understanding of family history from a biological perspective, with groups representing twins, siblings, and sperm within their collages. One group referenced evolution with a picture of an ape referring to it as "ancestors," and another group mentioned generations. Most of the references from this activity showed a common understanding of inheritance, with all groups mentioning traits and similarities within families, as well as genes and DNA. The collages illustrate that there was some understanding that physical traits are inherited and that health is also a factor in family history and inheritance with all four groups making some references to either disease or disability. Three of the four groups illustrated external factors as relevant to their understanding of family history and inheritance with sport, diet, fitness, and smoking depicted in their collages.

Individual portraits

Each participant produced a personal portrait highlighting inherited physical traits such as eye and hair color. Some participants included details of allergies and conditions, and the majority of portraits included nonbiological factors such as likes and dislikes.

Storytelling

Participants were asked to storyboard their ideas about genetics and cancer with a view to producing some sort of creative group output based on their learning from project activity. The emphasis for this activity was for participants to begin to focus on key messages that are most important to them and what they think other children their age need to know about cancer and genetics (Figure 9.3). Results from this activity capture not only learning from an individual perspective, but also attitudes relating to genetics and cancer.

Figure 9.3 Thinking about disease.

Figure 9.4 Capturing the discussions.

Of the stories analyzed, there were a myriad of references that illustrated participants' learning in genetics:

"Chromosome is a DNA strand
And it's the stuff
That makes your hand"

Participants demonstrated a confidence in their understanding when presenting information about genetics: "*Genes are the material from which you are made and DNA is like a recipe.*" Participants also demonstrated (Figure 9.4) an understanding of genetics as it relates to health mentioning disease and "faulty" genes: "*These genes aren't always perfect. They can have something called a mutation which can be passed down by your parents and grandparents.*"

Many of the scripted stories illustrated an understanding of risk and chance:

"40% of cancers are preventable
Take a look at how you live
Keep it sensible."

Risk described by participants referred to environmental factors and the role that individuals play in their own health and well-being and there were references to the fact that only 5% of cancers are inherited. The children's storyboards illustrated a good understanding of health promotion messages, highlighting preventative steps that can be taken by the individual including advising people to check themselves and the benefits of early diagnosis. The language used in this activity was proactive and engaging, with one participant developing his own rap (Figure 9.5). It was this rap that was used as a basis for the final creative group output.

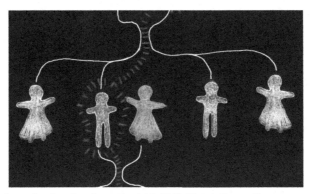

Figure 9.5 Screenshot of animation from the Let's Talk About Genes Rap.

THE LET'S TALK ABOUT GENES RAP

The lyrics for the rap were developed as a result of group activity that instructed participants to prioritize five key messages under the headings of genetics and family history, cancer, risk, and lifestyle. There was also a group discussion with an animator exploring images
they think would best convey key messages and in particular engage others their age with
the topic.

The Let's Talk About Genes Rap

Let's talk about genes
And I don't mean trousers
WOWZERS!
It's a lot to take in
From Stanwell's pupils
Let's begin.

Cancer is something that affects 1 in 3.
Many different types
Diet and genes
It also can contribute
To you getting ill,
But a healthy life choice
Helps a strong will.

You should always remember
To look after yourself
Stay fit
You can't put a price on your health.
Always check yourself,
Be body aware.
Make sure you check
Under here and over there

(Continued)

The Let's Talk About Genes Rap (Continued)

In a lifetime there can be many causes
Drinking and smoking
Even the sun can cause it.
Be careful of the summer,
Hat and sun cream
Keep protected with Factor 50.
A tan is visible proof
That your skin is damaged
If you lay off the sun bed
It can be managed.
Another effect is family history.
Cancer running in the family
Can be a mystery.
Small, tall, fat and thin.
How you look comes from within
GENETICS!

Knowing your family history
Is important.
To know if you have the genes
Are they dormant?

Chromosome is a DNA strand
And it's the stuff
That makes your hand.
Genotype is a recipe of you
Passed down from your ancestors to you.
Genetics is a study of inheritance
To learn about history
Could be relevant.

All humans have the same genes
So you're the same as me
It's a basic recipe.
Patterns of inheritance
Are hard to predict
Did you know?
Only 5% are genetic.

An equal number of traits
Are passed down
From parents to children
The circle turning around

(Continued)

The Let's Talk About Genes Rap (Continued)

In case you didn't know
Cancer can be fatal
Even your pets
Could feel the effects.

There are about 200 different cancers
You can use the internet
To find some answers.
It's definitely not cool
And I'm not joking
Cancer risk is heightened
If you're smoking.

Chemotherapy can make you lose your hair
Or you can wear a cap
To stop that.

40% of cancers are preventable
Take a look at how you live
Keep it sensible.

A healthy diet and exercise
Will stop you going up a size.
Thanks to medicine
Opening doors
Now oncologists
Are working on cures.

Let's talk about genes
And I don't mean trousers
WOWZERS!
It's a lot to take in
Stanwell's pupils gonna tell you something.
GENETICS!

ENGAGING CHILDREN: WAS IT SUCCESSFUL?

The data gathered from project activities and outputs produced by participants in this project indicate that it is both feasible and acceptable to discuss genetics and cancer with children. Children are aware of cancer and most informed about skin and lung cancer, however they hold a number of misconceptions about the nature and causes of the disease. At the start of project, participants demonstrated a simple understanding of inheritance and were able to identify traits. However, they were less aware of the science and language of genetics, such as how inherited diseases are passed on. By the end of the project, participants demonstrated a greater understanding of both cancer and genetics and evidenced an appreciation of the role of family history in cancer. Participants were able to identify genetic, environmental, and behavioral factors affecting health and engage in mature debate and discuss the issues confidently.

The methods and tools employed were well received by the participant group and allowed us to expose children to cancer and genetics learning in novel ways. The electronic voting gathered lots of data very quickly, but could be used to greater effect. Results from questions presented at both the start and end of project activity allowed us to gain an overview of participant understanding as well as observe any changes in learning and attitudes.

The group work was useful for generating a better understanding of participants' views and the interactive and kinesthetic activities were useful in generating interest and energy about a particular topic, as well as providing a novel way of illustrating connections. They provided space for participants to explore their understanding of a topic as well as make connections in their learning, and this was felt to be particularly useful for developing cancer genetic literacy for this age group.

Involving children in creative activities can be an effective method for both engagement and data collection. Thinking creatively about a topic can make it more accessible to a group of this age and provide a flexibility that may not be possible with interviews. The format of the creative activities allowed participants to stay in control of what they disclosed. The participant-led and collaborative approach to each day was not only well received by participants (and the school) but the approach enabled us to engage successfully with schoolchildren and collect data about how to talk to children about genetics and health. Genomics is increasingly being incorporated into mainstream health services and it is important to discuss the role that family history plays in health and disease as early as possible, including with children and young people. Of course, it is just as important for health professionals to become more genomically literate and in the next section, we describe one approach the GPU has taken to this, using stories.

SECTION 2—STORIES AS AN APPROACH TO ENGAGING AND EMPOWERING PEOPLE

About storytelling

On the face of it, Nancy's story is straightforward. Her ex-husband was diagnosed with Huntington's disease (HD) while he was in a nursing home, just 6 weeks before he died. Nancy's two adult sons then had to think about the implications of their father's diagnosis for themselves in relation to genetic testing. This is potentially a "textbook" clinical scenario, ideally suited to considering issues around inheritance, risk, genetic testing, and the condition itself, in the classroom. The reality was more complex however. As the eldest son, Dean received the letter from the neurologist who had seen his father, and the news made him very anxious. Both Dean and his 25-year-old brother Pete were referred to a genetics clinic; Pete decided to undergo testing but Dean decided to delay as he felt it would be "too much for everyone." Pete was tested positive and was in onset. Dean, who already had a diagnosis of schizophrenia, wished he could have had the diagnosis instead of Pete. The anxiety of the situation took its toll on the whole family and Dean took his own life (http://www.tellingstories.nhs.uk/index.php/18-stories/164-nancys-story).

Nancy's story is brief, yet incredibly rich as a learning resource for health professionals. It is powerful and illustrates all too tragically that life isn't like a textbook, but complex and messy. As Nancy said, *"HD takes its toll not just on the sufferer but family and friends as well; the pressure is immense."*

Such stories provide an invaluable vehicle for engaging people (public and health professionals) in the lives of others, drawing together the many threads into a memorable and coherent context. Patient stories are recognized as a valuable resource within health care and act as a powerful learning tool in health professional education [11,12].

Gidman [13] used a descriptive phenomenological approach to explore perceptions of learning from listening to digital stories among 12 prequalifying health and social care students during their practice placements. The data affirmed the value of service-user stories as an additional learning resource to promote learning and encourage reflection. The stories also helped students to understand alternative perspectives on care and this is an important aspect of using stories. The authentic voice of the storyteller can be heard, whether patient, carer, or other family member for example, and the learner has the opportunity to gain insights into the storyteller's world that might not be afforded through just reading clinical notes. With Nancy's story, the reader is taken beyond an account of a diagnosis of an inherited condition and the subsequent genetic risk for her two sons, to be able to see the whole picture from her perspective, and to connect to the psychological impact and tragic consequences for her and her family.

As well as offering alternative perspectives, stories can provide the opportunity to learn from situations which might otherwise be missed, because the learner or clinical supervisor may not recognize the nuances of a situation. This is particularly relevant to genomics, where knowledge, experience, and confidence may be limited and perhaps the situation relates to an individual with a rare condition whom the learner might not encounter directly. While the condition itself might be one that the learner is unlikely to come across, and although family experiences may differ, many of the issues faced by families affected by rare diseases are common. The knowledge and understanding gained can thus be transferred to other situations. The challenge of providing access for students to sufficient clinical placements with experienced supervisors who are able to maximize learning opportunities was noted by Paliadelis et al. [14]. To overcome this, they developed an online learning program using stories to engage and support learners and clinical supervisors in workplace-based learning. Evaluation clearly indicated that they identified with the authenticity of the stories, and that the stories enabled learning through emotional engagement and prompting reflection.

Developing an online story resource in genetics/genomics for health professionals

Telling Stories, Understanding Real Life Genetics (www.tellingstories.nhs.uk) is a freely available genetics/genomics education resource for healthcare professionals. It sets stories from patients, carers, family members, and professionals within an education framework that makes clear the links to professional practice. An accompanying toolkit of resources includes expert commentaries on the stories and supporting activities. Video clips can be downloaded and the story and toolkit can also be printed as a PDF document. Stories can be searched under a variety of themes, including genetic condition, professional role, and life-stage, to help the learner or teacher navigate according to their needs or interests (Figure 9.6). The site was launched in 2007, contains more than 100 stories and attracts tens of thousands of visitors from over 125 countries [15].

The idea of developing an online educational tool emerged from several experiences: the personal impact of hearing patients telling their stories at conferences;

Figure 9.6 Telling Stories screen shot.

observing the impact of a midwife recounting her experience of delivering a genetic test result on other midwives in a focus group; noting how "having a story to tell" enriched our own teaching; and the response of other educators to stories we had included in a report. Furthermore, work we had conducted in scoping the needs of educators to deliver genomics to student health professionals indicated that many had limited experience of this within a clinical setting and thus lacked stories to use in teaching. They wanted access to genetics specialists and genetics services users to support and enhance their teaching [16]. Providing "access" via an online digital story resource seemed to offer a practical alternative, at the same time providing access to multiple (and sometimes contrasting) experiences.

Involving the lay public in developing the resource through contributing their stories necessitated careful planning. We wanted to enable storytellers to contribute to education in a nonjudgmental and safe manner, empowering them to share their experiences. A framework was developed to ensure a consistent and rigorous approach was taken to gain informed consent for all stages from story collection to final publication, endorsed by the university Ethics Committee [11].

One of the key features of the resource is that the stories are set within an education framework that offers flexibility in how the stories can be used, for self-directed learning, learning within groups, or whole classroom teaching. Some of the approaches we have taken in using the stories to engage learners in genomics are outlined next.

Using stories to engage learners

In her review, Matthews reiterates the power of digital stories to engage learners emotionally and their value in enhancing communication skills and practice through reflection [17]. We find that the richness of stories enable us to help learners make connections across multiple layers of practice, with the lived experience of genomics at its core, as represented within the digital story (Figure 9.7).

While we use a variety of approaches to teach genomics, all must be underpinned by sound educational practice. We have chosen to draw on both adult learning theory recognizing the importance to the "student" of relevancy, goals, prior knowledge/experience, and active learning; and the work of Honey and Mumford [18] that uses a four-stage cycle of learning (having an experience, reviewing it, drawing conclusions, and planning the next steps). Through the real stories we are able to provide a tangible experience that can be reflected upon and developed further educationally, professionally, and personally.

Finding "the hook" to use at the start of a teaching session that will draw the student in and keep them engaged is essential. This is particularly important if learners arrive already feeling daunted or with negative perceptions of genetics/genomics, for

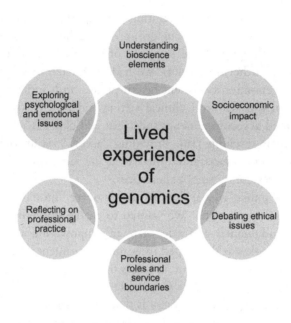

Figure 9.7 Lived experience of genomics as the core of education.

example, that it is too hard, all about science or not relevant. The use of a quote or video clip can be an excellent way to introduce a complex or sensitive subject and can offer a more accessible personal overlay to the topic.

Exercise 1

Our children can't live a normal life, and we can't be normal parents. We don't think the thoughts that normal parents would.

Rachel's story as a mother of two children with cystic fibrosis

Think about the different topics that this quote could be used to introduce; how many can you identify? Remember to think outside the area that you are familiar with and consider other occupations and population groups.

A selection identified by the authors are: enduring health and social care needs, living with uncertainty, role of the carer, personal relationships, parenting approaches, hopes and aspirations, work place/employment issues, choices, decision-making, mental health, and well-being.

Illustrating the breadth and complexity of issues affecting individuals and families can be readily achieved through stories. Using a structured, group-based learning approach, small groups (ideally no larger than five people) are provided with a story and worksheet (Figure 9.8). Individuals read the story; discuss with their peers

Figure 9.8 Students identifying themes within a story.

particular themes of interest (using the worksheet to capture their thoughts); reflect on the role of their profession (if appropriate) in providing information, support, etc. and then feed back to the whole group (Table 9.1). Additional information can be provided to help prompt discussion. For some health-focused sessions, we provide a list of bio-psycho-social "themes" that might be present within the stories; equally, discussions could, depending on the audience, be focused on public attitudes, social care, workplace/employment matters, ethical dilemmas, and so on. This activity also works well with self-directed learners.

The field of genomics is already vast and will only continue to expand. A challenge for those looking to engage the public and/or professionals of any discipline in genomics is selecting what to cover in a limited period of time. Learners can easily be alienated from a topic if too much information is squeezed into a teaching session. Using stories from the collection as the basis for scenarios, we have created role-plays that unfold during the course of a session and which are interspersed with formal "teaching moments." This approach can work well particularly for an all-day event where 6 h of formal teaching could quickly result in disengagement. With this type of "one-off" teaching event, there are often no opportunities for follow-up or reinforcement of learning and so sessions must be meaningful, memorable, and effective.

Based on many sudden adult death narratives, "Josh's scenario" follows the family of a teenager who has experienced symptoms that are later determined to be due to an inherited cardiac arrhythmia; the significant family history that is present has gone unnoticed over many years. In "The worried Andrews family," individuals are affected by or at risk of a familial form of bowel cancer. Participants are provided with limited information regarding their character (family member, school/work staff, friends/colleagues, and health professionals) and are encouraged to draw on personal and/or

Table 9.1 A selection of real-life quotes (from www.tellingstories.nhs.uk) and three example bio-psycho-social themes that they illustrate

Story quote	Theme
I have no other child to compare her with. The first couple of months were very difficult. It all seems a daze now. We were overcome with love for her, but also distraught that she had a disorder that it seemed we could do little about, and with not much hope for the future.... We asked them [the medical profession] how she would personally be affected and they told us they really could not say and we would have some idea at 1 year old. The uncertainty has been there from the start and still remains.	Living with uncertainty
Meriel's daughter has Down syndrome	Meriel has a diagnosis for her daughter but is still uncertain about her long-term future
I have suffered bullying both at school and in the workplace. I am never going to achieve any potential that I might have because of my shape and size which is due to Turner's. I have a degree and I am doing a low grade clerical job within the Civil Service. If you do not fit the mould, you may as well forget it and I am not going to get the reports to allow me to get on and am at a disadvantage in the workplace because people will not, when it comes down to it, take me seriously or believe that I am capable of managing staff. I feel that I am fighting a losing battle when it comes to walking in that interview room.	Stigmatization and discrimination
Collette has Turner syndrome	Collette feels that she will always be at a disadvantage in the workplace because of her condition
We knew that officially there was a 50/50 chance that your child could inherit this condition but even if the child did inherit it, it might be quite mild—as in my wife's case because she'd had it all her life and never known.	Implications of genotype
Mark's two children have tuberous sclerosis	While a diagnosis may provide an indication of the signs and symptoms to be expected, Mark makes the important point that for many genetic conditions the "expression" of the disease even within a family can be very variable. This can make decision-making very difficult

professional experience as well as envision and play out the possible concerns, questions, and consequences for the character.

Following broadly similar trajectories, both scenarios allow us, within the formal teaching moments, to introduce the skills of taking, drawing, and interpreting a family health history; deliver underpinning knowledge around core genetic concepts; convey practical information on specialist genetic services, testing, and screening; and provide opportunities to consider ethical, legal, and social issues (including prenatal testing and the testing of children, the right not to know one's genetic status, risk, communication, and employment and insurance concerns). Student evaluation is positive and indicates that the role-play helps them to develop confidence; they have a safe environment in which they can consider their interactions and responses to people and/or situations before they experience it for real. Their engagement as the story unfolds becomes evident as they become "caught up" in exploring the potential impact of genetic testing on family dynamics. As well as all-day events, we have used this approach in shorter (2-h) sessions to consolidate prior learning. On these occasions some of the formal teaching is replaced with individual and group activities.

In order to guide and support other users (educator or student) in accessing the most appropriate content within Telling Stories, a multifaceted education framework was established. Central to this are the professional competencies and learning outcomes. Stories have been mapped to the genetics/genomics frameworks for nursing [19], midwifery [20], general practice [21], and medical undergraduates [22], thus making it explicit how the stories can be used to teach or learn about the specific knowledge, skills, and attitudes required for clinical practice. Importantly the mapping information for each story is visible alongside the story, so for example, someone who is interested in *identifying clients* (nursing competence 1) and goes on to select Mike's story of living with sickle cell will be able to see that the story also relates to *sensitivity in tailoring information* (competence 2), *utility and limitations of testing* (competence 5), and *obtain and communicate information* (competence 7). This feature provides educators with the opportunity to use a single story as a common thread over a number of distinct sessions to address multiple competences/learning outcomes.

In addition the stories have also been organized in a number of different ways to further aid access to the right story(s). Users can use the theme search to select stories by genetic condition, mode of inheritance, genetic intervention, professional role, clinical specialism, or life-stage. This can be particularly useful when we consider the importance of "relevance" to the adult leaner. For example, someone working or interested in cardiology may not engage with or learn from a story about cancer, equally a nurse working solely in adult care may find it difficult to relate to a midwifery or child-focused story. As mentioned, supplementary to each story is a "toolbox" of resources (Figure 9.6) which can be utilized to further enhance the learning/teaching experience and meet the needs of different learning styles.

Evaluating utility and impact of a story resource

When developing any public engagement resource or activity it is good practice to integrate a robust evaluation plan from the outset. The benefits of evaluation are widely recognized and there are useful guidelines available [23,24]. Evaluation should include providing an assessment of value and impact to underpin continuous improvement, serving as a record of achievement and contributing to an evidence-base which may also help in securing funding for further work. Drawing on our practice and experiences of evaluating the Telling Stories project, we offer some guidance on what factors to consider.

Evaluation approach

Establishing clear project aims and objectives, defining what the evaluation needs to measure, and how the information collected will be used can help to inform the evaluation approach and allow it to be tailored to considerations such as who will be conducting it (participants, deliverers, independent observers, or a combination thereof), how it will be done (e.g., by a small number of participants over a long period of time, who may provide richer, more in-depth data, or a larger audience who may provide an instant gauge of any activity from one-off or intermittent participation), and what type of data will be collected—qualitative data (e.g., from interviews or open-text questionnaires); quantitative data (such as numerical questionnaire responses, web statistics); other creative outputs (e.g., photographs, videos, or artwork).

In the case of Telling Stories, the approach has been multifaceted and has entailed both formative and summative evaluation approaches using a range of methods:

- Storytellers are invited to evaluate their experience of participating in the project via a questionnaire. Several have also provided unsolicited feedback via e-mail.
- During development and prior to its online launch, storytellers and independent experts (genetics educators and e-learning specialists) were asked to provide feedback on web site design and content using a questionnaire with opportunity for open comment.
- At the launch of the web site, delegates were invited to participate in two evaluation approaches, using a written questionnaire and via anonymized electronic voting, to obtain their feedback on the design and content of the resource, and their attitudes to using it.
- Post-launch, follow-up telephone interviews were conducted with educators to explore the longer-term impact of the resource, the findings from which were published in a preliminary evaluation report [25], with interviewees describing the importance of patient stories within education:

You're starting with practice instead of starting with theory... You don't talk about the theory in isolation, you talk about it as it links to the patient (Ms A) and their impact on learning:

*Every time you read it you think of some other aspect that you might not have thought of...
I think it's particularly useful because in genetics... it can be quite a while between you seeing
one condition and you seeing another (Ms B)*

- Google Analytics software has been used to monitor and evaluate web site activity
  (e.g., type of visitor, such as unique vs. returning; number of hits on, or downloads
  of, a story or video clip; user demographics) and inform the expansion of the
  Telling Stories resource [15]. Google Analytics is a freely available web analytics
  service for tracking and reporting web site traffic, which allows web managers to
  analyze data ranging from the number of visitors from a particular geographical
  area and the content being viewed, to the technology (e.g., device, operating system, or browser) being used to access it. Using this software, we have identified
  that in 2009, almost 8000 users from 133 countries made over 62,000 page views
  on the Telling Stories web site, whereas by 2013 this had increased to over 18,000
  users viewing more than 75,000 pages.
- Visitors to the web site are invited to provide comment about all aspects of the
  resource, such as the stories themselves or technical issues, via a dedicated
  webpage.

Analysis, interpretation, and use of evaluation data

The approach to data analysis and interpretation of evaluation data will depend largely
on the type of information collected and how it is going to be used and presented
(e.g., as part of a journal article or report, or as an instant representation of feedback
about an activity at a public event such as a physical chart, an online graphic, or via
social media).

Evaluation data from Telling Stories have been used in numerous ways which
include:

- statistical analysis of web usage for use in journal articles [15], presentations, and
  impact statements for research purposes;
- thematic analysis of post-launch interviews in reports, to explore longer-term
  views of the resource and its impact on teaching practice, and to support funding
  applications;
- storyteller quotes and testimonials in education sessions, presentations, journal articles [11,15], and promotional material, to evidence the rationale for developing
  the resource and the motivation of storytellers to share their experiences.

In all cases, care to interpret evaluation data accurately is ensured by conducting
robust, critical analyses, preserving the authentic voice of the participant, drawing
evidence-based conclusions and working within an ethical framework at all times.

Ethical issues around evaluation

There should be transparency around collection and use of evaluation data; clear communication and, where appropriate, partnership-working with storytellers, which is particularly important when working with them over time. Good practice around involving storytellers can encourage further engagement and help ensure accurate representation of evaluation data. Their feedback is also valuable in promoting accessibility and sensitivity around difficult subjects which may arise during storytelling. Potential issues can be identified as part of the process of gaining ethics approval for which a sound, informed consent process is essential (Figure 9.9).

As has been described in more detail elsewhere, the ongoing involvement of storytellers as partners in Telling Stories and working with them within an ethical framework has been paramount to the success and impact of the project [11]. This includes making clear the purpose of the project and any associated risks, benefits, and outcomes, to ensure fully informed decision-making; being explicit about how all information, including evaluation data, will be used and providing options for its use; confidentiality of, and support for, participants if needed, both of which are

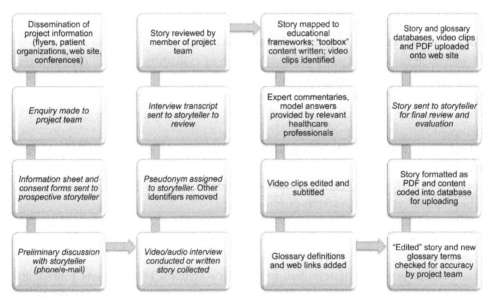

Figure 9.9 Overview of resource development process.
Overview of the story development process, beginning with the dissemination of project information to potential participants (top left) through to publication of the story and accompanying educational material on the web site (top right). Text in italics denotes where the storyteller is involved directly. Storytellers have the option to withdraw their story at any time, even after web site publication.

particularly important when disclosing sensitive information, as can often be the case in the field of genomics; preserving the authentic voice of the storyteller at all times through accurate representation of information. Others contributing to the evaluation, including educators and launch delegates, were provided with clear information about the intended use of their feedback.

An ongoing and multifaceted approach to evaluating Telling Stories from the outset has been integral to its continuing improvement. Feedback from storytellers and users has informed both the content and the usability of the resource, ensuring that it remains relevant, fit for purpose, and globally used [15].

Engagement of the storytellers

Involving the storytellers by soliciting their input and feedback throughout the story creation process and beyond has been important in engaging them as stakeholders in the project. Exploring their experiences of participating in the project has revealed that the motivation for sharing their stories and disclosing what can often be sensitive information (e.g., concerning genetic status, family history or attitudes, and beliefs) is largely altruistic, with the hope that sharing their own experiences will empower and inform others. While their individual reasons for engaging with the project vary, several common themes can be identified (see the following table):

To contribute to the education of health professionals, leading to improved care:

We feel that this website could make an important difference to the training of professionals and its application and use should be widened to include doctors at ALL levels. (Simon, Niemann—Pick disease)

I think having a training tool for professionals by the parents is so important. We are experts in our children's conditions. (Marianne, Edwards syndrome)

"Telling Stories" provided an ideal opportunity for me to help pass on my own experience not only to a consultant or GP but anyone who had an interest. Hopefully quite an audience. (Paul, Familial adenomatous polyposis)

I feel the more people that read about experiences such as mine, the more understanding and help is likely to become available... Reading the page "information and teaching" made me feel so pleased that I took part in your project as this is what I set out to do, tell the people who are able to help people such as myself. (Geraint, Brugada syndrome)

As long as you get some sort of educational understanding about it that would please me. (Tony, Sickle-cell anemia)

To promote awareness of often poorly understood genetic conditions and how they can impact on people's lives through providing information and support to others:

If it is possible to offer help, confidence and encouragement to other ataxians through this medium, then I will feel the whole experience to have been worthwhile. (Bob, Ataxia)

I am glad I have told my story and I really hope people will understand how hard it can be when faced with the news your child has a disability. (Paula, Prader—Willi syndrome)

If it helps anyone understand more about my conditions then that has to be a good thing. (Siobhan, Ankylosing spondylitis and type 2 diabetes)

Thank you for printing my story hopefully this way KS [Klinefelter syndrome] may be understood as sometimes I felt it was similar to a freak of nature and no-one was interested. (Andrew, Klinefelter syndrome)

I love my children and I love my husband and I didn't want it happening again and that's the reason I am doing this now. (Helen, Long QT syndrome)

Really enjoyed the experience. I think the information from sufferers out there is still quite poor and would love to get involved in more things in discussing NF with people doing research or wanting to inform others. (Amelia, Neurofibromatosis)

Thank you for the chance of raising awareness of my two conditions. If one person learns something then that would be enough to make it all worthwhile (Jenny, MODY and FSHMD)

As an opportunity for catharsis and an outlet for their voice to be heard:

I think that for me, it was a form of therapy—I needed to "shout out" and it's taken me 24 years! Thanks for giving me that release! (Brenda, Huntington's disease)

I'm very honoured to have done this in memory of our special little boy. It is his legacy to other children and families with T18 [trisomy 18]. (Marianne, Edwards syndrome)

Engaging and empowering storytellers

Working in partnership with storytellers within an ethical framework has been paramount in protecting their care and dignity. For those who have agreed to provide video clips, anonymity cannot be assured, but protecting privacy for all storytellers is important, such as through the use of pseudonyms and editing identifying details such as names of hospitals or of other family members. Feedback from their experiences of contributing to Telling Stories (Figure 9.10) has been extremely positive and is reflected in some of their endorsements:

Thank you very much for including my story and for handling it so respectfully and with sensitivity. I felt continuously included throughout this process.

(Amy, Hypermobility syndrome)

We feel that the story offers a very true portrayal of our experience... During the process, every effort was made by the researchers to make this a comfortable process and we found them very sensitive towards us and the difficulty in discussing such an emotive subject.

(Jeff & Helen, Long QT syndrome)

No pressure was put on me to include sensational bits or to include family members who might be concerned. At the end I felt worn out and yet really pleased at being given the opportunity to be part of a new way of training where patients' stories would really be appreciated and help fill in the gaps that text books are unable to do.

(Paul, Inherited bowel cancer)

Figure 9.10 Word cloud derived from participants' responses about their experience of being a storyteller.

CONCLUSION

Genomics is undoubtedly a complex field, not just because of its scientific and technological underpinnings, but also because of the psychological, social, and ethical implications of its applications for individuals and families. Promoting understanding and engagement of the public is crucial to maximizing the opportunities offered by advances in genomics yet overcoming its complexity is a challenge. The Let's Talk About Genes project demonstrated that with the right approaches, children can be successfully engaged with genomics. Engaging health professionals in genomics is equally important and equally challenging. Stories, from patients, family members, and carers can provide a flexible resource to aid learning and teaching that is relevant to the professional role, memorable and powerful. Importantly, storytellers themselves feel they derive some benefit from telling their stories and feel empowered by their role in educating others.

REFERENCES

[1] Kirk M, Calzone N, Arimori N, Tonkin E. Genetics—genomics competencies and nursing regulation. J Nurs Scholarsh 2011;43(2):107—16.
[2] Skirton H, O'Connor A, Humphreys A. Nurses' competence in genetics: a mixed method systematic review. J Adv Nurs 2012;68(11):2387—98.
[3] Ulph F, Townsend E, Glazebrook C. How should risk be communicated to children: a cross-sectional study comparing different formats of probability information. BMC Med Inform Decis Mak 2009;9:26.
[4] Ponder M, Lee J, Green J, Richards M. Family history and perceived vulnerability to some common diseases: a study of young people and their parents. J Med Genet 1996;33:485—92.

[5] Iredale R, Longley M, Thomas C, Shaw A. What choices should we be able to make about designer babies? A Citizens' Jury of Young People in South Wales. Health Expect 2006;9:207−17.

[6] Iredale R, Anderson C, Shaw A. The national DNA database on trial: project report. Pontypridd: University of Glamorgan; 2009. ISBN: 978-1-84054-220-2.

[7] Claassen L, et al. Using family history information to promote healthy lifestyles and prevent diseases; a discussion of the evidence. BMC Public Health 2010;10:248.

[8] Iredale R, Madden K, Taverner N, Yu J, McDonald K. The GAMY Project: young people's attitudes to genetics in the South Wales valleys. Hugo J 2010;4(1−4):49−60. Available from: http://dx.doi.org/10.1007/s11568-010-9148-8.

[9] Mayall B. Conversations with children. In: Christenson P, James A, editors. Research with children. London: Falmer Press; 2010.

[10] Let's Talk About Genes and I Don't Mean Trousers: encouraging cancer genetics literacy amongst children. ecancer 2014;8:408. Available from: <http://dx.doi.org/10.3332/ecancer.2014.408>.

[11] Kirk M, Tonkin E, Skirton H, McDonald K, Cope B, Morgan R. Storytellers as partners in developing a genetics education resource for health professionals. Nurse Educ Today 2013;33 (5):518−24.

[12] Christiansen A. Storytelling and professional learning: a phenomenographic study of students' experience of patient digital stories in nurse education. Nurse Educ Today 2011;31:289−93.

[13] Gidman J. Listening to stories: valuing knowledge from patient experience. Nurse Educ Pract 2013;13:192−6.

[14] Paliadelis P, Stupens I, Parker P, Piper D, Gillan P, Lea J, Jarrott H, Wilson R, Hudson J, Fagan A. The development and evaluation of online stories to enhance clinical learning experiences across health professions in rural Australia. Collegian; 2014. Available from: <http://dx.doi.org/10.1016/j.colegn.2014.08.003>.

[15] Kirk M, Morgan R, Tonkin E, McDonald K, Skirton H. An objective approach to evaluating an Internet-delivered genetics education resource developed for nurses: using Google Analytics to monitor global visitor engagement. J Res Nurs 2012;17(6):557−79.

[16] Kirk M, Tonkin E. Genetics education for nursing professional groups: survey of practice and needs of UK educators in delivering a genetics competence framework. Birmingham: NHS National Genetics Education & Development Centre; 2006.

[17] Matthews J. Voices from the heart: the use of digital storytelling in education. Community Pract 2014;87(1):28−30.

[18] Honey P, Mumford A. The Manual of Learning Styles Peter Honey, Maidenhead. 3rd. ed.; 1992. ISBN: 0-9508444-7-0.

[19] Kirk M, Tonkin E, Skirton H. An iterative consensus-building approach to revising a genetics/ genomics competency framework for nurse education in the UK. J Adv Nurs 2014;70(2):405−10.

[20] Tonkin E, Skirton H, Kirk M. Fit for Practice in the Genetics/Genomics Era: a revised competence based framework for midwifery education and training. <http://genomics.research.southwales.ac.uk/media/files/documents/2015-02-06/Midwifery_comps_2sides.pdf>; 2012.

[21] Farndon P, Martyn M, Stone A, Burke, S and Bennett C. Genetics in primary care. Royal College of General Practitioners Curriculum Statement 6; 2007.

[22] NHS National Genetics and Genomics Education Centre. Learning outcomes in genetics and genomics for medical students. <http://www.geneticseducation.nhs.uk/downloads/Learnng_Outcomes_in_Genetics_for_Medical_Students_2014.pdf>; 2006.

[23] Research Councils UK. Practical guidelines: a guide for evaluating public engagement activities: <http://www.rcuk.ac.uk/RCUK-prod/assets/documents/publications/evaluationguide.pdf>; 2011.

[24] Manchester Beacon for Public Engagement. Public engagement evaluation guide; 2011. <http://www.manchesterbeacon.org/files/manchester-beacon-pe-evaluation-guide.pdf>.

[25] Burke S. Telling Stories, Understanding Real Life Genetics. Educators' perceptions of relevance and usefulness: preliminary report. Birmingham: NHS National Genetics Education and Development Centre; 2008.

CHAPTER 10

The "Life Costs" of Living with Rare Genetic Diseases

Koichi Mikami[1], Alastair Kent[2] and Gill Haddow[1]
[1]Science, Technology and Innovation Studies, The University of Edinburgh, Edinburgh, Scotland, UK
[2]Genetic Alliance UK, London, UK

Contents

INTRODUCTION: THE LIFE COSTS OF RARE DISEASES?

Why should we pay *special* attention to individuals and their carers affected by rare diseases? Rare diseases by definition are *rare*, and supposedly each rare disease affects a small number of people. However, with rapid advances in medical genetics and a growing awareness of such diseases, the number of patients directly affected and also of the people living with them as caregivers, are estimated to be much larger than once believed [1]. The European Union recognized rare diseases as an important public health issue in the early 1990s and has been providing funding for projects addressing the issue for the last two decades. The issue of rare diseases is challenging particularly because it is multifactorial. Because they are rare, the information about each disease is often limited. Many clinicians have no experience of seeing a patient with a rare disease, making it difficult for them to locate information and make the diagnosis. Even if the right diagnosis is made, no established treatment might be available because little research may have been done on the disease or because its therapy is still in an early stage of development only offered as a clinical study.

Equal access to health care has therefore been a major challenge for those living with rare diseases. However, the situation seems to be improving partly owing to the success of rare disease reference portals, such as *Orphanet* [2], to the actions of various patient support organizations and to the progress in genetic approaches in medicine [3]. All these activities build capacity and knowledge to resolve the issue of rare

Genomics and Society
DOI: http://dx.doi.org/10.1016/B978-0-12-420195-8.00010-0

diseases and are of critical importance to improve the lives of those living with rare diseases. We discuss how such capacity and knowledge may be better utilized to improve their situation.

Central to our discussion is how decision-making about the provision of health care is increasingly inclined to the "economic" approach of a cost-benefit, or value-for-money, analysis. In this analysis, clinical benefit of therapy is weighted against its monetary opportunity "cost" [4]. However, we argue there are also nonmonetary costs that the patients and their carers bear if therapy is not made available and yet are hardly ever taken into account in such an analysis — in this chapter, we call such costs the "*life costs*." This issue of life costs potentially applies to all kinds of diseases — both rare and common — when therapies are not widely made available because evidence shows that their cost is higher than the estimated benefit. However, by virtue of the fact that conditions are rare and/or genetic, there are some additional reasons for life costs to be emphasized with respect to rare diseases. For example, the impact of a chronic disease on individuals and their carers living with, and possibly dying from, it may include a deterioration in social relationships, increased concern and constant worry, lack of time, alongside emotional and psychological burdens, such as anxiety and depression. This impact can be magnified if few people are available to talk with about their problems. Compared to common diseases, the patients and their carers of rare diseases are more likely to find themselves in such a situation struggling to find an expert who knows about their disease, let alone their friends or colleagues. Also, neglecting such nontangible costs has grave implications for rare diseases because the price of the drugs tends to be high, inevitably demanding the benefit to be equally substantial, if not more [5]. Hence, we suggest that the decisions on provision of health care for rare disease patients should they be available could be more balanced by considering the "life costs" to both the individuals and their caregivers.

In this chapter, we first sketch out the issues of applying a conventional cost-benefit analysis to rare diseases, and then argue that rare diseases have a significant impact not only on the lives of patients but also those of their families and significant others who often are the carers by default. The cases used here are two rare genetic diseases — Huntington's disease (HD) and phenylketonuria (PKU). In the case of HD, no curative therapy is currently available; we do however hypothesise the change should a cure become available, as one's situation may changes from dying from HD to living with it. The case of PKU where significant dietary changes have to be made from birth reveals the life costs this entails for parents; for PKU a therapy is now available but not widely offered in the United Kingdom and we examine why this is the case. In examining these cases, we do not mean to imply that having a particular rare genetic disease neces-sarily means the same life costs for individuals and their families. While not ignoring such diversity, patients' narratives and voices in addition to the findings from social science research with them demonstrates that the cost of living life can be high.

Hence, we conclude that more interdisciplinary exchange between health economists and other social scientists is required to explore how research from each field can inform the other and that this will offer added value to all involved, including healthcare providers.

As a final note, this chapter is grounded in our argument of promoting and recognizing the idea of life costs within an interdisciplinary approach. One of the coauthors (AK) has been involved for over 20 years representing some of the families living with rare genetic diseases. He is all too familiar with life costs, and his views and experience are therefore critical in which to locate the academic reviews of existing literature and research into both HD and PKU.

THE CHALLENGE OF "ADVERSE ECONOMIES OF SCALE"

Historically, the issue of rare diseases has been about the challenge of "adverse economies of scale." A rare disease is defined as a disease with the prevalence of less than 200,000 in the United States and as that of less than 5 in 10,000 in Europe. According to Huyard, "rare diseases" as a policy concept was first coined in the US Orphan Drug Act 1983 and became explicitly defined in its amendment in 1984 [6]. This Act was designed to mitigate the market disincentives, which discourage pharmaceutical companies to produce drugs for rare diseases, by offering counterincentives, including 7-year market exclusivity [7]. The coalition of patient organizations that played a major role in the passage of the Act then formed the National Organization for Rare Disorders [8] and is still acting as an umbrella group of rare disease organizations in the United States [9]. As Huyard points out, underlying its action is the belief that "number is power" [6]. The rarity of the diseases has therefore been perceived as a commercial and political challenge.

European adoption of the similar legislation came more than a decade later. Yet, even before the passage of the EC regulation on orphan medicinal products in 1999, the importance of addressing the issue of rare diseases as part of its public health program was already debated. For example, the Commission of the European Communities stated in 1993:

> . . .experiences show that by the very fact of their rareness and consequent lack of information available about them, they can produce significant problems for individuals' countries. There could thus be benefits in Community initiatives aimed at [rare] diseases to complement the efforts of Member States and maximise the exchange of information an experience, so that all Member States are assisted in taking appropriate action when confronted with an unfamiliar threat. With regard to rare illnesses, therefore, the Community will have to cooperate closely with Member States to draw up initiatives likely to generate economies of scale and create solidarity between Member States [10].

Successful cases of implementing this vision to achieve economies of scale in rare diseases by initiating European collaboration include *Orphanet* (the above-mentioned reference portal for rare diseases), *E-Rare* (a consortium of national agencies in Europe for funding rare diseases research), and *European Centres of Reference* (a coordinated effort to improve quality and accessibility of care for rare diseases patients) [1]. Thus, the European initiatives to build capability in and knowledge of rare diseases have reflected the challenge associated with their rarity in providing the patients with equal access to both research and care [11].

As a result of the provisions in orphan drug legislations and the more favorable conditions they create for developing and marketing orphan drugs targeting rare diseases, the number of such drugs approved has significantly increased [12]. However, it is also increasingly argued that the arrival of new orphan drugs will cause problems in the future for healthcare systems in developed countries [5,13]. The main concern underlying this argument is not the number of such drugs *per se* but the price attached to each of them. In the United States, lack of competition in the orphan drug market, as a result of the market exclusivity incentive, and that of a mechanism to lower the price of orphan drugs were suggested to be a potential problem already in the 1990s [14,15]. A similar concern was then raised in the United Kingdom in the mid-2000s when an enzyme replacement therapy for Gaucher's disease was introduced, and whether public healthcare providers, like the National Health Service (NHS) in the United Kingdom, should prescribe this expensive therapy to the patients despite its budget limits [16,17].

The basis of the UK debate was that because of the rarity of rare diseases, orphan drugs are inevitably inefficient [18]. In other words, orphan drug legislations have been successful in offering incentives for pharmaceutical companies to produce such drugs but not so in alleviating the challenge of "adverse economies of scale." As McCabe et al. put it:

> For efficiency's sake, the reimbursement authority must consider whether the benefit (or health) produced by the orphan drug is greater than the benefit (or health) foregone from spending the same resources on other health care interventions for the same or different patients. The high costs of orphan drugs mean that reimbursement authorities would normally need to move beyond simple market-based arguments to fund them [5].

Therefore, the debate is formed around the questions of whether there is any good reason for orphan drugs to be evaluated differently from other medicines and whether the decision to do so can result in the sustainability of healthcare systems. In the case of the above-mentioned enzyme replacement therapy, it did not pass the cost-benefit analysis by the National Institute of Health and Clinical Excellence (NICE) but in 2005 the national-level decision was made to place no restriction on its access for at least for 2 years. It was reported however that this decision caused problems of resource allocation at local levels [17].

In this debate, Drummond et al. argue that the benefit of orphan drugs are often difficult to measure because of the difficulties of conducting randomized clinical trials and also of the progressive nature of the diseases and that the assessment solely based on the cost-benefit analysis would be unfair for rare disease patients [18]. In contrast, while acknowledging that the uncertainty of clinical benefits would likely lead to the decision to await more evidence, in addition to the issue of high price of orphan drugs, which make it almost impossible for them to meet conventional criteria for cost-effectiveness, McCabe et al. suggest that:

> There seem to be no sustainable reasons why the cost effectiveness of drugs for rare diseases should be judged differently from that of other health care technologies [because giving them a special status and approving their use can] impose substantial and increasing costs on the healthcare system [and patients with common diseases] will [then] be unable to access effective and cost effective treatment as a result [16].
>
> On the point of "unfairness", applying the same analysis to all drugs, whether they target rare diseases or not, is suggested to be fairer and if one is to change the criteria this should be used for all drugs and may not allocate more resource to orphan drugs than it current receives [5].

As Lopez-Bastida and Oliva-Moreno point out, the issue of health resource allocation has to be based on what they call "the principle of scarcity" because there are, and never will be, sufficient resources [4]. The national healthcare systems in our society are increasingly finding it difficult to square the circle of increased potential for intervention, rising patient expectations, and static or declining resources. Without a doubt, financial calculations are an essential part of healthcare planning, and health economies in both developed and emerging economies have to find a way of matching the resources available to demand in ways that are not only fair but also transparent and robust. Decisions about heath resource allocation increasingly rely on a so-called cost-benefit, or value-for-money, analysis. Yet, the emerging issue of orphan drugs questions the appropriateness of developing resource allocation strategy solely on the basis of such an analysis.

How the analysis is actually done may vary across countries, and there are various attempts to be inclusive of nontangible costs of diseases. For example, NICE in the UK estimates the effectiveness of interventions using the measurement called quality-adjusted life years (QALYs), which it argues "to be the most appropriate generic measure of health benefit" [19].

However, some suggest that QALY measures do not necessarily reflect actual preferences of individuals or their families [20]. Similarly, use of self-rated measures, like patient reported outcomes, is also common, but such measures tend to focus on patients, not on their carers [21]. As we demonstrate in the following case studies, there are significant costs to patients and their carers living with rare diseases, which may be alleviated by giving them access to treatment — even though an economic approach may potentially suggest this would not be cost-effective. We also contend that the framings observed in the debates about the appropriateness of an economic approach for rare diseases are slightly misleading — as if lower price of orphan drugs would resolve all the problems. It may be true that it would reduce the cost for healthcare services, but there would still be the costs which patients and their carers still have to bear — the *life costs*.

CASE STUDY 1—HUNTINGTON'S DISEASE

Huntington's disease is a family disease. Every member of the family is affected — emotionally, physically, socially — whether patient, at risk, or spouse. And the disease occurs not only once, but over and over again in successive generations [22].

HD is a genetic disorder of the central nervous system. Its inheritance pattern is autosomal dominant, which means that a child of a parent with this disease has a 50% chance of inheriting the disorder. The onset of the disease is believed to be between the ages of 30 and 50, but some patients may not develop the symptoms until their 50s or 60s. There is also a juvenile type of this disease which affects young people before the age of 20. The precise age of onset is difficult to determine because the early signs of this disease are subtle changes to the mental and physical states of patients, including lack of concentration, short-term memory lapses, and stumbling and clumsiness. For significant others this is often the first signs that something may be amiss as is outlined in the following excerpt from a wife talking about finding out about husband's diagnosis [23]:

…Crunch time came about two weeks after I discovered that he was having an affair. That in itself shook my world to its foundations but it did give me the strength to suggest he go and see a doctor. He had been dropping things fairly regularly, his cigarettes, his coffee, car keys etc. As for his temper, what can I say, that was the reason I hadn't mentioned it before. He would have bitten my head off. But now he couldn't hurt me more than he had in the last couple of weeks, so when he tripped on the stairs for not the first time, I grasped the nettle and made the suggestion. To my utter amazement, he agreed without a fight…The initial diagnosis came roughly two weeks later. It had to be confirmed with a DNA test but the neurologist that his GP sent him to, recognised it as soon as he entered his room: Huntington disease. The affair paled into insignificance, this was far darker and too immense to get my head round.

As the disease progresses, HD patients experience dementia, involuntary movements, and difficulty in speech and swallowing as well as severe mental disturbances, often leading to depression and aggressiveness. This can cost in terms of increasing isolation not only for the individual but also for the family, as the wife of an HD patient continues:

The children soon stopped bringing friends round so the house became lifeless. Friends of mine stopped calling, as they just didn't know how to take him. His friends, people he had worked with for years, disappeared off the face of the earth. They can still be spotted now and again darting into shops to avoid me.

Currently, no cure is available for HD, but there are ways to treat or manage the symptoms, including medication, speech therapy, and high calorie diet [24,25]. Although medications can take time to prove effective in the management of the disease, other challenges remain:

The medication had no effect on the movements but his specialist assured me that it was probably the cause of his temper reducing. He sleeps till well after lunch these days, so I have a part time job to help protect my sanity. He finds it hard to swallow so every meal has to be chosen carefully. He has trouble getting to the toilet in time. Imagine that mess crossed with an extreme movement disorder.

When the Congressional Commission for the Control of HD and its Consequences in the United States published its report in 1977, the gene responsible for HD had not been discovered, and no early diagnostic test to distinguish carriers of the HD gene from noncarriers was available. Individuals knew they were at risk simply because they saw their parent suffering from the disease. Hence the Commission wrote "Not only must they bear witness to the painful decline of a parent; they must carry the burden of fear and anxiety that someday the same thing may happen to them" [22]. A life cost in terms of a "Damocles' Sword":

My first reaction was one of great sympathy, which rapidly changed to one of immense anger. We had done nothing wrong, especially our children but they each stood a 50/50 chance of inheriting the gene that would rob them slowly of their mind.

Because its early symptoms are so subtle, many children of HD patients could not spend their life without constantly watching for such symptoms. Nancy Wexler, who was executive director of the Commission and also is a trained clinical psychologist with the HD-affected mother, used the words genetic "Russian Roulette" to describe the experience of being at risk and explained that despite the reported 50% chance of developing the disease, "many at risk make themselves 100% miserable worrying while they are healthy" [26]. Wexler also reported that the idea of a presymptomatic test for HD, which would confirm the carrier or noncarrier status of the gene, received mixed reactions: some saw knowing their status as a way of planning for their future but others thought the knowledge of them carrying the gene was too difficult to bear. For those who saw their parent suffering from HD in their childhood, diagnosis was more like the termination of life, while actual death could mean "a welcome relief from life with symptoms" [26]. This view corresponds with the high suicide rate among HD patients at the time, noted in the 1977 Commission's report [22]. A test became possible after a team of researchers discovered in 1983 that a genetic marker called G-8 on chromosome 4 is linked to the disease [27]. However, this linkage-analysis test worked only when data were available from several generations in one's family, and with the possibility of errors in test result, there were ethical concerns about using the test for carrier screening, particularly given the deep anxiety of at-risk individuals [28,29].

In 1993 a consortium of geneticists studying HD identified the mutation causing the disease [30]. The mutation, named *Huntingtin*, was a prolonged repeat of CAG nucleotides on the chromosome, and it provided a highly accurate way of testing the carrier status [24]. Taylor argues that the availability of this test "generates challenges for at-risk individuals and their significant others" [31]. To begin with, some decided not to take the test, and among those who did some needed to adapt to a new "normal" life, whatever that might mean, while others had to accept the reality, as an individual or as a family, and prepare for the future. The preparation may include sharing of the knowledge with family members, seeking clinician advice and rearranging the living environment [32]. The high accuracy of this genetic test for HD also means that in the United Kingdom it is the only predictive genetic test approved by the Genetics and Insurance Committee to be used for the assessment of application [33]. The HD patients may also have to make advance decisions about life-sustaining treatment because they will lose the capacity to make such decisions as their disease progresses. Meantime, their family members may have to deal with the costs of often living in debt:

> My husband is now in a wheelchair, wears incontinence pads and has to be spoon-fed soft food like a baby. He rarely looses his temper, probably because there is no personality left to irritate. Carers come in to help twice a day. Like the bathroom adaptation, this is paid for by Social Care Services... I have reduced our debt a little as I now have full control of everything. I am still incredibly lonely.

And the disease ultimately can lead to life costs relating to the significant others' well-being:

> I now feel that we are noticed by Social Care Services but only because there is such a strong physical impairment. While he was causing all sorts of problems with his mental state, and probably doing our children far more harm, no one listened to us or offered any help. I was just told to increase my antidepressants.

In contrast to Gaucher's disease, HD has no specific therapies and no cure yet, other than to manage symptoms and attempt to delay its development. Although there is no immediate prospect of a drug that will halt the progress of this disease, much less provide a cure, should one become available it would shift the paradigm and may reduce the life costs to the family dramatically. Currently individuals who know themselves to be at risk because of their family history tend to choose not to have genetic testing because they would rather live with the uncertainty than run the risk of knowing for sure they were going to get HD. A drug that halted progression would change this situation, as there would be an incentive to know as soon as possible in order to arrest the trajectory before symptoms became too debilitating.

This would certainly put a premium on diagnostic services and genetic testing. It would also potentially raise difficult ethical questions. Assuming the drug to be effective and without unacceptably high levels of serious adverse events, presymptomatic or early intervention is an uncomplicated decision, but what about those in whom the disease is well advanced? How would it be possible to establish the wishes of an individual who is unable to engage with the world in ways we can reliably understand? Should we consider keeping someone in the advanced stages of their disease in that position, possibly for many years while we wait for a different disease to end their lives? And what about the impact on their family, physically, socially, and psychologically, especially in a context where the burden of care falls mainly on relatives due to constraints and shortages in the provision of publicly funded support. *Hence, over time, the availability of a treatment or cure that was previously unavailable changes the nature of the "life costs" for the family as well as the individual whose status changes from living with the prospect of a certain death from that condition, to living with the prospect of an alternative uncertain demise over a much longer extended time period.*

CASE STUDY 2—PHENYLKETONURIA

While some may argue that HD is a unique and extreme case of rare genetic diseases whereby patients and their family have to bear the considerable life costs, similar stories can be told about patients and families of many other diseases such as PKU where quality of life is severely affected and a drug treatment is now available but not offered widely. As demonstrated in the cases of enzyme replacement therapies for Gaucher's disease and other lysosomal storage disorders, the cost — benefit analysis presents significant challenges for rare disease patients, carers, and healthcare providers.

PKU is a rare congenital condition, affecting 1 in 10,000 in the United Kingdom, and is one of the rare diseases in which newborn mass screening has been implemented in many countries. The disease was discovered as early as in the 1930s and the high level of an amino acid called L-phenylalanine was then identified as the cause of intellectual disabilities. Special diet therapy with low phenylalanine was invented in the 1950s [34], and soon after Robert Guthrie developed a simple laboratory test to identify a new born with PKU in the 1960s, this test became used for mass screening worldwide [35]. The gene responsible for this metabolic disease, named *PAH*, was identified in the 1980s but this did not have much impact on the management of the disease. All newborns with positive results in the screening are placed on the strict dietary therapy to prevent intellectual disabilities. The parents of a PKU child are first told about the disease after the screening but they have to wait for the result of a follow-up confirmatory test to be certain about the illness [36]. Because it is rare, many parents may never have heard about the disease before, and the suggested possibility of their infant, despite no physical signs, being affected often causes tremendous stress. In such circumstances, many tend to search for information about the disease but what they find in books or on the Internet may not reduce their anxiety or concern.

Once the diagnosis is confirmed, the child will be on a dietary treatment for life. Adhering to this strict diet, however, poses a tremendous challenge for many PKU patients as well as their parents and carers [37]. First, it is the parents who have to master the knowledge of which foods are deemed acceptable for their child. For this reason, the management of PKU often places a special emphasis on education [38]. The diet also has to be supplemented with taking a medical product often referred as a "formula" but it is not only costly but also unappetizing. As the child grows, the parents may gradually lose control over their diet and have to pass their knowledge on so that the child becomes responsible for monitoring his or her own health. However, it is reported that "Adolescents with PKU are subject to peer pressures to eat everyday foods with their non-PKU peers in social situations" [38]. Paul and Brosco note, "Having to avoid common foods create profound barriers to sharing meals with others" and "in most contexts, the special diet is isolating" [37]. Many report experiences of being teased at school. In situations, like having dinner at restaurants, visiting friends' house, and traveling to other countries, adherence to the special diet often proves to be impractical as much as undesirable. Finally, having tried the taste of "normal" foods, many patients find it difficult to go back to the less tasteful diet. For all these reasons, despite the fact that any guideline for the management of PKU stresses the importance of adherence to the special diet, the actual practices of disease management among the patients significantly vary [38] and the level of compliance is often suboptimal [39].

As evident from the emphasis on education, clinicians tend to consider lack of knowledge as one of the major reasons for the low compliance. However, it is argued that to improve well-being of the patients, psychological aspects are also very important [40]. Some studies reveal "the primary obstacles to successful management of the disease are time constraints and stress related to preparation of the diet foods, the keeping records, and the restriction imposed on social life by PKU" [40]. From this point of view, not only adherence to special foods and a formula but also frequent visits to health professionals pose financial and psychological burdens to the patients and their caregivers. Some therefore express their desires to adopt a home monitor approach to test phenylalanine levels and to take a more active role in the disease management. A study in Europe has shown that such self-management approach can be fruitful [41].

For PKU, a pharmacologic treatment option of *sapropterin* was developed recently. This drug potentially frees some from dietary constraints and is becoming available in a growing number of countries [42]. However, it has not yet been made available to the majority of UK patients [43]. The drug was first approved by the US Food and Drug Administration in 2007 and then by the European Medicines Agency in 2008. It helps the body's metabolic system breakdown phenylalanine in foods. While any serious short-term adverse effect has not been reported about this product, its long-term impact on PKU patients is yet to be understood. In the United Kingdom,

NICE has not performed the technology appraisal on it yet but Horsely suggested even if it does it "is unlikely to meet generally accepted levels of cost-effectiveness in terms of cost per QALY" [39]. In 2013, the NHS made the decision to commission this drug only to the minority group of pregnant women with PKU who are unable to resume adequate dietary control and whose baby is at risk of developing severe intellectual disorders [43]. Despite the potential alleviation of the life costs for the patients and their carers living with PKU, further evidence is required before it becomes available to others.

CONCLUSION: TOWARD A MORE BALANCED APPROACH

Although personal stories may very well differ from one case to another, it is possible to draw out a number of common threads, which will illustrate our contention that, in addition to the financial costs to healthcare systems, in assessing the impact of life limiting rare genetic diseases, there are personal and societal costs that need to be taken into account. Healthcare systems are, to a greater or lesser extent, able to determine the price attached to the steps they take to respond to patient need. The price of a pathology test, a drug, and a surgical procedure can be worked out with varying degrees of precision by hospital administrators and from invoices submitted to the relevant authorities. However, it is our contention that such a narrow focus on the economic transactions in managing the clinical supervision of a patient with a chronic and life-limiting disease oversimplifies to the point of missing the point entirely. Ignoring the "elephant in the room" of all the costs associated with the presence of a life-limiting disease in the family is to negate consideration of the burden of these diseases in the round, and to exchange consideration of what matters for the measurement of that which can be counted, and so risk ending up in a place where we know the price of everything, but as the argument might go, the value of nothing.

Although there is a vast amount of literature on the economic and financial burdens of living with a rare disease, alongside a burgeoning body of research outlining the emotional, social, and psychological costs that individuals and their significant others bear (i.e., the "life costs") there appears to be little crossover between the two. Hence, although we are aware that there is a growing trend to integrate patient experiences as part of the evidences for making decisions on healthcare services [19,21], there is still uncertainty in who will be collecting the information about such costs (and indeed the type of costs and to whom) and how they will be used as evidences. Partly, the health economics' dominance relates to the relative simplicity of calculating the price and the value of treatment or medication as opposed to quantifying the life costs. Yet we would argue that more interdisciplinary exchange between economists and other social scientists is required to explore how research from each field can inform the other and therefore this will offer added value to all involved, including healthcare providers. There follows several beneficial outcomes beyond that of recognizing that health is not just disease but an illness suffered in a particular sociocultural context, that cost-effectiveness of clinical interventions does not necessarily enable robust health economics, and that taking account life costs seriously might offset economic ones. To have such benefits, a more balanced approach to healthcare planning is critical and the involvement of social scientists, who can offer their expertise in evaluating the life costs, is undeniably valuable.

In a sense it is a measure of our societal well-being that we can acknowledge the difficulties experienced by patients and families and respond appropriately to them in a timely and user-friendly manner. Publicly funded healthcare systems in developed economies are generally predicated on the notion of solidarity — the idea that access to care and support is based on need — rather than mutuality, where benefits are pooled across a group of individuals with similar risks and charges made accordingly. Well-being is about being healthy, and it requires more than the provision of interventions, such as surgery, therapeutics, and all other inputs traditionally seen as falling under the heading of "health care." This is an arena where the life costs, which can have significant impact on individuals and their families living with rare genetic diseases, thoroughly deserve our *special* attention.

REFERENCES

[1] Groft S, Posada de la Paz M. Rare diseases—avoiding misperceptions and establishing realities: the need for reliable epidemiological data. In: Posada de la Paz M, Groft S, editors. Rare diseases epidemiology. New York, NY: Springer; 2010. p. 3—14. [chapter 1].

[2] Orphanet. Homepage. <http://www.orpha.net/consor/cgi-bin/index.php?lng=EN> [accessed November 2014].

[3] Davies K. The era of genomic medicine. Clin Med 2013;13(6):594—601.

[4] Lopez-Bastida J, Oliva-Moreno J. Cost of illness and economic evaluation in rare diseases. In: Posada de la Paz M, Groft S, editors. Rare diseases epidemiology. New York, NY: Springer; 2010. p. 273—82. [chapter 16].

[5] McCabe C, Edlin R, Round J. Economic considerations in the provision of treatments for rare diseases. In: Posada de la Paz M, Groft S, editors. Rare diseases epidemiology. New York, NY: Springer; 2010. p. 211—22. [chapter 13].

[6] Huyard C. How did uncommon disorders become "rare diseases"? History of a boundary object. Sociol Health Illn 2009;31(4):463—77.

[7] Asbury CH. Orphan drugs: medical versus market value. Lexington: Health and Company,; 1985.

[8] National Organization for Rare Disorders. Homepage. <https://www.rarediseases.org> [accessed November 2014].

[9] Dunkle M, Pines W, Saltonstall PL. Advocacy groups and their role in rare diseases research. In: Posada de la Paz M, Groft S, editors. Rare diseases epidemiology. New York, NY: Springer; 2010. p. 515—25. [chapter 28].

[10] Commission of the European Communities. Commission communication on the framework for action in the field of public health, COM(93) 559 final. 1993.

[11] Kole A, Faurisson F. Rare diseases social epidemiology: analysis of inequalities. In: Posada de la Paz M, Groft S, editors. Rare diseases epidemiology. New York, NY: Springer; 2010. p. 223—50. [chapter 14].

[12] Linares J. A regulatory overview about rare diseases. In: Posada de la Paz M, Groft S, editors. Rare diseases epidemiology. New York, NY: Springer; 2010. p. 193—207. [chapter 12].

[13] Miles KA, Packer C, Stevens A. Qualifying emerging drugs for very rare conditions. QLM: Int J Med Law 2007;100(5):291—5.

[14] Gibbons A. Billion-dollar orphans: prescription for trouble. Science 1990;248(4956):678—9.

[15] Thoene JG. Curing the orphan drug act. Science 1991;251(4990):1158—9.

[16] McKabe C, Claxton K, Tsuchiya A. Orphan drugs and the NHS: should we value rarity? Br Med J 2005;331:1016—19.

[17] Burls A, Austin D, Moore D. Commissioning for rare diseases: view from the frontline. Br Med J 2005;331:1019—21.

[18] Drummond M, Wilson DA, Kanavos P, Ubel P, Rovira J. Assessing the economic challenges posed by orphan drugs. Int J Technol Assess Health Care 2007;23(1):36—42.

[19] National Institute for Health and Care Excellence. Guide to the methods of technology appraisal 2013. <http://publications.nice.org.uk/pmg9>; 2013 [accessed November 2014].

[20] Beresnial A, Medina-Lara A, Auray JP, De Wever A, Praet J-C, Tarricone R, Torbica A, Dupont D, Lamaure M, and Duru G. Validation of the underlying assumptions of the quality-adjusted life-years outcome: results from the ECHOUTCOME European Project, PharmacoEconomics, published online; 2014.

[21] Rajmil L, Perestelo-Rerez L, Herdman M. Quality of life and rare diseases. In: Posada de la Paz M, Groft S, editors. Rare diseases epidemiology. New York, NY: Springer; 2010. p. 251—72. [chapter 15].

[22] Commission for the Control of Huntington's Disease and its Consequences. Report: Commission for the Control of Huntington's Disease and its Consequences, US Dept. of Health Education, and Welfare; 1977.

[23] NHS National Genetics and Genomics Education Centre. Brenda's Story: Supporting a diagnosis of Huntington disease. <http://www.tellingstories.nhs.uk/index.php/joys-story?id=228> [accessed November 2014].

[24] Roos R. Huntington's disease: a clinical review. Orphanet J Rare Dis 2010;5(40).

[25] Huntington's Disease Association. Homepage. <http://hda.org.uk> [accessed November 2014].

[26] Wexler NS. Genetic "Russian Roulette": the experience of being "at risk" for Huntington's disease. In: Kessler S, editor. Genetic counseling: psychological dimensions. New York, NY: Academic Press, Inc.; 1979.

[27] Gusella JF, Wexler NS, Conneally PM, Naylor SL, Anderson MA, Tanzi RE, et al. A polymorphic DNA marker genetically linked to Huntington's disease. Nature 1983;306(5940):234—8.

[28] MacKay CR. Ethical issues in research design & conduct: developing a test to detect carriers of Huntington's disease. IRB 1984;6(4):1—5.

[29] Wexler A. Mapping fate: a memoir of family, risk, and genetic research. Berkley and Los Angeles: University of California Press; 1995.

[30] Huntington's Disease Collaborative Research Group. A novel gene containing a trinucleotide repeat that is expanded and unstable on Huntington's disease chromosomes. Cell 1993;72(6):971—83.

[31] Taylor SD. Predictive genetic test decisions for Huntington's disease: context, appraisal and new moral imperatives. Soc Sci Med 2004;58(1):137—49.

[32] Novas C, Rose N. Genetic risk and the birth of somatic individual. Econ Soc 2000;29 (4):485—513.

[33] Department of Health, and Association of British Insurers. Concordat and Moratorium on Genetics and Insurance. <https://www.gov.uk/government/uploads/system/uploads/attachment_data/file/216821/Concordat-and-Moratorium-on-Genetics-and-Insurance-20111.pdf> [accessed November 2014].

[34] Bickel H, Gerrard J, Hickmans EM. Influence of phenylalanine intake on phenylketonuria. Lancet 1953;265(6790):812—13.

[35] Scriver CR. The PAH gene, phenylketonuria, and a paradigm shift. Hum Mutat 2007;28 (9):831—45.

[36] Timmermans S, Buchbinder M. Patients-in-waiting: living between sickness and health in the genomic era. J Health Behav 2010;51(4):408—23.

[37] Paul DB, Brosco JP. The PKU paradox: a short history of a genetic disease. Baltimore, MD: Johns Hopkins University Press; 2013.

[38] Hagedorn TS, van Berkel P, Hammerschmidt G, Lhotakova M, Saludes RP. Requirements for a minimum standard of care for phenylketonuria: the patients' perspective. Orphanet J Rare Dis 2013;8(1).

[39] Horsley, W. Sapropterin (Kuvan®) in the management of phenylketonuria, NHS North East Treatment Advisory Group. <http://www.netag.nhs.uk/files/appraisal-reports/Sapropterin%20for %20PKU.%20NETAG%20Appraisal%20Report.%20April%202009.pdf> [accessed November 2014].

[40] Bilginsoy C, Waitzman N, Leonard CO, Ernst SL. Living with phenylketonuria: perspectives of patients and their families. J Inherit Metab Dis 2005;28(5):639—49

[41] Ten Hoedt AE, Hollak CEM, Boelen CCA, van der Herberg-van de Wetering NAP, ter Horst NM, Jonkers CF, et al. "MY PKU": increasing self-management in patients with phenylketonuria. A randomized controlled trial. Orphanet J Rare Dis 2011;6(1).

[42] Blau N, Belanger-Quintana A, Demirkol M, Feillet F, Giovannini M, MacDonald A, et al. Optimizing the use of sapropterin (BH4) in the management of phenylketonuria. Mol Genet Metab 2009;96(4):158—63.

[43] The NHS Commissioning Board. Clinical Commissioning Policy: Sapropterin (Kuvan®) for Phenylketonuria: USE in Pregnancy (Reference NHSCB/E12/p/a). <http://www.england.nhs.uk/wp-content/uploads/2013/04/e12-p-a.pdf>; 2013 [accessed November 2014].

CHAPTER 11

Genomics and the Bioeconomy: Opportunities to Meet Global Challenges

Gerardo Jiménez-Sánchez[1,2] and Jim Philp[3]
[1]Harvard School of Public Health, Department of Epidemiology, Harvard University, Boston, MA, USA
[2]Global Biotech Consulting Group (GBC Group), Mexico City, Mexico
[3]Organization for Economic Co-operation and Development (OECD), Directorate for Science, Technology and Innovation, Paris, France

Contents

Genomics and Society
DOI: http://dx.doi.org/10.1016/B978-0-12-420195-8.00011-2

INTRODUCTION

In the wake of the worst financial crisis in living memory, a new, sustainable economy must be created. This coincides with a time in our history when there are grand challenges such as climate change and energy security to be faced. By 2050 it is expected that there will be at least 9 billion people alive: food and water security are also increasingly important politically. Soil is being destroyed at unsafe rates. For most countries waste disposal is becoming increasingly difficult as suitable sites for properly engineered landfilling are becoming scarce.

Building a bioeconomy in which the relationship between economic growth and increasing greenhouse gas (GHG) emissions is decoupled is one way to start to tackle these challenges. Since the OECD published a policy agenda for a bioeconomy [1], interest has been growing. In 2012 the United States published its bioeconomy blueprint, and the European Union launched its bioeconomy strategy. Several nations have developed their own bioeconomy strategies, for example, Belgium, Denmark, Finland, Germany, Malaysia, the Netherlands, and South Africa. Several other countries, although lacking a formal bioeconomy strategy, have important policies consistent with the bioeconomy concept, for example, France, Italy, Japan, Korea, and the UK [2].

At roughly the same point in history, genome sequencing has been transformed into a readily available technology as the cost has plummeted. The grant scheme run by the US National Human Genome Research Institute, officially called the Advanced Sequencing Technology awards, is known more widely as the "$1000 dollar genome" program as it predicted that the cost of sequencing a human genome could be reduced to this cost. Started in 2004, the scheme has awarded grants to 97 groups of academic and industrial scientists to a value of some $230 million, including some at every major sequencing company [3].

This has demonstrated to policymakers and others that biotechnology is a foundation for economic growth, something that has been disputed for decades. The economic impact already being experienced from the Human Genome Project (HGP) has far-reaching consequences. The $3.8 billion investment of the HGP not only launched the genomics and DNA sequencing revolution but has also driven close to a trillion dollars in economic impact and generated over 300,000 jobs in the US economy [4]. It is estimated that the return on investment has been $178 for every public dollar [5].

In the present context, we hope to highlight early progress on genomics and associated –omics technologies in several of the key sectors that a bioeconomy will rely on. It is clear that there is much capacity for improvement as the age of practical application of genomics to societal problems other than human health is in its infancy. Because a bioeconomy will need to reconcile the needs of industry and agriculture, this chapter has a focus in applications of genomics in these areas.

THE GRAND CHALLENGES ECOSYSTEM

Certain grand challenges are either here or are approaching. It is argued that they are particularly challenging as policies to tackle one of them will have effects on one or more of the others, not always in a positive manner. For example, growing more crops on more land, or increasing the productivity of crops on the existing land addresses food security. This strategy is likely to negatively affect soil health, and will require more water, which is already stressed in many locations. It may also decrease biodiversity. Higher yields will require more artificial fertilizers, which mean more emissions and agriculture becoming even more dependent on the fossil industry. More agrochemicals can lead to further pollution. Bioenergy, biofuels, and bio-based materials produced from biomass instead of fossil resources address GHG emissions reductions, central to the mitigation of climate change. But this requires more biomass, which can impinge on food security, and can interfere in other ways.

Energy security

Most countries are plagued by energy insecurity. Nearly all European countries are net importers of oil and gas. India imports 80% of its domestic crude oil requirements [6]. The cost of crude oil imports account for more than 10% of the GDP of Thailand [7]. The United Kingdom became a significant net exporter of crude oil beginning in the early 1980s, but has been a net importer since 2005.[1]

No country illustrates the situation better than Japan, the world's third largest economy which is just 16% energy self-sufficient.[2] Japan is the world's largest importer of liquefied natural gas (LNG), the second largest importer of coal and the third largest net importer of oil. Japan relied on oil imports to meet about 42% of its energy needs in 2010 and to feed its vast oil refining capacity (some 4.7 million barrels per day at 30 facilities as of 2011) and relies on LNG imports for virtually all of its natural

---

[1] http://www.eia.gov/todayinenergy/detail.cfm?id=16971
[2] www.eia.gov/countries/cab.cfm?fips=JA

gas demand. Japan consumed an estimated 4.5 million barrels per day of oil in 2011, while it produced only about only 5000 barrels per day [8].

Meanwhile the stakes are rising. New oil discoveries globally have not kept up with annual consumption since at least 1980. Currently some specialist oil companies have high percentages of their potential capex over the next decade in high-cost, high-risk projects, especially deep water or oil sands, which may require a $95 per barrel price [9]. Currently almost a third of the oil consumed in the world comes from underwater reservoirs. And as offshore oil exploitation is moving into increasingly deep waters, the risks of accidents increases—for an average platform, each 30 m of added depth increases the incident probability by 8.5% [10].

Climate change

There is now overwhelming scientific consensus on the existence of anthropogenic global warming [11]. Over 80% of global emissions are caused by countries that have participated in the Copenhagen Accord,[3] which recognizes a need to limit the temperature rise though global warming to 2°C. And yet, the world seems on a trajectory consistent with a long-term average temperature increase, which is significantly higher [12]. Therefore most of the known and projected fossil fuel reserves may be unburnable [13]. This conclusion has been contested by key companies in the oil industry [14], although there is a gradual realization in the industry that climate change will affect how it does business [15].

McGlade and Ekins [16] have calculated that a third of oil reserves, half of gas reserves, and over 80% of current coal reserves should remain unused from 2010 to 2050 in order to meet the 2°C obligation. Critical work on climate change came from the Intergovernmental Panel on Climate Change (IPCC[4]) during 2014. In April 2014, the Working Group III contribution to the IPCC's Fifth Assessment Report Scenarios [17] showed that, to have a likely chance of limiting the increase in global mean temperature to 2°C, means lowering global GHG emissions to near-zero by the end of this century.

Food and water security

With 9.1—9.6 billion alive by 2050 as estimated in the medium variant option, food production will need to rise by 50—70% [18]. More arable land, or more efficient use of existing arable land, will be needed to meet the food demands, while less may be available because of changing climate conditions. Using more land for production also impacts biodiversity. The mantra will become "more from less."

[3] http://unfccc.int/resource/docs/2009/cop15/eng/l07.pdf
[4] www.ipcc.ch

As many as 2 billion people rely directly on aquifers for drinking water, and 40% of the food in the world is produced by irrigated agriculture that relies largely on groundwater. Globally, 70% of all freshwater use is for agriculture [19]. Vast territories of Asia rely on groundwater for 50–100% of the total drinking water [20] and groundwater depletion is accelerating worldwide. Some of the highest rates of depletion are in some of the world's major agricultural centers, including northwest India, northeast China, and northeast Pakistan [21]. Sophisticated modeling has suggested an 80% likelihood that at least one decades-long mega-drought will hit the Southwest and Midwest United States between 2050 and 2100 [22].

Soil destruction

Often overlooked in policy making, soil is the ultimate genetic resource; soils are the critical life-support surface on which all terrestrial biodiversity depends. More than 95% of all food is derived from cropland [23]. But soil is being destroyed at unprecedented rates due to soil erosion (e.g., through deforestation), pollution, and salination. About 2.5% of arable land in China is too contaminated for agricultural use [24].

It takes around 500 years to form 25 mm of soil under agricultural conditions, and about 1000 years to form the same amount in forest habitats.[5] Therefore soil should be treated as a nonrenewable resource. In the bioeconomy and sustainability context, soil accounts for some 20% of the capture of human CO_2 emissions [25].

FOOD PRODUCTION IN A BIOECONOMY

A primary focus of a bioeconomy is to reduce GHG emissions. That is a primary driver behind the bio-based industries replacing fossil resources exploitation to make fuels and materials. Agriculture is also enormously important in climate change mitigation strategies. Table 11.1 provides in broad terms the GHG emissions associated with the production of major protein sources. Although the sources of information use different methodologies, the table highlights that there are large differences, and that ruminant production is much worse in GHG emissions terms that chicken and fish. However, with the growth of the global middle class [26], demand for meat has increased tremendously.

In about the last 30 years meat consumption in developing countries has doubled, and egg consumption has quadrupled. The demand for more meat has significant environmental implications. Beef production is notoriously costly in resources such as water and land, and has been implicated in deforestation. For every kilogram of beef produced, 4–5 kg of high energy feed are required, and well over 10,000 l of water is consumed.

---

[5] Food and Agriculture Organization (FAO), www.fao.org/sd/epdirect/epre0045.htm

Table 11.1 GHG emissions for various food products and different studies. Variations within the same product category result from different farming methods, but also due to differences in life cycle assessment (LCA) calculations. Despite some large differences, there is general agreement that capture fish and salmon farming are much more efficient in terms of GHG emissions than beef or lamb (ruminant) production

Product	CO_2 (eq/kg)	Comments
Beef	44.8	Mainly a result of methane and N_2O, not CO_2
Belgian beef	14.5	
Idaho and Nebraska beef (average)	33.50	Farm-gate, quoted as 15.23 kg per pound of beef
Idaho lamb	44.96	Farm-gate, average of low and high productivity
Swedish pork	3.3–4.4	
Michigan pork	10.16	Farm-gate
Farmed trout	4.5	Raised in ponds. Frozen, leaving retailer
Cod	3.2	Frozen fillet, leaving retailer
Chicken	2.0	(Round weight, United States)
Poultry (United States)	1.4	
Chicken	4.6	(Round weight, United Kingdom)
Farmed salmon (sea-based, United Kingdom)	3.6	Including processing and transportation
Farmed salmon (sea-based, Canada)	4.2	Adjusted to fillet based on figures for live fish
Farmed salmon (sea-based, Norway)	3.0	Transportation to Paris
Farmed salmon (global average)	2.15	Farm-gate estimates
Capture fish (global average)	1.7	

Source: Adapted from Ref. [27–29].

What has genomics to offer? It is impossible to answer the question exhaustively in this chapter; rather a flavor of how genomics can influence food production is given. Moreover, the vast majority of applications of genomics to food and feed production have yet to be thought of. Nevertheless, at this early stage, it is quite clear that there are significant advances in the efficiency of food production to be made, even without genetic modification. Currently it is the alliance of genomics technologies with modern and conventional breeding technologies that is making greatest strides.

Beef production

Given the popularity of beef, genomics has started to be used to increase the efficiency of production, for example, to enhance meat quality during breeding or rearing. One of the greatest challenges to successful application of genomics, however, is the wide diversity of breeds used across the industry. Cattle functional genomics is in its infancy. Some approaches to beef cattle genomics are given.

Some economically relevant traits, such as early life growth traits, are easily measured in the field. Many of which are not easily measured are key to production efficiency factors, such as animal health and feed efficiency [30]. Partly this is the problem of the impracticality and/or cost of collecting field data, but the variety of breeds adds to the difficulties. To address this, sequencing efforts of important animals in the global beef industry are being made to identify variants and to associate those variants with the genetic variation observed across beef populations. In this way genomic selection tools may aid breeding programs.

Proven sire identification in breeding programs is extremely valuable, but parentage verification is a well-described problem, and is made more difficult by the large number of breeds. For example, McClure [31] employed microsatellite (MS) and single nucleotide polymorphism (SNP) from 39 breeds from *Bos taurus* and *Bos indicus*. An objective of this study was to develop a global SNP-MS reference panel that is inexpensive, easy to use, and can be used across the majority of these breeds.

When genomic approaches can routinely be combined with Artificial Reproductive Technologies, this will enable verification of the inheritance of favorable traits in *in vitro* embryos. This could make the whole selection and breeding process quicker and much more accurate [32].

Milk production

Genomics studies of milk have varied goals, underlining the significance of milk as a human food. Topics include the capacity of milk to manipulate the gut microbiota; manipulation of bovine milk fat; genetic selection for economically important traits, such as protein content; and diagnostics.

A very important economic trait is protein content, and in cattle the trait has a linkage to heritability. The application of genomics to breeding programs to improve protein yields would have obvious economic and societal benefits. For example, Raven et al. [33] produced evidence supporting a role for the RNASE5 (*angiogenin ribonuclease, RNase A family 5*) pathway in milk protein content, indicating that sequence polymorphisms associated with the genes involved explain some of the observed protein content differences. Their methodology contributes to the fundamental science of lactation, with practical implications for milk production with higher protein content.

Of particular relevance to milk is microbial spoilage [34]. Raw milk can harbor a variety of human pathogens and has been associated with serious foodborne illnesses such as diphtheria and brucellosis. Other, nonpathogenic species can produce off-flavors, unwanted acidification, and thereby can contribute to lower shelf life. With a rise in consumption of raw milk, the risks are obviously greater, but pasteurized milk is not completely risk-free. The indigenous strains found in the dairy vary significantly from type strains, thus necessitating genome sequencing of additional dairy isolates to better understand how they survive in milk, and to subsequently take measures to ensure their eradication from it.

Chicken as a bioeconomy food source

Chicken is a major source of protein in the world, with around 20 billion birds alive today, producing around 1.2 trillion eggs.[6] It was the first livestock species to be sequenced and so has led the way for others [35]. In parallel with the chicken genome sequencing project [36], a consortium set about identifying SNPs. When a large number of these are verified, the availability of a standard set of 10,000 or more SNPs holds much promise toward the identification of genes controlling quantitative trait loci (QTLs), including those of economic interest.

One of the key traits improved every year through selective breeding is feed efficiency (Figure 11.1)—the number of kilograms of animal feed needed to produce a kilogram of poultry meat [37]. Genomic technologies are expected to enhance this trend. Since animal breeding is cumulative, even small enhancements to the rate of improvement can multiply into huge differences for commercial customers over time

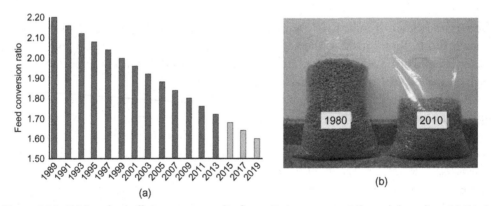

(a)   (b)

Figure 11.1 Chicken feed efficiency as a result of genetic improvement through breeding. (a) Feed conversion ratio over 20 years in meat-producing broilers at 42 days. (b) Broiler feed to produce a 2.5 kg chicken at 42 days. *Courtesy of Ref. [38].*

[6] http://www.bbsrc.ac.uk/news/food-security/2013/130404-f-what-lives-inside-a-chicken.aspx

and have very large impacts. The result of this is that more people can be fed from the same land resources.

The Aviagen[7] genomics project is concerned with identifying naturally occurring markers within the genome of elite birds and using those markers to help breed stronger and more productive birds through the current selective breeding program, a completely natural process. Aviagen became the first company to include genomic information as a critical additional source of information in a R&D breeding program.

Crop production

Feeding 9 billion people by 2050 is a major food security issue. Moreover, the demand for biomass for bio-based production of fuels, chemicals, and plastics will further stress land availability and productivity. The effects of climate change will exacerbate the difficulties facing conventional agriculture.

There are many applications of genomics to crop production that will be utilized in the future bioeconomy: pest resistance; more "efficient" plants that use less water; resistance to environmental stresses; and the development of crops that can fix nitrogen to replace synthetic fertilizers.

Heat and drought stress: an increasingly important problem

The 1988 drought in the Midwestern United States resulted in a 30% reduction in US corn production and cost about $39 billion [39]. The United States has just experienced its most widespread drought in more than half a century [40], and the drought in 2014 in California was perhaps the worst ever recorded [41]. Agriculture accounts for around 70% of all water use. Therefore measures that conserve water in agricultural use are of the utmost social and economic importance.

Drought is by far the most significant environmental stress in agriculture worldwide and improving yield under drought is a major goal of plant breeding. Despite much work on crop breeding for drought tolerance, there is still a large gap between yields in optimal and stress conditions. The complexity of drought tolerance mechanisms explains the slow progress in yield improvement.

Breeders now have much more genomics-related information available as new tools for breeding, such as markers for QTLs and single genes for plant transformation. Routine cloning of the genes underlying the QTLs is still a way off, but it will ultimately provide simple markers for an effective marker-assisted selection (MAS). Nevertheless MAS for drought tolerance will not be an easy task because dozens of QTLs for drought-related traits have been identified [42]. For many crop plants

---

[7] http://en.aviagen.com/research-development/

information on drought-related QTL findings have been collected in open source databases, such as GRAMENE[8] or GRAINGENES.[9]

Major food crops are targeted for genomics investigations into drought tolerance, for example, wheat and barley [43]. Whole genome re-sequencing of maize (corn) is being used to identify drought tolerance genes [44]. Literature mining of the tomato genome by Bolger et al. [45] and filtering against drought- or salt-related QTLs has resulted in the identification of 100 candidate genes. Potato is the fourth most important crop in the world. Potato yield losses due to climate change are expected to range between 18% and 32% up to 2030. Many more drought response genes have been identified in wheat, rice, and maize than in potato. The identification of genes controlling drought responses in potato only started in 2007, and some (limited) progress has been made over the last few years [46].

With the rise of cultivation of nonfood crops specifically for energy purposes (to reduce competition with food crops), there has also been a rise in interest in drought tolerance of these species. Examples are *Populus balsamifera* (balsam poplar), *Panicum virgatum* (switchgrass), and *Jatropha curas* (jatropha). Energy crops are often envisaged for growth on marginal land, where often the problem is lack of water. One of the principal portals to genomic information relevant to bioenergy crops is the phytozome portal.[10] Others are described by Ref. [47].

Evidence suggests that combined heat and drought stress can cause disproportionate damage to important crops compared with either stress individually (see Ref. [48]). Therefore, understanding dual stress tolerance such as heat and drought in crop plants in multiple locations over multiple seasons has become a top priority in agricultural research. Transgenic plant technologies derived from dual stress tolerance could enable farmers around the world to maintain higher yield and productivity over variable and adverse environmental conditions [49].

Too much water

Rice is a crop well adapted to wet, monsoon climates and allows farmers to produce food in flooded landscapes. Of the lowland rain-fed rice farms worldwide, over 22 million hectares are vulnerable to flash flooding, representing 18% of the global supply of rice. In total, some 30−40 million hectares get submerged, and this happens roughly every 3 years. Most rice varieties can tolerate only a few days of submergence and die after about a week. With traditional lowland rice, when flooded the plant reacts by spurring growth to get above the water, it continues to grow during the

---

[8] http://www.gramene.org/
[9] http://wheat.pw.usda.gov/GG2/
[10] www.phytozome.net

Immediately after flooding   Three months later

Figure 11.2 SUB1A enhances recovery after severe flooding damage. *Courtesy of Ref. [50].*

flooding, and finally runs out of nutrients and dies. Variety SUB1A does not grow while flooded and starts growing again after the flooding has subsided (Figure 11.2). In this case a single mutation is involved in tolerance.

SUB1A has been introduced into several mega-varieties of rice through MAS and backcrossing without apparent adverse effects on the plants [51]. Under submergence for 7−14 days, these tolerant cultivars have an average yield advantage of 1.5 tonnes per hectare over intolerant cultivars, with no reduction in yield under nonsubmerged conditions. SUB1 is gradually being introduced to all varieties developed for lowland ecosystems by the International Rice Research Institute, and several national programs are also introducing the gene into locally adapted varieties. To date, over 4 million farmers have been reached with seeds of SUB1.

Decoupling agriculture from fossil fuels

Nitrogenous compounds in fertilizers are major contributors to waterway eutrophication and GHG emissions, and the Haber−Bosch process for making fertilizers is very energy-intensive. It consumes 3−5% of the world's natural gas production and releases large quantities of CO_2 to the atmosphere [52]. When the price of Brent crude oil rose from around $50 per barrel to about $110 by January 2013, the prices for ammonia in Western Europe and the Mid-Western corn belt in the United States roughly tripled over the same period.[11]

Several efforts are ongoing in a tantalizing research area—creating plants that make their own fertilizer. A collaborative project with UK and US scientists aims to design and build a synthetic biological module that could work inside a cell to perform the function of fixing nitrogen [53]. This project aims to reengineer the cyanobacterial

[11] http://marketrealist.com/2013/02/brent-oil-moves-nitrogenous-fertilizer-prices/

machinery to fix nitrogen using solar energy as a first step toward transferring the machinery into plants themselves. This has the potential to revolutionize agriculture and significantly decouple it from the fossil fuels industry.

The oil palm: a classic bioeconomy quandary, and the power of genomics

Oil palm not only provides 45% of the world's edible oil, but the oil is well suited for use as biodiesel. By 2019, Indonesia and Malaysia are forecast to nearly double their production of biodiesel [54]. The central issue of food versus energy security is clearly a concern, and there are also social and environmental issues around its overexploitation in South East Asia [55]. Therefore, in this one crop, are illustrated some of the toughest bioeconomy issues.

The oil palm genome sequence was published by Singh et al. [56]. The sequence enables the discovery of genes for important traits as well as alterations that restrict the use of clones in commercial plantings. The oil palm is largely undomesticated and is an ideal candidate for genomic-based tools to harness the potential of this remarkably productive crop. The authors claim that the dense representation of sequenced scaffolds on the genetic map will facilitate identification of genes responsible for important yield and quality traits.

The modern oil palm tree *Elaeis guineensis* has three fruit forms: *dura* (thick-shelled); *pisifera* (shell-less); and *tenera* (thin-shelled) (Figure 11.3). The *tenera* palm yields far more oil than *dura* and is the basis for commercial palm oil production in all of South East Asia. In 2013, a remarkable discovery was made. The *Shell* gene has proven extremely challenging to identify in oil palm, given the large genome, long generation times and difficulty of phenotyping in experimental populations. Singh et al. [56] identified the gene and determined its central role in controlling oil yield. Regulation of the *Shell* gene will enable breeders to boost palm oil yields by nearly one-third, which is excellent news for the industry, the rainforest, and bioeconomy policymakers.

Seed producers can now use the genetic marker for the *Shell* gene to distinguish the three fruit forms in the nursery long before they are field-planted. Currently, it can take 6 years to identify whether an oil palm plantlet is a high-yielding palm. Even with selective breeding, 10−15% of plants are the low-yielding *dura* form due to uncontrollable wind and insect pollination, particularly in plantations without stringent quality control measures [57].

Accurate genotyping such as this has a critical implication for a bioeconomy. Enhanced oil yields can optimize and ultimately reduce the acreage devoted to oil palm plantations, providing an opportunity for conservation and restoration of dwindling rainforest reserves [58].

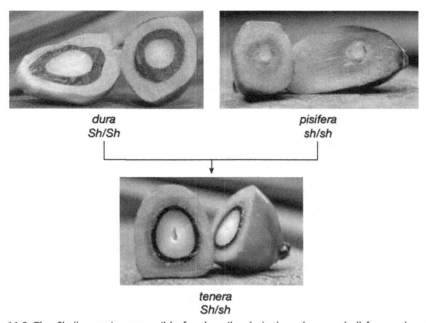

dura
Sh/Sh

pisifera
sh/sh

tenera
Sh/sh

Figure 11.3 The *Shell* gene is responsible for the oil palm's three known shell forms: *dura* (thick); *pisifera* (shell-less); and *tenera* (thin), a hybrid of *dura* and *pisifera* palms [56]. *Tenera* palms contain one mutant and one normal version, or allele, of *Shell*, an optimum combination that results in 30% more oil per land area than *dura* palms [57].

Fisheries and aquaculture

Stresses on land use could be alleviated by higher fish consumption (e.g., Ref. [27]). Seafood is already the highest value globally traded food commodity. There are many potential applications of genomics to sustainable wild and farmed fish production. Here a small selection of high-priority applications is outlined.

Genomics and the fishing industry

The social, economic, and environment contributions of fisheries are known to be under threat. Numerous wild fish populations are either overexploited or are in precipitous decline. Wild fisheries should therefore be regarded as "not necessarily renewable," and many are threatened. Many universal difficulties associated with wild fisheries are related to fish species misidentification. Incorrect identification can lead to errors in estimating numbers of fish species and the actual magnitude of fish stocks, with consequences for fisheries management and the fishing communities. Standardized DNA barcodes that are unambiguous, widely applicable, and globally accessible to the nonexpert can address such identification difficulties, along with other practical fishing industry problems, such as traceability, illegal fishing, and fish fraud [59].

Aquaculture and genomics

Aquaculture production has continued to grow annually at around 6—8%. Today, farmed seafood production (around 60 million tonnes) exceeds that of wild fisheries and has significant potential for future growth.

High-priority traits for farmed fish are the development of single-sex populations and improving disease resistance. Production of single-sex stocks (either male or female, depending on the species) is desirable in most commercial production to cut the cost of farming. In any commercial fish species, one sex usually reaches maturity and market size more quickly than the other; it is often the female (e.g., halibut), but not always, as in the case of Nile tilapia. Genomics is contributing to understanding the sex determination mechanism, as illustrated by two recent papers applying restriction-associated DNA (RAD) sequencing analysis to these two very important farmed species.

Palaiokostas et al. [60] described assays for sex-associated DNA markers developed from RAD sequencing analysis to help implement single-sex female halibut production. The same technique was used by Palaiokostas et al. [61] to identify a reduced candidate region for the sex-determining gene(s) and a set of tightly sex-linked SNP markers in male Nile tilapia, with no ambiguity is assigning sex.

Salmon genomics: a very special case

Salmonids, in particular Atlantic salmon, are among the most important aquaculture species. In 2010, approximately 1.5 million tonnes of Atlantic salmon were produced from farms worldwide, corresponding to a value of just over $7.8 billion [62]. It is an important bioeconomy species due to the low GHG emissions associated with farming salmon.

The genomic resources for Atlantic salmon are among the most extensive of all aquaculture species and include several genetic maps, a physical map, an extensive expressed sequence tag database of approximately 500,000 tags, and several microarrays [63]. In June 2014, the International Cooperation to Sequence the Atlantic Salmon Genome announced completion of a fully mapped and openly accessible salmon genome,[12] which is housed at its own web site.[13] Some of the expected outcomes of this research are: understanding the attacks by viruses and pathogens on salmon and to produce new vaccines to reduce losses through disease; applications for food security and traceability and brood-stock selection for commercially important traits; and better understanding of the interactions of farmed salmon with wild counterparts.

---

[12] http://www.genomebc.ca/news-events/news-releases/2014/scientific-breakthrough-international-collaboration-has-sequenced-atlantic-salmon-genome/

[13] http://www.icisb.org/atlantic-salmon-genome-sequence/

Selective breeding of salmon will be more targeted and efficient. This could, for example, select for individuals that are more resistant to disease and parasites, and select fish that grow more quickly while being adapted to new feed types (see the chicken example of "more from less"). In the longer term, genomic knowledge should help streamline the aquaculture industry while providing consumers with healthier farmed salmon, produced with as little environmental impact as possible.

In the case of salmon, then, the power of relatively small public research funding of genomics to transform an industry is illustrated. In this case, many of the problems of the industry can be addressed through a single tool, which makes genomics unique as a solution provider.

INDUSTRIAL BIOTECHNOLOGY—REPLACING THE OIL BARREL

The new industrial biotechnology is largely about the bio-based production of fuels, chemicals, and plastics. In most cases, the products already exist, made from fossil resources through oil refining and petrochemistry. There is no shortage of crude oil or natural gas, and replacing the oil barrel will take decades. However, increasing political pressure and societal awareness over climate change and energy security may necessitate its replacement long before crude oil becomes scarce.

Much of the focus on GHG emissions reductions has been in the energy industry and transport, which spurred the biofuels and bioenergy initiatives. However, the chemicals sector is the largest industrial energy user, accounting for one-tenth of global energy use [64] and is the third largest industrial source of emissions after the iron and steel and cement sectors [65]. This is one of the main drivers behind the development of a bio-based chemicals industry.

One of the questions asked frequently is: how much substitution of fossil-derived chemicals and plastics is possible using bio-based equivalents or new molecules? It has now been demonstrated that even completely unnatural compounds can be manufactured using microbial cells (e.g., Ref. [66]) and the technique of metabolic engineering. The open question regarding the ability to replace the entire oil barrel is becoming an increasingly important political question. Some of the classes of chemicals in the literature that can be made through metabolic engineering of microbial strains are shown in Figure 11.4.

The list of successes is increasing, and there is tentative evidence that the time from conception to production (the innovation cycle) is decreasing. However, what is achievable in the research laboratory may never see commercial mass production. Bio-based production of aromatics, an extremely important class of industrial chemicals, is problematic, due partly to their toxicity to microbial cells. However, prior to the global dominance of the petrochemicals industry, wood refineries existed in

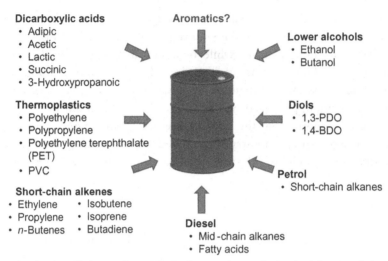

Figure 11.4 A selection of chemicals and fuels that can be made (in the laboratory) through metabolic engineering of microbes (from OECD research).

significant numbers (e.g., Re. [67]) and lignocellulose is also a potential source of aromatic compounds.

It is foreseeable that the combination of wood and "green" chemistry with microbial metabolic engineering could be a potent force in the eventual replacement of the oil barrel. Coal, although still vitally important as a feedstock for electricity generation, will need to be replaced in the long term to honor the 2°C obligation. However, as it becomes less popular as a fuel, it may achieve a renaissance as a source of aromatic chemicals.

The ultimate manufacturing model is the integrated biorefinery, where multiple classes of bio-based products are made from multiple sources of biomass. In these days of the infancy of bio-based production, there is a preponderance of single substrate, single product biorefineries. A great danger for such facilities is that if the price of the feedstock changes rapidly, the facilities can be put out of operation. This has already happened once to soy bean biodiesel production in Iowa. The greater the number of feedstocks that can be processed at a single facility, the greater the buffer against price volatility.

Biobased 1,3-propanediol: a metabolic engineering classic

Globally, the 1,3-propanediol (1,3-PDO) market is expected to grow from an estimated $157 million in 2012 to $560 million by 2019, with a compound annual growth rate of 19.9% during this period. It can be used in many synthetic reactions and has uses in solvents, adhesives, resins, detergents, and cosmetics. It is especially well known as a monomer for the synthesis of polytrimethylene terephthalate, a polyester with excellent properties for fibers, textiles, carpets, and coatings. The bio-based equivalent, now fully commercialized, has very significant environmental performance advantages compared to the petro-based counterpart.[14]

Its bio-based production was perfected in *Escherichia coli*. One of the considerations for working in *E. coli* is strains based on *E. coli* K12 are eligible for favorable regulatory status in the United States. The engineered strain relies on a carbon pathway that diverts carbon from dihydroxyacetone phosphate, a major pipeline in central carbon metabolism, to 1,3-PDO.

During the metabolic engineering, the two most fundamental changes described were (Figure 11.5):

1. To remove a theoretical yield limitation, the phosphotransferase system was replaced with a synthetic system comprising galactose permease (*galP*) and glucokinase (*glk*); both genes are endogenous to *E. coli*.

2. Triosephosphate isomerase (*tpi*) was deleted in an early construct (part (a) in Figure 11.5). But this also imposed a yield limitation. To overcome this, *gap* (glyceraldehydes 3-phosphate dehydrogenase) was down-regulated, which, along with the reinstatement of *tpi* (part (b) in Figure 11.5), provided an improved flux control point.

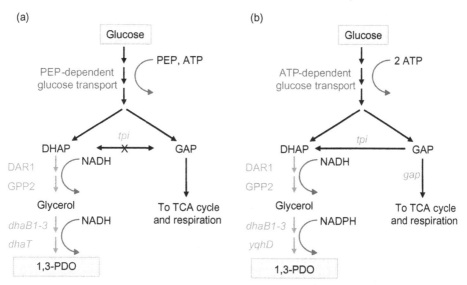

Figure 11.5 Metabolic engineering for the production of 1,3-propanediol [68]. (a) An early construct; (b) a later construct with improved yield. GAP, glyceraldehyde 3-phosphate; DHAP, dihydroxyacetone phosphate.

[14] http://www2.dupont.com/Bio-based_Propanediol/en_CN/

Along with other changes, the end result was a metabolically engineered organism that produced 1,3-PDO at a titer of 135 g/l.

Sugar to plastic through metabolic engineering and fermentation

Polylactic acid (PLA) has been considered a good alternative to petroleum-based plastic because it possesses several desirable properties such as biodegradability and biocompatibility. The major driver for its production is for large-scale use in fibers and fabrics. For example, it is being used in car interiors, replacing plastics with greater GHG emissions. Current manufacturing consists of fermentation to produce lactic acid followed by one of two major chemical routes to the polymer, both of which are difficult and either use high temperatures and solvents or heavy metal catalysts [69]. But there is no existing natural bacterial route to PLA. However, Jung and Lee [70] described efficient production of PLA by a direct fermentation of glucose without a chemical step (Figure 11.6) in a metabolically engineered *E. coli* chassis strain.

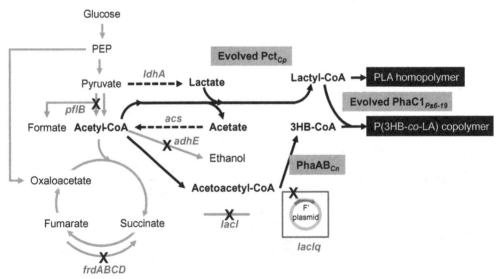

Figure 11.6 Direct fermentation of glucose to PLA in *E. coli*, replacing the chemical polymerization step [70]. The overall metabolic network is shown in blue together with the introduced metabolic pathways shown in black for the production of the PLA homopolymer and the P(3HB-*co*-LA) copolymer in *E. coli*. The genes with cross marks shown in black represent the chromosomal gene inactivation and the elimination of F′ plasmid shown in the box, and the genes with dashed arrows shown in black represent the overexpression of the genes by chromosomal promoter replacement. For interpretation of the references to color in this figure legend, the reader is referred to the web version of this book.

Consolidated bioprocessing

There are many areas of research required to bring about this revolution in manufacturing, especially in both genomics and bioinformatics. One area receiving attention is to combine several bioprocess functions into a single biocatalyst (the so-called consolidated bioprocessing, CBP), rather than to rely on multiple, expensive hydrolytic enzymes or thermal processes for pretreatment of cellulosic biomass to fermentable sugars. The United States Department of Energy (US DoE) opined that CBP technology will lessen the complexity, cost, and energy intensity of the cellulosic biorefinery [71], endorsing the view that this technology may offer the ultimate low-cost configuration for cellulose hydrolysis and fermentation. Biomass conversion allied to synthetic biology has been termed "a match made in heaven" [72]. An example of the potential was demonstrated by Bokinsky et al. [73], in which an engineered *E. coli* was able to utilize pretreated switchgrass to produce three advanced biofuels (Figure 11.7).

The food versus fuel controversy, and avoiding it in future

The rapid expansion of the bio-ethanol industry based on corn (maize) as a feedstock (first-generation biofuels) was accompanied by a debate concerning the role of bio-fuels in food prices increases around 2008, the so-called food versus fuel debate (e.g., Ref. [74]). Evidence linked first-generation biofuels to the price spike, some of it

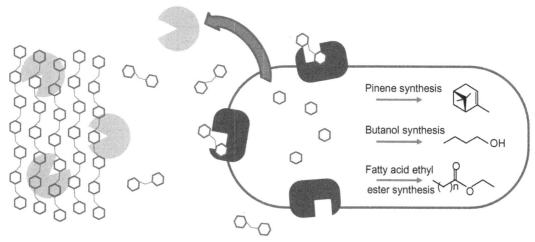

Figure 11.7 Engineering *E. coli* for use in consolidated bioprocessing. Cellulose and hemicellulose were hydrolyzed by secreted cellulase and hemicellulose enzymes into soluble oligosaccharides (blue). β-Glucosidase enzymes (red) further hydrolyzed the oligosaccharides into monosaccharides, which were metabolized into biofuels via heterologous pathways (from Ref. [73]). For interpretation of the references to color in this figure legend, the reader is referred to the web version of this book.

showing a marginal effect among a host of factors, but the actual extent of the linkage will probably never be known. Many studies (e.g., Refs [75–78]) have arrived at the view that there were several causes, interacting in a complex way, and that biofuels were only a part of the cause. However, the quest was already under way to use organic waste sources as feedstocks in future biorefineries. This avoids the use of food crops for biorefinery operations. The waste materials that are available in large quantities are the "cellulosic" agricultural and forestry wastes, municipal solid waste (MSW), including food wastes, and waste industrial gases.

Cellulosic biorefineries

The arrival, albeit in small scale, of cellulosic biorefining heralds a landmark achievement. The first commercial scale cellulosic ethanol (second-generation ethanol) plant in Europe is now open in Crescentino, Italy,[15] with the aid of public finance. At least three plants will be open in the United States in 2015, all built with public support. The success of these plants could be critical to the future of cellulosic ethanol [79] and further expansion of the biofuels industry.

One of the more significant challenges in utilizing the vast global lignocellulose resource is the need for large quantities of glycoside hydrolase enzymes to efficiently convert lignocellulose, hemicellulose, and cellulose into fermentable sugars (Table 11.2).

Table 11.2 Composition of different lignocellulosic materials

	Glucose	Xylose	Arabinose	Mannose	Lignin
Hardwood					
Birch	38.2	18.5	–	1.2	22.8
Willow	43.0	24.9	1.2	3.2	24.2
Softwood					
Spruce	43.4	4.9	1.1	12.0	28.1
Pine	46.4	8.8	2.4	11.7	29.4
Grasses					
Wheat straw	38.2	21.2	2.5	0.3	23.4
Rice straw	34.2	24.5	n/d	n/d	11.9
Corn stover	35.6	18.9	2.9	0.3	12.3

Figures are percentage of total dry weight. Glucose is mainly derived from cellulose, xylose, arabinose, and mannose from hemicellulose. Lignin is comprised mainly of phenolics.
n/d, not determined; –, below detection limit.
Source: From Ref. [80].

[15] http://www.novozymes.com/en/news/news-archive/Pages/World%E2%80%99s-first-advanced-bio-fuels-facility-opens.aspx

The harsh conditions used in the pretreatment of the raw material release fermentation inhibitors including weak organic acids (particularly acetic and formic acids), furan derivatives, and phenolic compounds (there are several reviews, e.g., Ref. [81]). To improve the fermentation ability of industrial *Saccharomyces cerevisiae* yeast strains for ethanol production, several strategies have been applied to overcome the effect of inhibitors. These approaches include: controlling inhibitor concentrations during the fermentation [82]; a mutagenesis and genome shuffling approach [83]; and the overexpression of genes encoding enzymes that confer resistance toward specific inhibitors [84].

The ability of *S. cerevisiae* to convert glucose to ethanol is the basis of much of the biofuels industry, with a theoretical ethanol yield of over 160 g/l [85]. It cannot, however, ferment C5 (pentose) sugars efficiently to ethanol, and xylose and arabinose may constitute up to 30% of the sugars that can be derived from lignocellulose. And yet it has a full xylose metabolic pathway [86]. There are other yeasts that can utilize C5 sugars, but their yields and productivity of ethanol are low by comparison. *S. cerevisiae* is also a GRAS organism (generally regarded as safe), with favorable regulatory status. As a result there is high interest in metabolic engineering of *S. cerevisiae* to improve C5 fermentation to ethanol (e.g., Ref. [87]).

Arguably more is known about *E. coli* than any other organism, and K12 strains also have favorable regulatory status. However, despite its flexibility and its very low risk level, it is not always possible to ensure efficient transcription/translation of a heterologous gene in *E. coli*, and post-translational protein modification does not occur in prokaryotic production systems. The *E. coli* catabolite repression limits simultaneous utilization of multiple carbohydrates [88], making industrial bioprocessing less efficient. Hence the development of specialized eukaryotic production hosts such as yeast. Some limited progress has been made in the co-utilization of mixed sugars for industrial purposes, both by metabolic pathway implantation and by isolation of catabolite repression-negative microorganisms (reviewed by Ref. [89]).

MSW as a feedstock for biorefining

Much of the organic matter in MSW is suitable as a fermentation feedstock. It is of a very mixed nature, and in its solid form would be extremely difficult to maintain as an on-specification feedstock for biorefining. However, virtually any form of organic matter can be converted into syngas through high-temperature gasification, and this can be done with the organic fraction of MSW. The first stage in MSW biorefining, then, is gasification, followed by syngas fermentation.

Although globally there are 1.3 billion tonnes of MSW, about 420 million tonnes are suitable for use in the current first generation of MSW biorefineries. That is the equivalent of 160 billion liters of renewable fuels from one sector alone, more than doubling the addressable market for biofuels with just the one feedstock.

MSW biorefining addresses two other societal challenges, the diminishing supply of geologically suitable landfill sites, and the ever-increasing headache for cities of how to deal with waste (Box 11.1). The future products include those of lower value, such as heat and power, through to higher value materials such as chemicals and pharmaceuticals. This highlights that environmental and economic aspirations need not contradict each other.

BOX 11.1 Edmonton's solution to landfilling

The City of Edmonton is moving from 60% waste recovery for recycling into other materials, already a high proportion, to 90%, arguably the best in the world. This is to be achieved via a MSW biorefinery that converts residuals from the City of Edmonton's composting, recycling, and processing facilities, waste that would otherwise be landfilled, into biofuels. The annual amount of this refuse-derived fuel is 100,000 tonnes.

From the Edmonton municipality point of view, it costs roughly CAD 70 per tonne, in fully loaded costs, to open up a new landfill. When a combustion technology to generate some power and slow the rate at which the site is filled to capacity is added in, that rises to around CAD 90 per tonne of waste. The commercial deal with Edmonton calls for a 25-year, CAD 45 per tonne deal that ultimately converts 30% of the city's waste stream to liquid fuels and chemicals. The first products are ethanol and methanol.

Beyond Edmonton, cities that have expressed strong interest in finding solutions sooner rather than later to landfilling problems include Philadelphia, Toronto, and Los Angeles. In 2015 there is a commitment to complete a commercial-scale facility in Varennes, Quebec, an option to double the capacity in Edmonton which the city and company are now mutually exploring, and a DoE-sponsored commercial-scale project in Pontotoc, Mississippi, that was conceived out of funds from the Recovery Act.

Source: Various, and Biofuels Digest [90].

Food waste, a major component of MSW

A huge amount of food is wasted unnecessarily. In particular, bread is wasted in large amounts, with a fraction that falls between 12% and 39% of MSW among different countries. In the United Kingdom, it is the largest contributor to food waste; 32% of all bread purchased is dumped when it could be eaten.[16]

Leung et al. [91] investigated the feasibility of fermenting waste bread to succinic acid. The resultant succinic acid production bioprocess gave an overall yield of 0.55 g succinic acid per gram of bread, at the time the highest yield among other food waste-derived media reported. Succinic acid is a precursor for many chemicals, with a production capacity of 30,000 tonnes per year and a corresponding market value of $225 million [92].

---

[16] http://www.bbc.com/news/magazine-17353707

Fermenting industrial waste gases to bio-based products

Industrial waste gases, such as CO_2 and CO, are starting to be used as feedstocks for fermentation. This has the enormous advantage of decoupling bio-based production from food production. A good example is the use of steel mill off-gases, especially highly toxic CO, to produce valuable chemicals [93], and in the future, a jet fuel. Photosynthetic and nonphotosynthetic biocatalysts are being developed that rely on metabolic engineering to improve yields and titers. Similarly, Calysta[17] of Norway uses natural gas-fed fermentation to produce feed-quality protein with high nutritional value for use in aquaculture. There are two means of carbon capture from waste gases for industrial purposes to consider, photosynthetic and nonphotosynthetic.

Photosynthetic carbon capture from waste gases

Much more attention has been given to photosynthetic processes and marine algal applications in particular. If successful, algae could deliver six to ten times more energy per hectare than conventional cropland biofuels while reducing carbon emissions by up to 80% relative to fossil fuels [94].

Cyanobacteria and algae grow faster than terrestrial plants and have simpler genetic backgrounds, which are easier to manipulate [95]. Despite the availability of a relatively large number of completed genome sequences, applications of synthetic biology in cyanobacteria and algae have significantly lagged behind those in *E. coli* and yeast.

Despite the obvious potential, there are several technical barriers and many clear targets for synthetic biology studies [96]. There is a serious lack of chassis strains. There is a lack of cyanobacterial standardized parts, and it cannot be assumed that *E. coli* or yeast parts will perform the same way in cyanobacteria (or vice versa). Indeed, performance will differ across different cyanobacterial species.

Transformation efficiencies need to be improved. *In vivo* restriction activities are an important barrier to introducing foreign DNA into cyanobacterial cells. A horizontal barrier is that solar conversion efficiencies are low, with yields around 5—7% during the growing season and around 3% in bioreactors on an annual basis [97]. In photo-bioreactors, excessive photon capture by the cells in the surface layer can block the light availability to the cells underneath [98]. Ribulose bisphosphate carboxylase is an essential enzyme in photosynthetic carbon fixation, but the reaction is slow. However, the carbon fixation efficiency can be greatly increased [99]. Despite early progress, synthetic biology in cyanobacteria and algae is in its infancy.

Nonphotosynthetic carbon capture from waste gases

Microorganisms capable of fermenting syngas are ubiquitous. They have diverse metabolic capabilities, resulting in the formation of a variety of desirable native products

---

[17] http://calystanutrition.com/nutrition/

such as acetate, ethanol, butanol, butyrate, formate, and H_2 [100] but not at industrial-scale efficiency. The vast majority of syngas fermenting organisms are anaerobic acetogens, which have a chemoautotrophic mode of metabolism [101].

Developing technologies based on purely chemoautotrophic organisms that utilize CO_2 and other waste gases for producing bio-based chemicals and fuels is attractive but technically very challenging. Six natural carbon fixation pathways are known so far, of which five are found to some extent in chemoautotrophs. Of these, the Wood—Ljungdahl pathway seems to be the most efficient in bio-based production conditions [102]. What may turn out to be critical is that it also can operate under heterotrophic conditions [100].

The choice is similar to the choice in other bio-based production technologies—if contemplating the introduction of a complete carbon fixation pathway into a prokaryotic host, whether to introduce a natural pathway or a synthetic one. And the dilemma is also the same—natural pathways have been optimized for the survival and reproduction of the native organisms in their natural environments, not for the survival in the artificial, extreme environment of a bioreactor, using high substrate concentrations to make high titers and yields of desired industrial products.

However, the task represents another classic for synthetic biology. Carbon fixation requires a relatively large set of genes, most of which involve complex, largely unexplored regulation [103]. Then there are the familiar tasks, the creation and insertion of the genes necessary to make the industrial product and the removal of competing or interfering genes or pathways. Additional complications arise from: anaerobic or microaerobic conditions; suitable redox environments; specialized metals chaperones; and membrane systems for ATP coupling [102]. All this also requires a synthetic biology strategy that minimizes the complex regulatory systems. As with many other bioeconomy applications of synthetic biology, the promise is great but the tasks ahead gargantuan.

As examples of the promise, ethanol production from CO is considered to be a viable approach to low-carbon fuel production, and at least three companies are seeking to develop the technology as a commercial process. The biochemistry and metabolic engineering of gas fermentation for biofuels was reviewed recently [104]. Köpke et al. [105] demonstrated the production of 2,3-butanediol from waste gases. A recent patent filing described the production of one or more terpenes by recombinant, acetogenic fermentation of CO [106].

The short-chain alkenes: the powerhouse of the petrochemicals industry

The six short chain alkenes are the main building blocks in modern petrochemistry. They are used in the plastics industry, for example, for producing polypropylene or polyethylene (Table 11.3). These short chain alkenes are currently produced by catalytic cracking of petroleum products and natural gas.

Table 11.3 Short chain alkenes and their uses

Olefin	Existing market $ billion	Main applications
Ethylene	144	Polyethylene (60%)
Propylene	88	Polypropylene (65%)
Linear butenes	37–74	Co-monomers in various plastics
Isobutene	29	Tires, organic glass, PET, fuels
Butadiene	14.6	Tires, nylon, coating polymers
Isoprene	2	Tires, adhesives

A patented method [107], using techniques of synthetic biology, describes the production of alkenes from a 3-hydroxyalkanoate by enzymatically converting it into the corresponding 3-phosphonoxyalkanoate, and using a second enzyme to convert 3-phosphonoxyalkanoate into an alkene.

In 2010, Global Bioenergies (Evry, France) generated an artificial metabolic pathway to isobutene from glucose. Isobutene is a gas, and its recovery from a fermenter is therefore simplified. There is no product toxicity, no feedback inhibition, and the downstream processing is inexpensive. Because of the central role of these compounds in the modern petrochemicals industry, perfecting bio-based versions of the short-chain alkenes would give enormous impetus to replacing the oil barrel.

CHALLENGES ON MANY FRONTS

If we are truly at the start of the era of bio-manufacturing and a long journey to a new energy order, then there is also an extremely complex web of new policy to be developed. Not only does this cover the entire supply and value chains, it brings in other aspects such as education and training, and public opinion. Many of the technologies remain to be proven at full-scale, and investors find scale-up very risky. A constant message from industry is the need for policy consistency and stability: most of the investments for this new future will have to come from the private sector, but public policy stability is needed to de-risk the investments. There will be a need for years to come for public policy that supports these developments if society feels that it is necessary to go down this path. In sustainability terms, there are very few options, if any, other than the bioeconomy that makes full use of all forms of biomass and waste materials in a post-fossil era.

Biomass sustainability

At the very heart of all this is biomass sustainability [108], given the competition for its use between food and industrial production. At present there is no consensus on what the metrics for biomass sustainability should be [109]. Therefore it cannot be accurately assessed how much biomass can actually be grown sustainably. The very first international biomass sustainability disputes have already arisen. Currently the regulatory framework is a confusing patchwork.

Public opinion

One of the toughest issues in Europe relating to the roles of -omics technologies will be gaining public support, as negative reaction to genetic modification testifies. Something highlighted here is that application of genomics technologies without genetic modification or synthetic biology can drive progress toward a bioeconomy. In industrial production, contained use (in bioreactors) regulation pertains. However, the greatest difficulties are related to deliberate release of genetically modified crops.

Education and training

To take these aspirations to realities will need a workforce qualified in skills that are not currently in profusion. Bioeconomy poses a long-recognized conundrum for higher education—the need for breadth and depth, necessitating multidisciplinary higher education. Most governments face this difficulty as science and engineering tend to be taught by discipline. Biologists who understand engineering are few but increasing in number. Adding business skills may also be necessary. Bioinformatics is rapidly becoming the -omics bottleneck, given the progress in high-throughput sequencing. And yet, too rapid educational progress could create oversupply (e.g., Ref. [110]). Not only do new educational programs need careful design, but also careful monitoring.

CONCLUDING REMARKS

In the wake of the HGP and the rush to high-throughput genome sequencing, the impact of genomics outside of human health is comparatively underreported. We do not ignore human health as a factor in a bioeconomy—part of the problem is, in fact, the ageing population and the associated challenges to health care. By 2050, the International Monetary Fund has estimated that the European Union will move from 4:1 working people-to-elderly (currently) to only 2:1 [111], with serious economic consequences.

For this chapter, however, we chose to focus on other sectors that are vital to the future bioeconomy: the trinity of food, energy, and chemicals. Much of the world has embarked on energy reform. It is clearly a long, expensive process. It is harder to foresee a society with drastic reductions in chemical products without large lifestyle changes. All three currently are highly dependent on fossil fuel resources, and that will ultimately change. Nevertheless, the futures of food, energy, and chemicals are interlinked, which is essentially the overarching challenge for policymakers and society. With sufficient political and societal will the -omics technologies could help make the necessary transitions easier to manage.

Gestalt or laissez faire?

This generation is the most privileged of all time. Decades of refinement to petrochemical processes has led to the current "way of life," with a seemingly endless stream of new products that make life more convenient. Less people are hungry now than ever before. A result, however, is a set of grand challenges that threaten this way of life. The history of climate change negotiations shows just how difficult it has been to get acceptance of the fact that our planet is heading in the wrong direction as the result of our own hand. To correct matters requires Draconian action over long timescales; history tells us that society strongly resists a change in the existing order [112]. Nevertheless, all the evidence being accumulated points to a need for action. The bioeconomy concept holds some of the answers.

So we conclude with a question:

"What would you want the high school history text books of 2115 to say about our generation?"

ACKNOWLEDGMENTS

The opinions expressed and arguments employed herein are those of the author(s) and do not necessarily reflect the official views of the Organization for Economic Cooperation and Development (OECD), or of the governments of its member countries.

*Funded in part by CONACYT Grants 224031 and 249698 to GJS (GBC Group).

REFERENCES

[1] OECD. The bioeconomy to 2030—designing a policy agenda. Paris: OECD Publishing; 2009. ISBN: 978-92-64-03853-0.
[2] Bioökonomierat. Bioeconomy policy synopsis and analysis of strategies in the G7. Berlin, Germany: The Office of the Bioeconomy Council, German Federal Government; 2015.
[3] Hayden EC. The $1,000 genome. Nature 2014;507:294—5.
[4] Battelle Technology Partnership Practice. The economic impact of the Human Genome Project. How a $3.8 billion investment drove $796 billion in economic impact, created 310,000 jobs and launched the genomic revolution. Battelle, Columbus, OH; 2011.
[5] Wadman M. Economic return from Human Genome Project grows. Report finds genomics effort has added US$1 trillion to US economy. Nature 2013. News, June 12, 2013. <http://dx.doi.org/10.1038/nature.2013.13187>.
[6] Ministry of Petroleum and Natural Gas, Government of India. Basic statistics on Indian petroleum and natural gas, Economic Division. <http://petroleum.nic.in/total.pdf>; 2009.
[7] Siriwardhana M, Opathella GKC, Jha MH. Bio-diesel: initiatives, potential and prospects in Thailand: a review. Energy Policy 2009;37:554—9.

[8] OECD. Biobased chemicals and plastics. Finding the right policy balance. OECD Science, Technology and Industry Policy Papers No. 17. OECD Publishing, Paris; 2014.

[9] Tracker C. Carbon supply cost curves: evaluating financial risk to oil capital expenditures. Carbon Tracker Initiative; 2014.

[10] Muehlenbachs L, Cohen MA, Gerarden T. The impact of water depth on safety and environmental performance in offshore oil and gas production. Energy Policy 2013;55:699−705.

[11] Cook J., Nuccitelli D., Green S.A., Richardson M., Winkler B., Painting R., et al. Quantifying the consensus on anthropogenic global warming in the scientific literature. Environmental Research Letters 2013;8. <http://dx.doi.org/10.1088/1748-9326/8/2/024024>.

[12] IEA. World Energy Outlook 2013. Paris: International Energy Agency; 2013. ISBN: 978-92-64-20130-9.

[13] Tracker C. Unburnable Carbon 2013: wasted capital and stranded assets. In conjunction with the Grantham Research Institute of Climate Change and the Environment. 40 pp. <www.carbontracker.org>; 2013.

[14] The Economist. The elephant in the atmosphere; 2014, p. 59.

[15] Financial Times. How oil and gas majors are rethinking on climate change; 2015.

[16] McGlade C, Ekins P. The geographical distribution of fossil fuels unused when limiting global warming to 2°C. Nature 2015;517:187−203.

[17] IPCC WGIII. Climate Change 2014: mitigation of climate change. <http://www.mitigation2014.org/>; 2014.

[18] UN FAO. The state of food and agriculture. Livestock in the balance. Rome: FAO; 2009. ISBN: 978-92-5-106215-9.

[19] Sophocleous M. Global and regional water availability and demand: prospects for the future. Nat Resour Res 2004;13:61−75.

[20] UNEP. Groundwater and its susceptibility to degradation: a global assessment of the problem and options for management. Early Warning and Assessment Report Series RS 03-3, United Nations Environment Programme; 2003.

[21] Wada Y, van Beek LPH, van Kempen CM, Reckman JWTM, Vasak S, Bierkens MFP. Global depletion of groundwater resources. Geophys Res Lett 2010;37:L20402. Available from: http://dx.doi.org/10.1029/2010GL044571.

[22] Underwood E. Models predict longer, deeper U.S. droughts. Future western "megadroughts" could be worse than ever. Science 2015;347:707.

[23] Gore A. The future: six drivers of global change. New York, US: Pub. Random House; 2013. ISBN: 9780753540503.

[24] Chen R, Ye C. Land management: resolving soil pollution in China. Nature 2014;505:483.

[25] European Commission. Environment fact sheet: soil protection—a new policy for the EU. KH-15-04-014-EN-C. <http://ec.europa.eu/environment/pubs/pdf/factsheets/soil.pdf>;2007.

[26] OECD. The emerging middle class in developing countries. Paris: OECD Publishing; 2010. <http://www2.oecd.org/oecdinfo/info.aspx?app = OLIScoteENandRef = DEV/DOC(2010)2>.

[27] Ellingsen H, Olaussen JO, Utne IB. Environmental analysis of the Norwegian fishery and aquaculture industry—a preliminary study focusing on farmed salmon. Mar Policy 2009;33:479−88.

[28] Pelletier N, Tyedmers P, Sonesson U, Scholz A, Ziegler F, Flysjo A, et al. Not all salmon are created equal: life cycle assessment (LCA) of global salmon farming systems. Environ Sci Technol 2009; 43:8730−6.

[29] Hamerschlag K, Venkat K. Meat eaters guide to climate change and health: life cycle assessments methodology and results. Environmental Working Group. <http://static.ewg.org/reports/2011/meateaters/pdf/report_ewg_meat_eaters_guide_to_health_and_climate_2011.pdf?_ga = 1.82716050.502837435.1429633963>; 2011.

[30] Pollak EJ. Genomics and the global beef cattle industry. Abstract from the International Forum: Genomics, Innovation and Economic Growth. Harnessing Science, Technology and Industry for Sustainable Development. Mexico City; 2013.

[31] McClure MC, Sonstegard TS, Wiggans GR, Van Eenennaam AL, Weber KL, Penedo CT, et al. Imputation of microsatellite alleles from dense SNP genotypes for parentage verification across multiple *Bos taurus* and *Bos indicus* breeds. Front Genet 2013;18. Available from: http://dx.doi.org/10.3389/fgene.2013.00176.

[32] Garcia JF. Genomics innovation for the cattle industry. Abstract from the International Forum: Genomics, Innovation and Economic Growth. Harnessing Science, Technology and Industry for Sustainable Development. Mexico City; 2013.

[33] Raven LA, Cocks BG, Pryce JE, Cottrell JJ, Hayes BJ. Genes of the RNASE5 pathway contain SNP associated with milk production traits in dairy cattle. Genet Sel Evol 2013;45:25. Available from: http://dx.doi.org/10.1186/1297-9686-45-25.

[34] Marco ML, Wells-Bennik MHJ. Impact of bacterial genomics on determining quality and safety in the dairy production chain. Int Dairy J 2008;18:486−95.

[35] Burt DW. Chicken genome: current status and future opportunities. Genome Res 2005;15:1692−8.

[36] Hillier LW, Miller W, Birney E, Warren W, Hardison RC, et al. Sequence and comparative analysis of the chicken genome provide unique perspectives on vertebrate evolution. Nature 2004;432:695−716.

[37] Technology Strategy Board. Boosting global food security. <https://connect.innovateuk.org/documents/3285671/6079410/Boosting+Global+Food+Security.pdf/b0ee6ed1-2e98-41f2-b4e2-2b594db4a035>; 2010.

[38] Jiménez-Sánchez G. Presentation at the OECD workshop Genomics for sustainable development in emerging economies: food, environment, and industry. Kuala Lumpur, March 14, 2015.

[39] Mishra V, Cherkauer KA. Retrospective droughts in the crop growing season: implications to corn and soybean yield in the Midwestern United States. Agric Forest Meteorol 2010;150:1030−45.

[40] Reardon S, Hodson H. Water wars loom as US runs dry. New Sci 2013;217:8−9.

[41] National Post. California to impose water-use fines amid "worst drought we have ever seen"; 2014.

[42] Cattivelli L, Rizz F, Badeck F-W, Mazzucotelli E, Mastrangelo AM, Francia E, et al. Drought tolerance improvement in crop plants: an integrated view from breeding to genomics. Field Crops Research 2008;105:1−14.

[43] Fleury D, Jefferies S, Kuchel H, Langridge P. Genetic and genomic tools to improve drought tolerance in wheat. J Exp Bot 2010;61:3211−22.

[44] Xu J, Yuan Y, Xu Y, Zhang G, Guo X, Wu F, et al. Identification of candidate genes for drought tolerance by whole-genome resequencing in maize. BMC Plant Biol 2014;14:83. Available from: http://dx.doi.org/10.1186/1471-2229-14-83.

[45] Bolger A, Scossa F, Bolger ME, Lanz C, Maumus F, Tohge T, et al. The genome of the stress-tolerant wild tomato species *Solanum pennellii*. Nat Genet 2014;46:1034−8.

[46] Monneveux P, Ramírez DA, Pino M-T. Drought tolerance in potato (*S. tuberosum* L.): can we learn from drought tolerance research in cereals? Plant Sci 2013;205−206:76−86.

[47] Yin H, Chen CJ, Yang J, Weston DJ, Chen J-G, Muchero W, et al. Functional genomics of drought tolerance in bioenergy crops. CRC Crit Rev Plant Sci 2014;33:205−24.

[48] Atkinson NJ, Urwin PE. The interaction of plant biotic and abiotic stresses: from genes to the field. J Exp Bot 2012;63:3523−43.

[49] Yang S, Vanderbeld B, Wan J, Huang Y. Narrowing down the targets: towards successful genetic engineering of drought-tolerant crops. Mol Plant 2010;3:469−90.

[50] Ismail A. Presentation at the OECD workshop Genomics for sustainable development in emerging economies: food, environment, and industry. Kuala Lumpur, March 14, 2015.

[51] Bailey-Serres J, Fukao T, Ronald P, Ismail A, Heuer S, Mackill D. Submergence tolerant rice: SUB1's journey from landrace to modern cultivar. Rice 2010;3:138−47.

[52] Licht S, Cui B, Wang B, Li F-F, Lau J, Liu S. Ammonia synthesis by N_2 and steam electrolysis in molten hydroxide suspensions of nanoscale Fe_2O_3. Science 2014;345:637−40.

[53] National Science Foundation Press Release. Press Release 13-147: US and UK scientists collaborate to design crops of the future. <https://www.nsf.gov/news/news_summ.jsp?org=NSF&cntn_id=128878&preview=false>; 2013.

[54] USDA Foreign Agricultural Service. Malaysia: biofuels annual report 2009. USDA Foreign Agricultural Service, Washington, DC. <http://gain.fas.usda.gov/Recent%20GAIN%20Publications/General%20Report_Kuala%20Lumpur_Malaysia_6-12-2009.pdf>; 2009.

[55] Obidzinski K, Andriani R, Komarudin H, Andrianto A. Environmental and social impacts of oil palm plantations and their implications for biofuel production in Indonesia. Ecol Soc 2012;17:25. Available from: http://dx.doi.org/10.5751/ES-04775-170125.

[56] Singh R, Low ET, Ooi LC, Ong-Abdullah M, Ting NC, Nagappan J, et al. The oil palm SHELL gene controls oil yield and encodes a homologue of SEEDSTICK. Nature 2013;500:340—4.

[57] Cold Spring Harbor Laboratory News. Full genome map of oil palm indicates a way to raise yields and protect rainforest. <http://www.cshl.edu/News-Features/full-genome-map-of-oil-palm-indicates-a-way-to-raise-yields-and-protect-rainforest.html>;2013.

[58] Danielsen F, Beukema H, Burgess ND, Parish F, Bruhl CA, Donald PF, et al. Biofuel plantations on forested lands: double jeopardy for biodiversity and climate. Conserv Biol 2009;23:348—58.

[59] Costa FO, Landi M, Martins R, Costa MH, Costa ME, Carneiro M, et al. A ranking system for reference libraries of DNA Barcodes: application to marine fish species from Portugal. PLoS ONE 2012;7:e35858. Available from: http://dx.doi.org/10.1371/journal.pone.0035858.

[60] Palaiokostas C, Bekaert M, Davie A, Cowan ME, Oral M, Taggart JB, et al. Mapping the sex determination locus in the Atlantic halibut (Hippoglossus hippoglossus) using RAD sequencing. BMC Genomics 2013;14:566. Available from: http://dx.doi.org/10.1186/1471-2164-14-566.

[61] Palaiokostas C, Bekaert M, Khan MGQ, Taggart JB, Gharbi K, McAndrew BJ, et al. Mapping and validation of the major sex-determining region in Nile Tilapia (Oreochromis niloticus L.) using RAD sequencing. PLoS ONE 2013;8(7):e68389. Available from: http://dx.doi.org/10.1371/journal.pone.0068389.

[62] UN FAO. Species fact sheets Salmo salar. <http://www.fao.org/fishery/species/2929/en>; 2010.

[63] Gonen S, Lowe NR, Cezard T, Gharbi K, Bishop SC, Houston RD. Linkage maps of the Atlantic salmon (Salmo salar) genome derived from RAD sequencing. BMC Genomics 2014;15:166. Available from: http://dx.doi.org/10.1186/1471-2164-15-166.

[64] Broeren MLM, Saygin D, Patel MK. Forecasting global developments in the basic chemical industry for environmental policy analysis. Energy Policy 2014;64:273—87.

[65] IEA. Energy technology perspectives 2012—pathways to a clean energy system. Paris: International Energy Agency; 2012.

[66] Yim H, Haselbeck R, Niu W, Pujol-Baxley C, Burgard A, Boldt J, et al. Metabolic engineering of Escherichia coli for direct production of 1,4-butanediol. Nat Chem Biol 2011;7:445—52.

[67] USDA. Hardwood-distillation industry. Forest Products Laboratory, Madison, Wisconsin. No. 738; 1956.

[68] Nakamura CE, Whited GM. Metabolic engineering for the microbial production of 1,3-propanediol. Curr Opin Biotechnol 2003;14:454—9.

[69] Mehta R, Kumar V, Bhunia H, Upadhyay SN. Synthesis of poly(lactic acid): a review. J Macromol Sci Part C: Polym Rev 2005;45:325—49.

[70] Jung YK, Lee SY. Efficient production of polylactic acid and its copolymers by metabolically engineered Escherichia coli. J Biotechnol 2011;151:94—101.

[71] US DoE. Breaking the biological barriers to cellulosic ethanol: a joint research agenda. DoE/SC-0095. US Department of Energy Office of Science and Office of Energy Efficiency and Renewable Energy. <www.doegenomestolife.org/biofuels/ >; 2006.

[72] French CE. Synthetic biology and biomass conversion: a match made in heaven? J R Soc Interface 2009;6:S547—58.

[73] Bokinsky G, Peralta-Yahya PP, George A, Holmes BM, Steen EJ, Dietrich J, et al. Synthesis of three advanced biofuels from ionic liquid-pretreated switchgrass using engineered Escherichia coli. Proc Natl Acad Sci 2011;108:19949—54.

[74] Mueller SA, Anderson JE, Wallington TJ. Impact of biofuel production and other supply and demand factors on food price increases in 2008. Biomass Bioenergy 2011;35:1623—32.

[75] Abbott P, Hurt C, Tyner WE. What's driving food prices? Farm Foundation Issue Report. Farm Foundation, Oak Brook, IL; 2008.

[76] Timmer PC. Causes of high food price plus technical annexes. ADB Economics Working Paper Series No. 128 (October), Asian Development Bank, Manila; 2008.

[77] IFPRI. Reflections on the global food crisis. How did it happen? How has it hurt? And how can we prevent the next one? International Food Policy Research Institute Washington, DC; 2010. ISBN: 978-0-89629-178-2.

[78] De Gorter H, Drabik D, Just DR. How biofuels policies affect the level of grains and oilseed prices: theory, models and evidence. Global Food Secur 2013;2:82−8.

[79] Peplow M. Cellulosic ethanol fights for life. Pioneering biofuel producers hope that US government largesse will ease their way into a tough market. Nature 2014;507:152.

[80] Jørgensen H, Kristensen JB, Felby C. Enzymatic conversion of lignocellulose into fermentable sugars: challenges and opportunities. Biofuels, Bioprod Biorefining 2007;1:119−34.

[81] Almeida JRM, Modig T, Petersson A, Hahn-Hägerdal B, Lidén G, Gorwa-Grauslund MF. Increased tolerance and conversion of inhibitors in lignocellulosic hydrolysates by *Saccharomyces cerevisiae*. J Chem Technol Biotechnol 2007;82:340−9.

[82] Martin C, Alriksson B, Sjöde A, Nilvebrant NO, Jönsson LJ. Dilute sulfuric acid pretreatment of agricultural and agro-industrial residues for ethanol production. Appl Biochem Biotechnol 2007;137:339−52.

[83] Zheng DQ, Wu XC, Tao XL, Wang PM, Li P, Chi XQ, et al. Screening and construction of *Saccharomyces cerevisiae* strains with improved multi-tolerance and bioethanol fermentation performance. Bioresour Technol 2011;102:3020−7.

[84] Hasunuma T, Kondo A. Development of yeast cell factories for consolidated bioprocessing of lignocellulose to bioethanol through cell surface engineering. Biotechnol Adv 2012;30:1207−18.

[85] Dai W, Word DP, Hahn J. Modeling and dynamic optimization of fuel-grade ethanol fermentation using fed-batch process. Control Eng Pract 2014;22:231−41.

[86] Batt CA, Carvallo S, Easson DD, Akedo M, Sinskey AJ. Direct evidence for a xylose metabolic pathway in *Saccharomyces cerevisiae*. Biotechnol Bioeng 1986;28:549−53.

[87] Chu BCH, Lee H. Genetic improvement of *Saccharomyces cerevisiae* for xylose fermentation. Biotechnol Adv 2007;25:425−41.

[88] Vinuselvi P, Kim MK, Lee SK, Ghim CM. Rewiring carbon catabolite repression for microbial cell factory. BMB Rep 2012;45:59−70.

[89] Kim J-H, Block DE, Mills DA. Simultaneous consumption of pentose and hexose sugars: an optimal microbial phenotype for efficient fermentation of lignocellulosic biomass. Appl Microbiol Biotechnol 2010;88:1077−85.

[90] Biofuels Digest. Enerkem: Alberta's municipal waste to fuels juggernaut, in pictures; 2014.

[91] Leung CCJ, Cheung ASY, Zhang AY-Z, Lam KF, Lin CSK. Utilisation of waste bread for fermentative succinic acid production. Biochem Eng J 2012;65:10−15.

[92] Taylor P. Biosuccinic acid ready for take off? Royal Society of Chemistry. <http://www.rsc.org/chemistryworld/News/2010/January/21011003.asp>; 2010.

[93] Bomgardner MM. Biobased chemicals without biomass. Chem Eng News 2012;90:25.

[94] Schmidt M. Introduction. In: Schmidt M, editor. Synthetic biology: industrial and environmental applications. Weinheim, Germany: Pub. Wiley-Blackwell; 2012.

[95] Koksharova O, Wolk C. Genetic tools for cyanobacteria. Appl Microbiol Biotechnol 2002;58:123−37.

[96] Wang B, Wang J, Zhang W, Meldrum DR. Application of synthetic biology in cyanobacteria and algae. Front Microbiol 2012;3: article 344, http://dx.doi.org/10.3389/fmicb.2012.00344.

[97] Blankenship RE, Tiede DM, Barber J, Brudvig GW, Fleming G, Ghirardi M, et al. Comparing photosynthetic and photovoltaic efficiencies and recognizing the potential for improvement. Science 2011;332:805−9.

[98] Melis A. Solar energy conversion efficiencies in photosynthesis: minimizing the chlorophyll antennae to maximize efficiency. Plant Sci 2009;177:272−80.

[99] Espie GS, Kimber MS. Carboxysomes: cyanobacterial RuBisCO comes in small packages. Photosynth Res 2011;109:7−20.

[100] Latif H, Zeidan AA, Nielsen AT, Zengler K. Trash to treasure: production of biofuels and commodity chemicals via syngas fermenting microorganisms. Curr Opin Biotechnol 2014;27:79–87.

[101] Schiel-Bengelsdorf B, Dürre P. Pathway engineering and synthetic biology using acetogens. FEBS Lett 2012;586:2191–8.

[102] Fast AG, Papoutsakis ET. Stoichiometric and energetic analyses of non-photosynthetic CO_2-fixation pathways to support synthetic biology strategies for production of fuels and chemicals. Curr Opin Chem Eng 2012;1:380–95.

[103] Kusian B, Bowien B. Organization and regulation of cbb CO_2 assimilation genes in autotrophic bacteria. FEMS Microbiol Rev 1997;21:135–55.

[104] Liew FM, Köpke M, Simpson SD. Gas fermentation for commercial biofuels production. In: Fang Z, editor. Liquid, gaseous and solid biofuels—conversion techniques. Pub. Intech; 2013. Chapter 5, ISBN: 978-953-51-1050-7.

[105] Köpke M, Mihalcea C, Liew F, Tizard JH, Ali MS, Conolly JJ, et al. 2,3-Butanediol production by acetogenic bacteria, an alternative route to chemical synthesis, using industrial waste gas. Appl Environ Microbiol 2011;77:5467–75.

[106] US20130323820 A1. Recombinant microorganisms and uses therefor. Original assignee: Lanzatech New Zealand Limited; 2013.

[107] WO2012052427. 2012. Production of alkenes by combined enzymatic conversion of 3-hydroxyalkanoic acids. International patent application number: PCT/EP2011/068174, Publication date: 26.04.2012.

[108] Pavanan KC, Bosch RA, Cornelissen R, Philp JC. Biomass sustainability and certification. Trends Biotechnol 2013;31:385–7.

[109] van Dam J, Junginger M. Striving to further harmonization of sustainability criteria for bioenergy in Europe: recommendations from a stakeholder questionnaire. Energy Policy 2011;39:4051–66.

[110] Watkinson II, Bridgwater AV, Luxmore C. Advanced education and training in bioenergy in Europe. Biomass Bioenergy 2012;38:128–43.

[111] Carone G, Costello D. Can Europe afford to grow old? Finance and Development 43. <http://www.imf.org/external/pubs/ft/fandd/2006/09/carone.htm>; 2006.

[112] Machiavelli N. 1532. The Prince. (Currently available as a Penguin Classic from 2003, ISBN-13: 978-0140449150).

CHAPTER 12

Socioeconomic Outcomes of Genomics in the Developing World

Dhavendra Kumar
Genomic Policy Unit, Faculty of Life Sciences and Education, The University of South Wales, Pontypridd, Wales, UK

Contents

Health of any nation depends upon its sound economic base, political stability, healthy agriculture (crops/animal stock), a safe environment (clean water, sanitation, and climate control), and the general well-being of its people. Among several biological and ecological factors, the role of genetics and genomics is now widely accepted as a major determinant for the health of any nation. There are diverse and far-reaching applications of genetics and genomics that we are beginning to utilize in all walks of life, including genetically engineered vaccines and biological products (insulin and growth hormone), new drugs (cancer and heart disease), genetically modified crops (enhanced produce and biofortification), developing bioengineered vectors for controlling malaria and other parasitic diseases, and the preservation of animal and plant life, specifically rare and extinct species.

All new scientific developments lead to many applications that require significant socioeconomic adjustments in major areas, for example, the practice of medicine and health. The developments in genomics have had an impact and provoked a sociocultural shift in the values and practice of medicine in most developed nations and rapidly expanding in developing countries. This is echoed much more in the *emerging economies*, such as BRICS nations (Brazil, Russia, India, China, and South Africa) and

Genomics and Society
DOI: http://dx.doi.org/10.1016/B978-0-12-420195-8.00012-4

Mexico, where developments in science and technology including genomics have contributed significantly to economic growth and enhanced social status. This chapter provides a broad overview on socioeconomic outcomes of wider genomic applications in the developing world.

INTRODUCTION

Developments in science and biotechnology are reliable denominators that can be used to address health and socioeconomic progress in any nation including developing countries [1]. This largely depends upon the development of infrastructure and adopting a diverse set of policies aimed at translating scientific discoveries, such as in genetics and genomics, into products and services. These technologies include molecular diagnostics, recombinant vaccines, techniques of vaccine and drug delivery, bioremediation (use of living organisms to degrade hazardous matter), sequencing pathogen genomes, female-controlled protection against sexually transmitted infections, bioinformatics, nutritionally enriched genetically modified crops, recombinant therapeutic proteins, combinatorial chemistry, and probably many more [1]. A full review and discussion on these technologies is beyond the scope of this chapter; however, outcomes of major developments are briefly discussed.

Most developed nations in the West have invested and continue to do so, in harnessing the benefits of genetics and genomics. However, large populations in developing and underdeveloped countries (Figure 12.1), comprising more than two-thirds of the global population, continue to be deprived of potential health and socioeconomic benefits. It is estimated that the current global population of around 7 billion will rise to around 10 billion by 2050 (Figure 12.2). There is alarming disproportion—currently around 5 billion people in developing countries are deprived of the full benefits of modern science and technology, including genetics and genomics.

Conventionally most physical, chemical, and biological sciences have contributed in raising the socioeconomic status, improved health, education, and to some extent political stability. Regrettably, these were, and continued to be, applied in negative and destructive uses, for example, developing and use of weapons of mass destruction. Further, inequity and monopoly in science undeniably creates a huge gulf between nations thus creating the artificial and unclear boundaries of developed, developing, and underdeveloped nations.

Apart from the emerging economies of the recognized developing world, for example, BRICS nations, most countries in Central and Eastern Europe are called *economies* or *nations in transition*, where the growth of science and technology is comparatively slower. Some extremely poor parts of Africa, Asia, and the Pacific are called *least-developed countries* (LDCs), where basic healthcare provision is questionable [2].

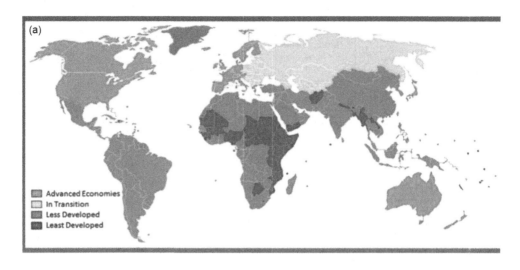

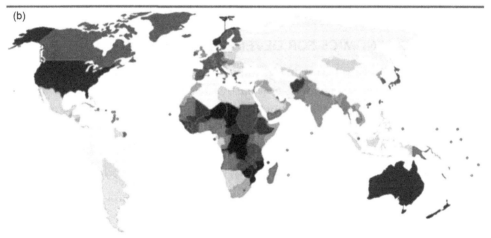

Figure 12.1 The global map of the developing world showing both developed and underdeveloped nations.

However, boundaries are arbitrary and could not be defined by any geopolitical dimensions. Nevertheless, LDCs, economics in transition, and the merging economies collectively represent "the developing world," with unquestionable huge potential for socioeconomic and biotechnological/biomedical growth and future developments [3]. However, unlike most developed countries, where there is relative homogeneity of policies governing science and biotechnology, there are huge geographical, political, economic, and sociocultural variations within the developing world [4]. Invariably

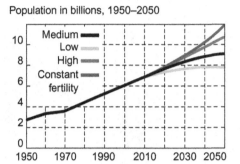

Figure 12.2 World population forecast to year 2050 (United Nations/WHO).

these have hampered the scientific and technological progress in most developing nations. This chapter reviews potential fields of genomic applications in developing nations with measurable socioeconomic outcomes.

HARNESSING GENOMICS FOR DEVELOPING COUNTRIES

There are six key areas for harnessing genomics [5]: *demography and population variation; food and nutrition; medicine and health; biotechnology and bioindustry; education, training; and research and development (R&D)*. Socioeconomic outcomes of genomics are broadly discussed in the respective area.

Demography and population variation

Mapping individual genome is now possible and increasingly at progressively low cost. This powerful method allows delineation of any genomic variation that could be used for establishing individual identity, parental links, kinship, and ultimately extending to community and population levels (Figure 12.3) [6]. Establishing the genome map of people and populations has helped in defining the demographic landscape and database. Several genome-based databases now exist [7,8], including India [9,10], China [11], Asia-Pacific [12], Arab and Middle East [13,14], Russia and Siberia [15], and Latin America [16,17] (Figure 12.4). Similar efforts are under way for genome mapping the indigenous and tribal populations in Tamil Nadu and Andman and Nicobar [18], Aborigines and Polynesian people (Australasia) [19], selected areas of Africa, and rare tribes in Latin America [20]. Undeniably this could be a very expensive approach for sorting out the demography and population structure. Nevertheless, it is extremely important to have a reliable basis of human population variation that is acceptable, replicable, and reflect the ancestral, past and present human migration. The genome mapping is without any question the best solution to this challenging issue. Perhaps

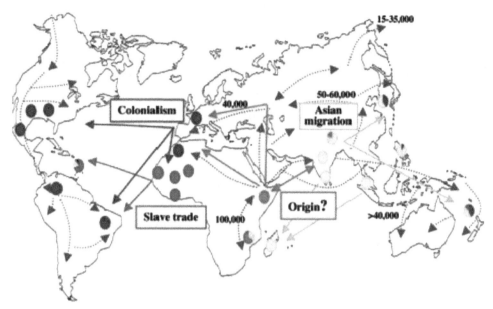

Figure 12.3 Movement and migration of man verified by population-level genomic analysis.

Figure 12.4 The linguistic population of India based on the genomic structure of geophysical and socio-cultural heterogeneous people of India (Indian Genome Variation Project, CSIR, Government of India).

the widely prevalent caste system in India could be sorted out through the genome mapping.

Apart from clarifying the ethnic and population diversity, the human genome variation databases (single nucleotide polymorphisms; specific gene polymorphisms; copy number variation, and sequence number variation) have the potential of applying in a number of situations [21]. Examples include:

- Forensic identification of people in major natural and man-made disasters, for example, tsunami, earthquakes, terrorist attacks, and major air/train accidents. The DNA Fingerprinting Centre in Hyderabad is one of the very few facilities in the world that regularly undertake such tasks.
- Tracking global migration and immigration of people across several countries.
- Managing large-scale endemics and pandemics through identification of people moving out of high-risk areas. Examples include the SARS virus from China/Far East, HIV from Africa, and the HIN1 Pandemic influenza [22].

Food and nutrition

Enhancing and improving the food production and nutritional status of the nation is of paramount importance to any country, whether developed, developing, or struggling to emerge from the stigma of being underdeveloped. Developed nations in the West are faced with growing epidemic of Obesity [23]. On the contrary malnutrition or undernutrition remains a big challenge for both developing and underdeveloped countries. This is further compounded by widespread poverty. It is natural to ask how genetics and genomics could be applied for improving the food supply and positively change the nutritional status of people?

Translational use of genetics and genomics is now increasingly used in agriculture (improving/enhancing crops), horticulture (household plants for vegetables), and animal/poultry/fish industry (improved/enhanced dairy production and quantitative/qualitative production of meat/chicken/fish products) [24,25].

Achievements so far in this area include the creation of the brown (golden) rice facilitated by the sequencing of the rice genome fortified with beta-carotene gene sequence (Figure 12.5) [26]. This variety of rice crop can be grown in less favorable conditions with relatively higher per hectare yield. Extensive research and field work have helped in developing efficient and cost-effective variety of Casava (Figure 12.6) [27], the major staple food source in Africa. Similarly, the genome modified rice [28] and wheat crop [29] can provide a significant greater yield in relatively smaller field and natural surroundings. Developments in the field of genome modified maize crop are also promising. A notable example is the creation of special variety of maize enriched with carotenoids providing a natural and enriched source of vitamin A, essential for normal cornea and retina functions [30].

(a)

(b)

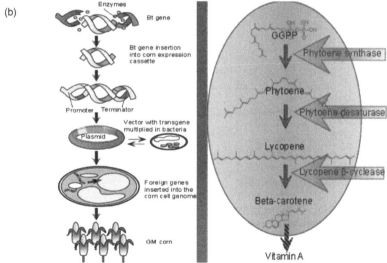

Figure 12.5 The creation of genetically modified rice—golden rice (a) and corn (orange variety) with beta-carotene gene (b) maize crop enhancement for mass supply of vitamin A to corn eating populations for the prevention of vitamin-A-deficient blindness—an example of genomic applications in population health.

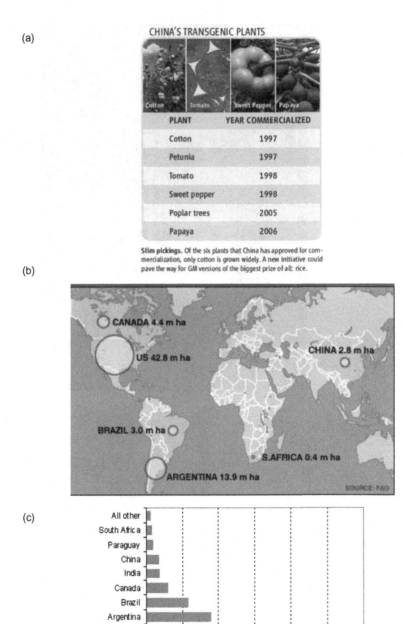

Figure 12.6 Genomic applications in enhanced plant (horticulture) and crop (agriculture) production: (a) China's transgenic plants; (b) global map of GM crops—United Nations Food and Agricultural Organization (www.fao.org); and (c) global GM production per hectares.

This is a promising application of genomics in combating vitamin A deficiency related keratopathy, a leading preventable cause of blindness.

There are several examples where applied and translational genomics has shown promising results in horticulture, animal breeding, and marine farming. Interested reader can find this information from the Food and Agricultural Organization, a leading United Nations establishment in Rome [31]. It is important that developing nations, like India, include genomics as a high priority in the agricultural research and invest in developing newer improved and high-yielding genome-based enhance varieties of common crops.

Genomics came into the public limelight, with issues related to genetic engineering in agriculture in the developing countries, when debates about genetically modified (GM) crops (e.g., BT cotton and Golden rice) raised huge controversies about biosafety, benefit sharing, and intellectual property rights for farmers, along with health and safety concerns for the ordinary public. Some countries—for example, India, Kenya, Brazil, and China—adopted national policies for GM crops, and in some respects these policies were actually more cautious than those adopted in Europe [32]. The degree of caution is interesting, given the conspicuous unmet food production needs in some of these countries today, with basic food and nutrition being a fundamental priority in public policy.

Medicine and health

Applications of genetics in medicine and health have been around for over 50 years. A number of medical conditions and developmental anomalies detected during pregnancy and identified at birth (congenital) have established and detectable genetic causes. Several other medical conditions manifest later in childhood, adolescence, and later in life where a genetic cause might be confirmed. In many early and late onset conditions, genetic factors are coexistent with environmental (often avoidable or modifiable) factors. Genetic and new genomic-led laboratory diagnostic methods are regularly used and are now an integral part of the specialist laboratory medicine service in all developed and to some extent in some developing countries. Special aspects of this intervention include prenatal genetic diagnosis and screening, genetic counseling, diagnostic and predictive genetic testing, preimplantation genetic diagnosis and testing, new born screening and cascade genetic testing in high-risk families, and ethnic minority communities.

The clinical use of medical genetics and human genetic research is concentrated in developed nations of Europe and North America that cover less than 1 billion people. In contrast, India alone has over 1 billion people equally at risk for being affected with a wide range of genetic diseases! The volume and burden of genetic disease afflicting those in other, developing or less developed or "low-income" countries are beyond imagination and comprehension. It is extremely important that developing nations develop and implement strategies for utilizing genomic technologies in health care and facilitate provision of genetic and genomic services focusing on selected and cost-effective, highly sensitive, and specific diagnostic methods, targeted therapeutic and preventive applications. It is encouraging to note some positive indications in this regard, for example, diagnostic and therapeutic options are now available in some selected centers across India for beta-thalassemia and rare genetic metabolic diseases.

Traditional medicine to modern medicine: relevance of genomics

Most regions of the developing world historically have followed traditional practices of medicine, such as Arabic, Ayurveda, and Chinese that primarily focus on the holistic nature of a being, and were based on a relationship with nature and drawing upon medicinal plants from nature [33]. Initially associated with a variety of negative terms during colonial times, more recently the "traditional medicine" has acquired largely positive connotations in the West [34]. It needs to be acknowledged that a variety of practices and epistemologies fall under the rubric of "traditional," and most medical traditions are intrinsically plural in nature. The plurality of traditional medical practices depends on how the tradition is interpreted and codified, and viewed in its practical applications; for example, there are several versions of oriental/Chinese medicine seen in Buddhist countries, ranging from Tibet to Thailand.

The distinctions between traditional and modern medicine are seen in the ways in which treatment and therapy is viewed, and the ethical codes that guide them. Most traditional practices, including some African traditions, take a holistic approach and focus on balancing the body—mind with nature [35]. The body and mind (spirit) are not separate, and exist in harmony with each other. Ancient Indian physicians asserted that both nature and man are made up of the same matter, the five gross elements: earth, fire, air, water, and space; additionally, human beings have consciousness. The human body is a combination of body, mind, and self; physical and psychological processes are inseparable and interact, and both are expressions of the life force. Similar concepts are seen in African traditions focusing on harmony with nature, and most rural communities have relied on the spiritual and practical skills of traditional medical practitioners, whose botanical knowledge of plant species and their ecology are invaluable.

Recent progress in the field of environmental sciences and immunology has led researchers to appreciate the capacity and rationality of various traditional taxonomies, and the effectiveness of their treatments. The globalization of Chinese medicine and forms of Ayurveda has, in turn, led to the modernization of traditional medicine [33]. A multitude of these traditional products and practices is increasingly in evidence in

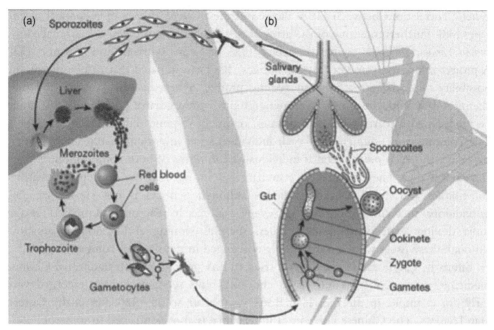

Figure 12.7 The potential of discovery and development of new genomics-led drugs and vaccines targeted at multiple genomes involved in Malaria.

the European and American markets. Modernization is seen at the level of economics, as well as the ways in which these traditional approaches are subjected to clinical trials and then standardized, commodified, and mass distributed—such as the curative use of artemesinin (*qing hao*) for malaria, and the controversial patent of the Neem tree (*Azadirachta indica*) for its medicinal properties (Figure 12.7) [36]. Nevertheless, despite the modernization of traditional medicine, the traditional practices continue to thrive by default where provision of new biomedicine is not accessible. The rapid growth of traditional medicine also demonstrates that modern medicine may not always be the preferred treatment option.

Traditionally, the concept of "inheritance" didn't exist in medicine in the Asian medical system and physical traits and diseases that were not believed to be transmitted from one generation to other. The type and the severity of the condition were based on one's *karma* in the past. However, with the developments in genomics, attempts are now being made to find out if there are any correlations between genetics and traditional medical systems. Ayurveda classifies human populations into three major constituents of *prakriti* (phenotypes); namely, *vata* (motion), *pitta* (metabolism), and *kapha* (structure) as discreet phenotype groupings [37]. This is generally independent of racial, geographic, and ethnic considerations, and focuses only on the individual's diseases and physiological differences. However, researchers have now postulated that there are genetic correlations between HLA alleles and the *tri-dosha* theory of individual *prakriti* types [38]. Furthermore, one of the arms of new biobanking project in India, called the *Indian Variome Project*, is trying to establish links between Ayurveda and genomics [39]. A project under the name of "Ayurgenomics" has been initiated that is focusing on the possibility of correlations between genetic profile, Ayurvedic classification, and biochemical data, with the aim of identifying groups within normal populations that could be predisposed to certain kinds of diseases, or might respond differently to drugs [40]. The constitutional uniqueness of each individual is an important feature in Ayurvedic medicine, and each prescription is individualized in terms of both composition of drugs and suitable diet. These features have similarities with modern-day concepts of pharmacogenomics and individualized medicine, although genetics is not clearly specified in traditional medicine. Also, there are several examples in religious and medical texts in India dealing with contemporary issues, such as cloning and xenotransplantation, although these issues were never ethically challenged in the medical context.

Similarly, Chinese researchers have tried to link genomics with traditional Chinese medicine, where concepts related to choosing traits seemed to have emerged very early; for example, in the East Han Book written in AD 25–220 during the Eastern Han Dynasty. The Chinese concept of inheritance is also mentioned in some old literature; for example, the book *Zhu Shi Yi Shu*. In it, Jing Xue Ben says that children are similar to parents because their *jing* (essence) and *xue* (blood) are from parents, although genes are not specifically mentioned. Although Chinese medicine and modern medicine are based on totally different fundamentals, in recent years researchers have hypothesized using the concept of *chi* (electromagnetic waves) from Chinese medicine in the process of cellular treatment, through changing the abnormal electrocharged cell to normal, stimulating the cell function and activating gene functions. However, the challenge is how to detect, measure, and control the flow of *chi*. These are considered to be the basic directions in which genetics in Chinese medicine is likely to go forward [41].

Genomic medicine and health for developing countries

Genomics and the genomic era have been accompanied by an impressive array of new and powerful technologies that have a direct impact on the diagnosis, control, and prevention of infectious diseases. An analysis of the top 10 biotechnologies for improving health in developing countries identified *diagnostics, vaccines, and drug and vaccine delivery* as the top three technologies [1]. In broader terms, the key genomic technologies for the developing world include:

1. The rapid identification of pathogens, viral, bacterial, or parasitic.
2. Rapid, cheap, easy to use point-of-care diagnostics—ideally, these tests should be able to detect multiple pathogens and be usable in least-developed settings.

3. Genomics for better, more effective, easily deployable vaccines in the future, for example, against TB, malaria, HIV/AIDS, the three major causes of mortality in developing countries.
4. Genomics in drug discovery—utilizing pathogen genome sequences, directed at neglected diseases affecting developing countries; for example, the African Network for Drugs and Diagnostics Innovation.
5. Applications to vector control—particularly in Malaria, applying the recent availability of complete genome sequences *Aedes* and *Anopheles* mosquitoes.
6. Monitoring antimicrobial (antibiotic) resistance—viral, bacterial, protozoal, using genomic/molecular markers for surveillance and appropriate therapeutic responses.
7. Translating new knowledge of host—pathogen interactions into better interventions leading to individualized diagnostics and therapeutics—pharmacogenomics and personalized medicine.

It is extremely important for health planners in developing countries to ensure that infectious disease control programs utilize the above areas where applications of genomic and molecular diagnostics and therapeutics are highly likely to be beneficial and cost-effective [42].

BIOTECHNOLOGY AND BIOINDUSTRY

In addition to medicine and health, whole nations and their people are being deprived of the socioeconomic benefits from investment and infrastructure developments in areas like biomedical and agriculture technology. The current and future applications of genome science and technology in pharmaceutical, bioengineering, and the food and agriculture industries hold great promise for the new emerging economies of the developing world. This challenge offers new opportunities for promoting international cooperation in relevant biomedical research in developing countries as recommended by the United Nations' Millennium Project Task Force on Science, Technology, and Innovation [43]. These recommendations are included in the UN Millennium Development Goals.

It would be necessary to introduce major organizational and administrative changes for transforming or creating relevant biotechnology and bioengineering infrastructure and facilities to adapt new genomic-led applications. Examples include new drug/vaccine discovery and development, biofuels from sugars and seaweeds, biofortification for sanitation and clear water, information and technology equipment/service, laboratory/industry hardware and equipment, and many more such applications [44]. These efforts would require full support of genetic and genomic education, training, research facilities to cater the rapidly growing demand at all levels and in all countries.

EDUCATION AND TRAINING

Strategic planning and investing in effective and targeted education, training, and research are undoubtedly and unquestionably vital for the progress of science and technology. This is evident from several leading institutions across the developing world including India. However, much needs to be undertaken and achieved in the field of biological sciences, particularly genetics and genomics. It is extremely important to engage with professionals, consumers (the public), and regulators (politicians) in working out the strategy, identifying potential areas and carefully selecting locations [45,46]. The focus needs to be on raising public awareness and encouraging young, bright, and enthusiastic students, for example, trainee nurses for choosing a course/career in genetic/genomics [47]. Lead genetics and genomics organizations would require strategy for convincing education and training planners, organizations, and institutions for setting up relevant courses and training programs in different areas of genomic applications; state-led or sponsored research programs for strengthening support areas of bioethics, bioeconomy, bioinformatics, bioengineering and setting standards and guidelines for intellectual property rights, patents, and commercial freedom and protection.

RESEARCH AND DEVELOPMENT

Investments in R&D are universally acknowledged as an essential component of any scientific and professional organization or institution. In a fast and dynamic field of genetics and genomics, it is undeniably important that a set proportion of annual budget and capital expenditure is allocated for R&D activities [42]. Apart from focusing and supporting specific research projects, emphasis should be made for creating a positive R&D environment and culture [48]. This should include support for developing/making applications for research grants, information and technology aspects, adequate training for good clinical practice, mechanisms for research ethics approval, infrastructure for research governance, support for dissemination of research outcomes (conference presentations and publishing in leading/high-impact journals), and recognizing and supporting potential patenting and commercial opportunities [49].

It is difficult to comment on what should be the R&D strategy in terms of specific areas as outlined above. This should be clearly the responsibility of respective developing nation. In this respect it is important to draw attention to public—private partnership. There are several examples like the Wellcome Sanger Genome Centre in United Kingdom (www.sanger.ac.uk), Howard Hughes Medical Institutes in the United States (www.hhmi.org), The Gates Foundation (www.gatesfoundation.org), and the Wellcome—DBT India collaboration (www.wellcomedbt.org). In author's opinion, developing nations should avoid R&D investments in basic genetics and

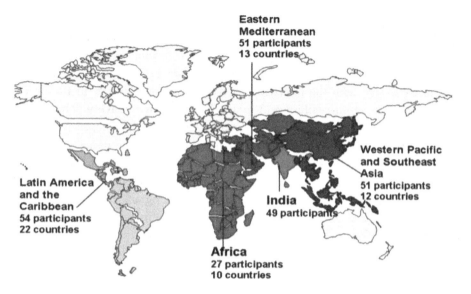

Figure 12.8 Investments in developing countries for harnessing benefits of biotechnology and bioindustry.

genomics that are less likely to be productive and contributory to health care, industrial applications, and socioeconomic development (Figure 12.8). Basic science research is nevertheless important and should be left to developed nations and probably undertaken by few developing nations with sufficient resources. India's investment in the chain of CSIR (www.csir.res.in) and ICMR (www.icmr.res.in) institutes are good examples. However, the evidence from other developing nations point to higher rewards from applied research, for example, Japan's electronic and automobile industries, China's massive industrial boom, and India's prosperous IT and service sectors. Similar approach for applied and translational genetic and genomic research should be worthwhile for the developing world. The future looks bright.

It is important to acknowledge, however, that active efforts and being made in several developing nations, particularly India, China, Brazil, and South Africa, for promoting and supporting genetic/genomic education, training, and research. This is reflected in huge sums invested and outcomes measured against number of students graduating in genetics/genomics, successful research grants, publications in leading peer-reviewed scientific, medical and industrial journals, patents and property rights secured and establishing strong, and viable public—private and international collaborations.

There is huge potential of growth. For example, United Kingdom has over 100 fully trained medically qualified geneticists serving only 60 million people compared to just around 20 trained medical geneticists in India for over 1 billion people. This is also evident in other branches of scientific and industrial professions. The challenges are huge and impossible to meet, however, achievable. We would need to rewrite the curriculum for secondary and graduate courses. We should ensure that sufficient time and resources are allocated at all levels for genetic/genomic education and training. We should be enterprising and look for avenues and opportunities. It is extremely important to engage with the international community, particularly in the developed world, for developing curriculum and methods of effective delivery. The Indo—UK Genetic Education Forum is one such example (www.walesgenepark.cardiff.ac.uk). Several organizations and institutions across India welcome the concept and offered, and continue to do so, to host dedicated educational seminars and symposia. It was not too difficult, although a bit challenging, for me to convince my fellow UK and European colleagues to join me in this endeavor. This program has attracted the attention of other countries and is likely to be expanded to include China, Asia-Pacific, Arab world, and Africa. The Afro—UK Genetic Education Forum is being set up in collaboration with the African Society of Human Genetics (www.afshg.org).

This chapter sets out the basis and highlights the potential areas of genomic applications in the developing world (Figure 12.9). While it is necessary to prioritize the current limited resources in dealing with the most common socioeconomic and health problems faced by developing countries, especially the LDCs, resulting from poverty, malnutrition, unsafe water supply, poor sanitation, and communicable diseases, it is nevertheless important to invest in new science and technology, such as genetics and genomics, to bridge the gap and prepare some ground for future developments [50]. Unfortunately, geographic, economic, and political challenges in India and other developing countries often restrict investment and improvement in the infrastructure development necessary to sustain progress in any area. Health is the prime area to focus. There is ample evidence to argue that genetic and genomic factors play an important role in the causation of the common health problems affecting developing and least-developed nations and consequently their people. This issue was examined and highlighted in the WHO expert committee report on role of genomics for global health. This has been followed by several reports and publications that lend further support to the view that the developing nations in South Asia (www.saarc-sec.org) and other parts of the world (Brazil—Russia—India—China—South Africa, BRICS nations) should not be left behind in acquiring relevant genetic and genomic technologies for the betterment of health, industry, and socioeconomic status (https://community.oecd.org/docs/DOC-42174). High-level R&D support in developing countries has increased considerably. The future of genetics and genomics in developing countries is promising, particularly India, Latin America, the Arab world, and parts of Africa. The importance of genetics and genomics is acknowledged in the recent WHO statement on the global prevention of birth defects endorsed by the Sixty-Third World Assembly (WHO, April 2010). The WHO "Grand Challenges" project on applications of genomics in the public health in the developing world

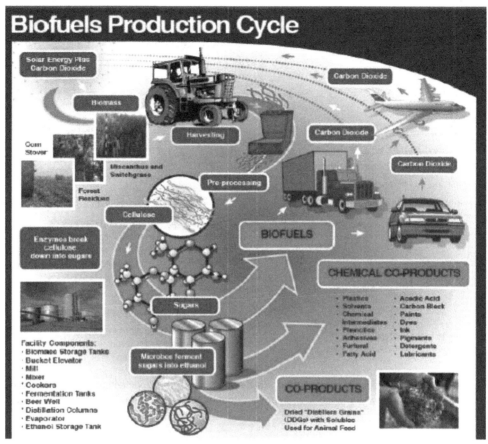

Figure 12.9 Genomics applications in biofuel and bioenergy development program (www. energy.gov).

(2011−2013) is expected to identify potential areas and offer strategic guidance to member nations in the developing world (www.who.int/rpc/grand_challenges.pdf). The Department of Energy, United States, has launched an ambitious project covering wide-ranging ecological and socioeconomic applications of genome sciences, the success of these objectives would set the direction and pace of massive R&D efforts in major aspects of genome science and technology (Figure 12.10; www.genomicscience. energy.gov/userfacilities/jgi.shtml).

In conclusion, it is now widely believed that the rapid development and inclusion of genetic and genomic technologies will be crucial to any nation's socioeconomic well-being, and the health of its population. It is important that this technology and expertise should not remain beyond the reach of the developing world, confined to the global West and North. I hope that this supported by few examples generate discussion and debate among wide range of professionals in genetic/genomic sciences,

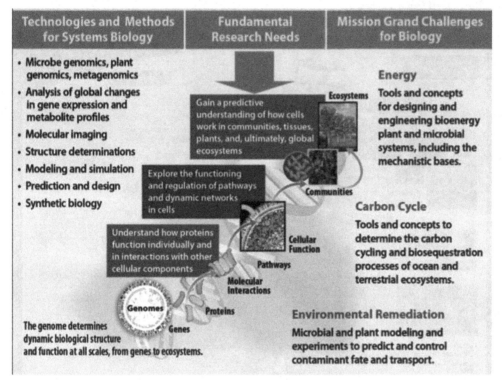

Figure 12.10 The model of genomics-led R&D applicable to both developed and developing nations based on the US Department of Energy's Genomic Science Program (www.genomicscience. energy.gov).

anthropology, medicine, public and population health, the biotechnology industries, media, and public services. It is important that we engage with lay people and more importantly politicians in this debate for setting up the desired direction and making strategic decisions for harnessing the huge potential of genomics for enhancing science, generating wealth and improving the health of our great nation.

Finally, I would like to conclude this chapter by including the famous quotation of Mahatma Gandhi, *"Science to be science must afford the fullest scope for satisfying the hunger of body, mind and soul."* Genomics has the potential to offer this outcome.

Thus all of us need to ensure that genomics, like other major scientific disciplines, is continued to be nurtured, applied, and translated for the benefit of man irrespective of geographic, racial, linguistic, ethnic, political, or economic variation.

ACKNOWLEDGMENTS

The author gratefully acknowledges permission from Oxford University Press to use material from chapters 3, 10, 15, 20, 30, and 31 "Genomics and Health in the Developing World" (New York, 2012).

This chapter is based on the Public Lecture "Harnessing Genomics for the Developing World" delivered by the Author at the Ranbaxy Science Foundation (India) 19th Annual Symposium "Gains of genomic research in biology and medicine" (www.ranbaxysciencefoundation.net).

REFERENCES

[1] Daar AS, et al. Top ten biotechnologies for improving health in developing countries. Nat Genet 2002;32(2):229—32.
[2] Bhutta ZA. Ethics in international health research: a perspective from the developing world. Bull World Health Organ 2002;80(2):114—20.
[3] Cantrell RP, Reeves TG. The cereal of the world's poor takes center stage. Science 2002;53—53
[4] Naylor RL, et al. Biotechnology in the developing world: a case for increased investments in orphan crops. Food Policy 2004;29(1):15—44.
[5] Weatherall DJ. Genomics and global health: time for a reappraisal. Science 2003;302(5645):597—9.
[6] Bustamante CD, Francisco M, Burchard EG. *Genomics for the world*. Nature 475(7355):163—165.
[7] Watkins WS, et al. Genetic variation among world populations: inferences from 100 Alu insertion polymorphisms. Genome Res 2003;13(7):1607—18.
[8] International HapMap, C. Integrating ethics and science in the International HapMap Project. Nat Rev Genet 2004;5(6):467.
[9] Kivisild T, et al. Deep common ancestry of Indian and western-Eurasian mitochondrial DNA lineages. Curr Biol 1999;9(22):1331—4.
[10] Basu A, et al. Ethnic India: a genomic view, with special reference to peopling and structure. Genome Res 2003;13(10):2277—90.
[11] Chen J, et al. Genetic structure of the Han Chinese population revealed by genome-wide SNP variation. Am J Hum Genet 2009;85(6):775—85.
[12] Friedlaender JS, et al. The genetic structure of Pacific Islanders. PLoS Genet 2008;4(1):e19.
[13] Behar DM, et al. The genome-wide structure of the Jewish people. Nature 466(7303):238—242.
[14] Hunter-Zinck H, et al. Population genetic structure of the people of Qatar. Am J Hum Genet. 87 (1):17—25.
[15] Torroni A, et al. mtDNA variation of aboriginal Siberians reveals distinct genetic affinities with Native Americans. Am J Hum Genet 1993;53(3):591.
[16] Sans M. Admixture studies in Latin America: from the 20th to the 21st century. Hum Biol 2000;155—77.
[17] Salzano FM. Interethnic variability and admixture in Latin America-social implications. Rev Biol Trop 2004;52(3):405—15.
[18] Majumder PP. People of India; biological diversity and affinities. The Indian human heritage 1998;45—59.
[19] Ingman M, Gyllensten U. Mitochondrial genome variation and evolutionary history of Australian and New Guinean aborigines. Genome Res 2003;13(7):1600—6.
[20] Wang S, et al. Geographic patterns of genome admixture in Latin American Mestizos. PLoS Genet 2008;4(3):e1000037.
[21] Conrad DF, et al. Origins and functional impact of copy number variation in the human genome. Nature 2009;464(7289):704—12.
[22] Rambaut A, et al. The genomic and epidemiological dynamics of human influenza A virus. Nature 2008;453(7195):615—19.
[23] World Health Organization. Obesity: preventing and managing the global epidemic. Geneva: World Health Organization; 2000.
[24] Khush GS. What it will take to feed 5.0 billion rice consumers in 2030. Plant Mol Biol 2005;59 (1):1—6.
[25] Lack G. Clinical risk assessment of GM foods. Toxicol Lett 2002;127(1):337—40.
[26] Paine JA, et al. Improving the nutritional value of Golden rice through increased pro-vitamin A content. Nat Biotechnol 2005;23(4):482—7.

[27] González C, Johnson NL, Qaim M. Consumer acceptance of second generation GM foods: The case of biofortified cassava in the Northeast of Brazil [Approved article]; 2009.

[28] Kim HY, et al. Growth and nitrogen uptake of CO_2-enriched rice under field conditions. New Phytol 2001;150(2):223—9.

[29] Kirigwi FM, et al. Markers associated with a QTL for grain yield in wheat under drought. Mol Breed 2007;20(4):401—13.

[30] Li S, et al. Vitamin A equivalence of the ß-carotene in ß-carotene-biofortified maize porridge consumed by women. Am J Clin Nutr. 92(5):1105—1112.

[31] Fao J, Foods MHI. URL: http://faostat.fao.org Food and Agriculture organization of the United Nations. Rome; 2004

[32] Mielniczuk F. Food security, biotechnology, and the BRICS: a necessary relationship. In International affairs forum. Taylor & Francis, London.

[33] Patwardhan B, et al. Ayurveda and traditional Chinese medicine: a comparative overview. Evid Based Complement Alternat Med 2005;2(4):465—73.

[34] Corson TW, Crews CM. Molecular understanding and modern application of traditional medicines: triumphs and trials. Cell 2007;130(5):769—74.

[35] Fokunang CN, et al. Traditional medicine: past, present and future research and development prospects and integration in the National Health System of Cameroon. Afr J Tradit Complement Altern Med 8(3).

[36] Kumar D. Various chapters in this book have discussed key. Genomics and Health in the Developing World, (62):66.

[37] Chopra A, Doiphode VV. Ayurvedic medicine: core concept, therapeutic principles, and current relevance. Med Clin North Am 2002;86(1):75—89.

[38] Patwardhan B, Bodeker G. Ayurvedic genomics: establishing a genetic basis for mind-body typologies. J Altern Complement Med 2008;14(5):571—6.

[39] Joshi K, Ghodke Y, Shintre P. Traditional medicine and genomics. J Ayurveda Integr Med. 1(1): 26.

[40] Mukerji M, Prasher B. Ayurgenomics: a new approach in personalized and preventive medicine. Sci Cult 77:10—17.

[41] Efferth T, et al. From traditional Chinese medicine to rational cancer therapy. Trends Mol Med 2007;13(8):353—61.

[42] Collins FS, et al. A vision for the future of genomics research. Nature 2003;422(6934):835—47.

[43] Sachs JD, McArthur JW. The millennium project: a plan for meeting the millennium development goals. Lancet 2005;365(9456):347—53.

[44] Aguilar A, et al. Biotechnology and applied genomics for health: initiatives of the European Union. Eur J Med Chem 2003;38(4):329—37.

[45] Guttmacher AE, Porteous ME, McInerney JD. Educating health-care professionals about genetics and genomics. Nat Rev Genet 2007;8(2):151—7.

[46] Skirton H, et al. Genetic education and the challenge of genomic medicine: development of core competences to support preparation of health professionals in Europe. Eur J Hum Genet. 18(9): 972—977.

[47] Prows CA, et al. Genomics in nursing education. J Nurs Scholarsh 2005;37(3):196—202.

[48] Green ED, Guyer MS, NHGR Institute. Charting a course for genomic medicine from base pairs to bedside. Nature. 470(7333):204—213.

[49] Khoury MJ, et al. The continuum of translation research in genomic medicine: how can we accelerate the appropriate integration of human genome discoveries into health care and disease prevention? Genet Med 2007;9(10):665—74.

[50] Thomas MA, Klaper R. Genomics for the ecological toolbox. Trends Ecol Evol 2004;19 (8):439—45.

CHAPTER 13

Roles of Genomics in Addressing Global Food Security

Denis J Murphy
Genomics and Computational Biology Research Group, University of South Wales, Pontypridd, Wales, UK

Contents

INTRODUCTION

Food security is one of the most important social and scientific issues facing humanity in the twenty-first century. According to current projections by the United Nations and other agencies, the global population of about 7.5 billion in 2015 is likely to increase to at least 9 billion by 2050, and possibly to as much as 11−13 billion by the end of the century [1]. This near-certain increase in population will require the provision of 20−90% additional calories just to keep people adequately fed, never mind improving the often-poor quality of many diets.

However, the actual demand for agricultural products is likely to be much higher than is implied by the figure of an additional 20−90% calories of global food production. This is because populations in those developing countries with rapidly expanding economies, such as India, China, and Brazil, are increasingly switching away from diets based mainly on crops, such as rice and cassava, and toward the inclusion of more livestock products such as beef, pork, and poultry [2]. The intensive production of such animal-based foods normally involves increased use of crop-derived rations as livestock feed. Some of the most common types of animal feed are grains such as maize and soybeans that would otherwise be consumed by people. The production of animal-based foods also involves the use of much more water and land than traditional crop-based diets [3].

Genomics and Society
DOI: http://dx.doi.org/10.1016/B978-0-12-420195-8.00013-6

Because animal-derived foods provide a much lower calorie return per hectare than those from plants, this will result in an effective decrease in the overall availability of food calories unless more land is farmed or higher yielding crops and/or more efficient livestock animals are developed. It will, therefore, be necessary to generate higher yields from existing crops both to feed people directly and to feed the ever-increasing numbers of livestock animals.

It is also recognized that many diets in poorer countries are deficient in micronutrients such as vitamins and minerals and that the development of crops with improved nutrient profiles will be necessary in order to optimize human health [4,5]. Examples range from increasing the vitamin A content of staples such as rice, eggplant, and maize to enhancing levels of unsaturated fats in oil crops. The world is also facing ever greater uncertainties due to climate change and resource depletion that are predicted to impact with particular severity in many developing countries [6–8]. Modern genomics has great potential to contribute to improving many aspects of crop and livestock performance, including increased yield and quality, better disease resistance, and reduced levels of inputs and greater resilience in the face of increasingly variable environmental factors [9–11].

THE CHALLENGE OF GLOBAL FOOD SECURITY

At present, most population projections involve the growth to about 9 billion people by 2050 followed by a plateauing off to about 10 billion by 2100. However, a recent detailed study suggests a much higher rate of population increase than previously predicted, with a possible sustained rise to as much as 13 billion people by 2100. No less than 4 billion of these people will be living in Africa [12]. This analysis, by United Nations and University of Washington, Seattle, is the first of its kind to use modern statistical methods rather than expert opinions to estimate future birth rates, one of the determining factors in population forecasts. One statistic that particularly emphasizes the discrepancy in birth rates between industrialized and developing regions is the comparison of birth rates between Europe and Nigeria. Between 2008 and 2011, 5.4 million babies were born in the whole continent of Europe while in a single African country, Nigeria, no less than 23.7 million babies were born over the same period.

Nigeria already imports much of its food despite having the capacity to grow most of its requirements locally. For example, the region of West Africa that includes Nigeria is the center of origin of a major food crop, the oil palm, which is a rich source of calories and vitamins. Until the 1950s, Nigeria produced a surplus of palm oil that not only fed its own population but was also exported around the world. By 2015, the neglect of indigenous oil palm cultivation meant that Nigeria was reduced to importing this oil from Indonesia and Malaysia. This imposes a considerable cost to the Nigerian economy and means that the country cannot control the supply of a

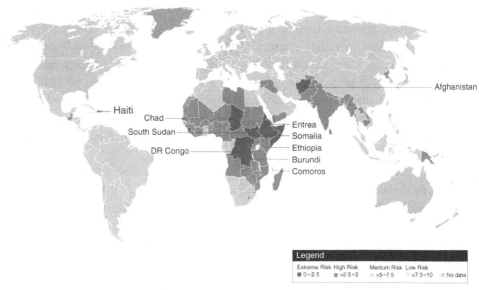

Figure 13.1 Food insecurity is a large and growing problem in many countries, especially in the tropics. (© Maplecroft 2012 www.maplecroft.com). *Image available at: http://www.sciencedirect. com/cache/MiamiImageURL/1-s2.0-S0960982212014911-f13-01-9780124201958_lrg.jpg/0? wchp = dGLbVlk-zSkWb&pii = S0960982212014911.*

traditional staple food that has been part of the local diet for millennia. Overall, therefore, areas of highest risk for food insecurity, both now and in the future, are decisively concentrated in Sub-Saharan Africa and in some parts of South Asia [13].

As shown in Figure 13.1, some of the countries at particular risk include South Sudan, Chad, DR Congo, Somalia, Ethiopia, and Afghanistan. It is no coincidence that each of these countries has witnessed prolonged civil conflict, which have significantly disrupted economic activity, and especially the production and distribution of food. However, in these cases, civil conflict has merely exacerbated underlying structural problems that include inadequate networks of crop breeding, management, and distribution. Hence, we find that even relatively peaceful countries such as India, Tanzania, and Zambia are also assessed as being at high risk in terms of their food security. One of the principal factors behind the chronic deficits in food production in Sub-Saharan Africa is the failure of the crop breeding and management advances that were behind the Green Revolution in Asia and the Americas to be fully implemented in Africa. In the future it will be important for such countries to establish secure and well-resourced scientific and management structures in order to take advantage of improvements in agronomy via new technologies such as genomics [10].

Clearly, an increase in the global population to as much as 13 billion would only be possible if crop yields were to be increased on a continuous basis for the rest of this century. The alternative is likely to be mass starvation and the accompanying civil strife that would probably have a global impact as increasingly desperate people attempted to migrate to richer countries. Efforts to increase food yields on such a scale are likely to be a serious challenge to current plant breeding efforts where it is clear that the yields of some major commercial crops are now leveling off [14]. These crop yield plateaus mainly affect major commercial crops such as maize, wheat, and soybeans. Luckily there is still scope for yield increases in several of the more important developing country crop staples such as cassava and sorghum, especially via genomics.

Another factor that is limiting food production is the increase in the consumption of animal products, most notably meat. This trend is especially marked in developing countries such as India and China where increased affluence is associated with rapid rises in the use of animal products, especially in the diet of their burgeoning middle classes. The OECD [15] estimates that the developing world will account for 80% of the growth in meat consumption over the next decade. This means that considerably more grain crops are now being diverted from human food consumption to serve as livestock fodder with the likelihood that such trends will continue at least into the medium-term future. In addition to increasing crop yields, it will therefore be important to use genomics to improve the yields of livestock animals.

THE IMPACT OF CLIMATE CHANGE

Several recent reports have concluded that climate change is beginning to constrain where and what crops we can grow [7,16,17]. Rising temperatures in southern Africa and tropical highlands worldwide could be particularly hazardous [18]. Some developing country crops, such as cassava and sorghum, are at particular risk from climate change. Another staple crop, the potato, is especially vulnerable to heat stress, which reduces growth and starch formation. Developing and distributing heat-tolerant potato varieties could reduce climate-related damage for about 65% (7.7 million hectares) of the world's potato crop. Therefore, one of the priority areas for crop genomics is to understand the genetic basis of plant responses to climate so that better adapted varieties can be developed in the future. It is encouraging that in many crops there appears to be considerable genetic variation in the response to climatic factors but in other cases where inherent genetic variation is lacking it may be necessary to introduce new genes via breeding methods such as assisted hybridization or by transgenesis [11].

Crop production also faces new and uncertain risks due to pests and pathogens. Changing climates are resulting in the emergence of new crop diseases that are often spread by humans via increasingly globalized transport and trade networks. A recent example is the spread of new strains of the wheat rust fungus, UG99, from east Africa into Asia and beyond [11]. The value of maintaining effective surveillance and intervention systems that include genome analysis of emerging disease organisms is shown by the estimate in 2014 that global research to control stem rust disease saves wheat farmers losses worth US\$1.12 billion per year [19]. This study found that, had there not been investment in stem rust research and ensuing effective global control during 1961—2009, losses in wheat production would have amounted to 6.2 million tonnes annually, or 1.3% of the total harvest. This equates to losses of US\$1.12 billion per year at 2010 prices, or enough wheat to satisfy almost the entire annual calorie deficit of Sub-Saharan Africa's undernourished population.

Unfortunately, however, more broadly speaking the investments in breeding for resistance to crop diseases have declined in recent decades so it is vital that better vigilance and monitoring systems, including DNA-based assays, are developed.

SO, HOW CAN GENOMICS HELP?

The application of genomics in agriculture is already having direct impacts on crop breeding and thereby in addressing overall global food security [20,21]. For example, one of the most dramatic example of technology improvements in the twenty-first century has been in DNA sequencing where the cost per base has decreased by 100,000-fold since 2000 [22,23]. The first plant genome to be fully sequenced was the model species, *Arabidopsis thaliana*, published in 2001, while the first crop genome was rice, where a high-quality sequence was published in 2005. The sequencing of the much larger maize genome required a massive effort by private and public laboratories and the results were published in a series of papers in 2009. Other large-scale projects are currently under way for developing country crops such as sorghum and foxtail millet and sequence data are now being publicly released at an increasingly rapid pace.

Advances in next generation sequencing technologies are now enabling the genomes of even comparatively minor crops to be characterized [24]. Over the past decade, the genomes of virtually all of the major commercial crops have been sequenced and a new initiative involving sequencing of no fewer than 1000 rice genomes has been undertaken in China. These advances are now driving new discoveries relating to crop yield/quality performance and responses to the abiotic environment and to biotic threats such as pests and diseases. In turn, this will underpin identification and manipulation of the key genes that regulate such traits in crops.

The construction of detailed genetic maps and now the availability of full genome sequence data has enabled breeders to develop DNA-based molecular markers that are reducing the time required for bringing new varieties to market by many years

[25–29]. In particular the use of single nucleotide polymorphisms for marker-assisted selection can reduce the time required to develop a new crop variety by as much as 3–5 years. These advances will be essential in enabling farmers to increase overall crop yields and to address sustainability criteria, while also being able to respond to environmental threats such as climate change and more immediate anthropogenic issues such as salinization and lack of moisture [30–34]. Sustainability issues include the use of energy-intensive and environmentally impacting chemical inputs such as fertilizers, pesticides, fungicides, and herbicides. Biological versions of many of these inputs are now being developed, for example, inbuilt insecticides such as Bt proteins and biofertilizers in the form of nitrogen-fixing soil bacteria [10,11].

Most of the environmental threats to crops that result from climate change relate to traits involved in the regulation of heat/cold/drought/flooding tolerance. Unfortunately, however, most of these traits are highly complex and are controlled by large numbers of genes. Also, the tolerance mechanisms are often specific to one crop group and are therefore not necessarily transferrable to other crops, for example, via transgenesis. More immediate anthropogenic climatic factors that affect crops include effects of pollution, especially ozone, as well as salinization due to poor irrigation management, and a lack of moisture due to upstream water extraction. These man-made problems are particularly acute in Asia and in many cases the solutions will require a combination of improved environmental legislation and advances in the breeding of more resilient crops.

WHAT ARE THE KEY FUTURE TARGETS FOR GENOMICS-ASSISTED CROP IMPROVEMENT?

The two most important short-term target groups of crop traits that need to be addressed immediately are (1) overall yield of the crop product(s) in question and (2) the quality of such products. These are the key traits for short/medium-term priority over the next 5–10 years. Improving these traits has the potential both to increase food production and to reduce requirements for further expansion of farmland. Therefore they can reduce the overall environmental footprint of agriculture. They can also be implemented quickly in conjunction with improved input strategies [10].

In the medium term, that is, by the 2020s, some of the most important target traits are as follows:

- Abiotic stress tolerance: drought, flooding, heat, cold, salinity.
- Pathogen tolerance: especially new fungal, viral and bacterial diseases.
- Pest tolerance: especially pesticide-resistant insects.

These are mostly complex multigenic traits that require further investment in R&D. The breeding of inherent tolerance to these external threats in crops has the potential to greatly reduce use of biocide chemicals and hence to reduce costs and environmental footprint and thereby improve sustainability [11,35].

Over the longer-term target traits, that is, by 2040 and beyond, our increasing knowledge of fundamental plant developmental processes has the potential for breeders to make more radical alterations to the structure and performance of major crops. Some of the most important target traits include:

- Modifying crop architecture to maximize yield and harvestability.
- Increasing nitrogen efficiency, including the introduction of nitrogen fixation into nonlegume crop species that currently require supplementation by external nitrogen in the form of fertilizers.
- Increasing the efficiency of CO_2 uptake during photosynthesis, for example, engineering Rubisco and the C4 pathway of photosynthesis.
- Apomixis, that is, the ability of plants to produce fertile seeds without pollination.
- Orphan crops (see later).
- Domesticating new crops (see later).

These can be considered as "blue skies" targets that are high risk with timescales measured in decades but they are still worth pursuing as long-term public sector research programs and/or via public—private partnerships (PPPs) [11,35,36].

Orphan crops: There are large numbers of the so-called orphan crops that have yet to benefit from many aspects of modern breeding technologies, including genomics [37]. Many of these crops are grown in regions such as Sub-Saharan Africa where a combination of poverty, poor infrastructure, corruption, and lack of security has seen crop yields stagnate or even decline. While some countries such as Kenya and Uganda are now making good progress, the overall picture is patchy and there is still a large gulf between those relatively disadvantaged areas of the developing world that are most exposed to food insecurity and the rest of the world where there is scope for cautious optimism. Orphan crops are a weakness—but also an opportunity. They suffer from very low yields and may be severely impacted by environmental changes. However, they have been relatively neglected by modern breeders and therefore still have considerable scope for significant genetic improvement for yield, quality, and resilience using new genomic approaches that include bioinformatics [38].

The African Orphan Crops Consortium (http://www.mars.com/global/african-orphan-crops.aspx) is an international effort to improve the nutrition, productivity, and climatic adaptability of some of Africa's most important food crops, helping to decrease the malnutrition and stunting rife among the continent's rural children. The goal is to sequence, assemble, and annotate the genomes of 100 traditional African food crops, which will enable higher nutritional content for society over the decades to come. The resulting information will be put into the public domain, with the endorsement of the African Union. These "orphan crops" are species that have been neglected by researchers and industry because they are not economically important on the global market.

Table 13.1 Sub-Saharan African cassava production in 2007

Sub-Saharan Africa	
Country	Cassava yield (kg/ha)
Niger	21818
Reunion	20000
Malawi	16539
Mali	16400
Mauritius	15000
Sudan	1667
Burkina Faso	2000
Gambia	3000
Sudan	1667
Burkina Faso	2000

The productivity gap further increases among Sub-Saharan countries: a comparison between Sub-Saharan countries with high (top 5) and low productivity (bottom 5) reveals that the top 5 productivity is between three and 13 times higher. Source: www.gapminder.org (*Gapminder Agriculture*).

An example of the potential to improve "orphan crops" in Africa is given in the case of cassava in Table 13.1. This table provides the variation in the average yield of cassava crops in major producer countries. Note that most of the countries with the lowest yields are in Sub-Saharan Africa. There are many factors behind such a wide variation in yield but poor agronomic practice and lack of access to improved varieties are some of the most important reasons. With modern genomics-assisted breeding it is possible to increase yields to as much as 30 ton/ha, which is more than 10 times the yields currently obtained in countries such as Sudan and Burkina Faso.

Domesticating new crops: Why should we domesticate new crops? The reason is that over half of global food calories are provided by only four major crops, namely rice, wheat, maize, and potato (Table 13.2). Each of these four crops was domesticated over 10,000 years ago and, because they provide such a high proportion of human food intake, any setback in their production due to disease, climate, or other factors could spell disaster for large numbers of the global population. The best way to increase resilience in our food supply is to embark on a program aimed at domesticating new crops.

There are many noncrop plants that could potentially yield considerable amounts of food providing they are domesticated into forms that are amenable to agriculture. This process is analogous to the taming of selected wild animals into livestock species such as cattle and sheep. New discoveries have recently revealed the genetic basis of the key domestication processes for crops [39,40]. Using this knowledge we can embark on a systematic program of domestication of new crops designed for the twenty-first century [11].

Table 13.2 Global food production of selected major crops

Crop	Area		Output		Yield (t/ha)
	Mha	%	Mt	%	
Wheat	241	29.4	606	20.9	2.8
Maize	158	19.3	792	27.3	5.0
Rice	156	19.0	660	22.8	4.2
Soybean	90	11.0	221	7.6	2.4
Sorghum	47	5.7	63	2.2	1.4
Rapeseed	31	3.8	51	1.8	1.6
Dry beans	27	3.3	18	0.6	0.7
Sugarcane	23	2.8	107	3.7	4.7
Sunflower	22	2.7	27	0.9	1.2
Potato	19	2.3	309	10.7	16.7

Source: Data reproduced with permission from Ref. [11].

IMPACTS OF GENOMICS IN DEVELOPING COUNTRIES

At present the impact of genomic technologies on agricultural systems in developing countries has been mixed. In some of the more rapidly expanding economies such as India, China, Brazil, Malaysia, and South Africa, uptake of these technologies has been relatively rapid. For example, in 2013, the Malaysian Palm Oil Board (MPOB) (a major government research center) published the sequence of the oil palm genome which has already enabled its scientists to develop methods to select new yield-related traits in one of the world's most important food and industrial crops [41,42,43]. We and others are now working with MPOB to use their genomic data for gene discovery and manipulation of key yield and quality traits [44,45].

Oil palm is a uniquely productive tropical crop with a potential yield capacity well in excess of 10 tonnes of oil per hectare (t/ha)—for comparison temperate oil crops such as rapeseed and sunflower yield about 0.5−1.0 t/ha. However, current oil palm yields are well below the 10 t/ha figure and are typically about 4−6 t/ha for the best commercial plantations and 3−4 t/ha for smallholder farmers. Palm oil is mainly used as an edible product and is an important dietary component for well over 1 billion people worldwide. If we can more than double the yield of oil palm, as is feasible using existing breeding approaches, there will be much less pressure to convert tropical forests to plantations—with all the resultant unfortunate ecological and environmental consequences [44].

The construction of detailed genetic maps and the availability of full sequence data have enabled breeders to develop DNA-based molecular markers that are reducing the time required for bringing new varieties to market by many years. These advances will be essential to enable farmers to increase overall yield and to address sustainability criteria while also being able to respond to environmental threats such as climate change and more immediate anthropogenic issues such as salinization and lack of moisture. Other major genomics-related breeding tools include TILLING (Targeting Local Lesion IN Genomes—a high-tech form of mutagenesis), and the many forma of transgenesis (often popularly referred to as GM) [11,35,46].

Despite recent advances, however, most of these genomics-related technologies are still relatively resource intensive and require rather high skill levels and sustained funding and infrastructure. For these reasons they have not been taken up widely in many developing country crop systems where food insecurity is most acute. This is especially true for parts of Sub-Saharan Africa where there are potentially great gains to be won in terms of increasing crop yield and quality. In future it would be beneficial if there could be improved knowledge sharing and the targeting of genomic tools on selected crops in developing countries. One of the major tools to achieve this ambitious objective could be a more imaginative use of PPPs [9,11,37,47–49].

CONCLUSIONS

- Due to population increase, environmental degradation, and climate change, the guaranteeing of an acceptable level of food security across the world will become increasingly difficult in the future [9].
- We need to increase food production from crops—but in the most sustainable possible manner.
- There are many nonbiological challenges to achieving this goal—for example, poverty, poor governance, and unequal distribution.
- However, crop breeding is one of the most important tools at our disposal to increase food security to endangered populations in developing countries.
- Crop breeding has already benefited greatly from modern genomic technologies—but this has yet to be achieved in most of the regions of greatest need.
- One of the major future challenges is to enable all of humankind to share in the increasing rewards from modern scientific breeding, including genomics [49].
- PPPs are one of the most effective strategies to address this challenge.

REFERENCES

[1] Gerland P, Raftery AE, Sevčíková H, Li N, Gu D, Spoorenberg T, et al. World population stabilization unlikely this century. Science 2014;346:234–7.
[2] Eshel G, Shepon A, Makov T, Milo R. Land, irrigation water, greenhouse gas, and reactive nitrogen burdens of meat, eggs, and dairy production in the United States. Proc Natl Acad Sci USA 110:11996–12001.
[3] Bebber DP, Holmes T, Gurr SJ. The global spread of crop pests and pathogens. Global Ecology and Biogeography 2014;23:1–10.
[4] Cakmak I, Graham RD, Welch RM. Agricultural and molecular genetic approaches to improving nutrition and preventing micronutrient malnutrition globally. In: Cakmak I, Welch RM, editors. Impacts of agriculture on human health and nutrition. Oxford, UK: UNESCO-EOLSS. Encyclopedia of Life Support Systems; 2004. [Chapter 13.7].

[5] Mayer JE, Pfeiffer WH, Beyer P. Biofortified crops to alleviate micronutrient malnutrition. Curr Opin Plant Biol 2008;11:166—70.

[6] Alliance for a Green Revolution in Africa (AGRA) Africa agriculture status report: climate change and smallholder agriculture in Sub-Saharan Africa, Nairobi, Kenya 2014.

[7] Nelson GN, Rosegrant MW, Koo J, Robertson R, Sulser T, Zhu T, Ringler C, Msangi S Palazzo A, Batka M, Magalhaes M, Valmonte-Santos R, Ewing M, Lee D. Climate change: impact on agriculture and costs of adaptation. Food policy report. International Food Policy Research Institute, Washington, DC; 2009. <http://www.ifpri.org/sites/default/files/publications/pr21.pdf>.

[8] Rosegrant M, Koo J, Cenacchi N, Ringler C, Robertson R, Fisher M, et al. Food security in a world of growing natural resource scarcity, the role of agricultural technologies. International Food Policy Research Institute (IFPRI); 2014. <http://www.ifpri.org/publication/food-security-world-growing-natural-resource-scarcity>.

[9] West PC, Gerber JS, Mueller ND, Brauman KA, Carlson KM, Cassidy ES, et al. Leverage points for improving global food security and the environment. Science, 345. 2014.

[10] Murphy DJ. Current status and options for crop biotechnologies in developing countries. Biotechnologies for agricultural development. Rome: FAO; 2011. p. 6—24.

[11] Murphy DJ. Plants, biotechnology, and agriculture. UK: CABI Press; 2011.

[12] Gerland P, Raftery AE. Ševčíková H, Li N, Gu D, Spoorenberg T, Alkema L, Fosdick BK, Chunn J, Lalic N, Bay G, Buettner T, Heilig GK. Wilmoth J. World population stabilization unlikely this century, Science 346:234-7.

[13] Jobbins G, Pillot D. EU and IFAD Review of CGIAR Research Programme 7: Climate Change, Agriculture and Food Security (CCAFS). IFAD—AGRINATURA-EEIG Institutional Contract, <https://cgspace.cgiar.org/handle/10568/34270>; 2013.

[14] Grassini P, Eskridge K, Cassman KG. Distinguishing between yield advances and yield plateaus in historical crop production trends. Nat Commun 2014;4:2918.

[15] OECD/FAO. OECD-FAO Agricultural Outlook 2014. Paris: OECD Publishing; 2014. <http://dx.doi.org/10.1787/agr_outlook-2014-en>.

[16] FAO Climate change adaptation and mitigation: challenges and opportunities for food security. Information document prepared for the High-level Conference on World Food Security: the Challenges of Climate Change and Bioenergy, Rome. Available from: <ftp://ftp.fao.org/docrep/fao/meeting/013/k2545e.pdf>; 2008.

[17] Rosenzweig C, Elliott J, Deryng D, Ruane AC, Müller C, Arneth A, et al. Assessing agricultural risks of climate change in the 21st century in a global gridded crop model intercomparison. Proc Natl Acad Sci USA 2014;111(9):3268—73. Available from: <http://dx.doi.org/10.1073/pnas.1222463110>.

[18] Müller C, Waha K, Bondeau A, Heinke J. Hotspots of climate change impacts in Sub-Saharan Africa and implications for adaptation and development. Global Change Biology 2014;20 (8):2505—17. Available from: <http://dx.doi.org/10.1111/gcb.12586>.

[19] Pardey PG, Beddow JM, Kriticos DJ, Hurley TM, Park RF, Duveiller E, et al. Right-sizing stem-rust research. Science 2013;340:147—8.

[20] Varshney RK, Tuberosa R. Genomics-assisted crop improvement. Genomics approaches and plat-forms, vol. 1. Berlin: Springer; 2007.

[21] Varshney RK, Tuberosa R. Genomics-assisted crop improvement. Genomics applications in crops, vol. 2. Berlin: Springer; 2007.

[22] Mardis E. The impact of next-generation sequencing technology on genetics. Trends Genet 2008;24:133—41.

[23] Shendure J, Ji H. Next-generation DNA sequencing. Nat Biotechnol 2008;26:1135—45.

[24] Edwards D, Batley J. Plant genome sequencing: applications for crop improvement. Plant Biotechnol J 2010;8:2—9.

[25] Bernardo R. Molecular markers and selection for complex traits in plants: learning from the past 20 years. Crop Sci 2008;48:1649—64.

[26] Dargie J. Marker-assisted selection: policy considerations and options for developing countries. In: Guimaraes EP, Ruane J, Scherf B, Sonnino A, editors. Marker-Assisted Selection (MAS) in crops, livestock, forestry and fish: current status and the way forward. Rome: FAO; 2007.

[27] FAO. Molecular marker-assisted selection Current status and future perspectives in crops, livestock, forestry and fish, Guimarães EP, Ruane J, Scherf BD, Sonnino A, Dargie JD, eds. Available from: <http://www.fao.org/docrep/010/a1120e/a1120e00.HTM>; 2007.

[28] Stafford W. Marker assisted selection (MAS) key issues for Africa, The African Centre for Biosafety. Available from: <http://www.biosafetyafrica.org.za/index.php/20090521225/MAS-Key-Issues-for-Africa.-Author-William-Stafford/menu-id-100025.html>; 2009.

[29] Xu Y, Crouch JH. Marker-assisted selection in plant breeding: from publications to practice. Crop Sci 2008;48:391—407.

[30] Yamaguchi T, Blumwald E. Developing salt-tolerant crop plants: challenges and opportunities. Trends Plant Sci 2005;10:615—20.

[31] Zhang H, Kim MS, Sun Y, Dowd SE, Shi H, Paré PW. Soil bacteria confer plant salt tolerance by tissue-specific regulation of the sodium transporter HKT1. Mol Plant Microbe Interact 2008;21:737—44.

[32] Ashraf M, Athar HR, Harris PJC, Kwon TR. Some prospective strategies for improving crop salt tolerance. Adv Agron 2008;97:45—110.

[33] Flowers TJ. Improving crop salt tolerance. J Exp Bot 2004;55:307—19.

[34] Munns R. Genes and salt tolerance: bringing them together. New Phytol 2005;167:645—63.

[35] Xu Y. Molecular plant breeding. Oxford, UK: CABI; 2010.

[36] Cook DR, Varshney RK. From genome studies to agricultural biotechnology: closing the gap between basic plant science and applied agriculture. Curr Opin Plant Biol 2010;13:115—18.

[37] Dawson IK, Jaenicke H. Underutilised plant species: the role of biotechnology. International Centre for Underutilised Crops Position Paper No. 1. International Centre for Underutilised Crops (ICUC), Colombo, Sri Lanka. Available from: <http://www.icuc-iwmi.org/>; 2006.

[38] Armstead I, Huang L, Ravagnani A, Robson P, Ougham H. Bioinformatics in the orphan crops. Brief Bioinform 2009;10:645—53.

[39] Sang S. Genes and mutations underlying domestication transitions in grasses. Plant Physiol 2009;149:63—70.

[40] Weeden NF. Genetic changes accompanying the domestication of pisum sativum: is there a common genetic basis to the "domestication syndrome" for legumes? Ann Bot 2007;100:1017—25.

[41] Singh R, et al. Oil palm genome sequence reveals divergence of interfertile species in Old and New worlds. Nature 2013. Available from: <http://dx.doi.org/10.1038/nature12309>.

[42] Singh R, et al. The Shell gene of the oil palm (Elaeis guineensis) controls oil yield and encodes a homologue of SEEDSTICK. Nature 2013. Available from: <http://dx.doi.org/10.1038/nature12356>.

[43] Soh AC. Genomics and plant breeding. J Oil Palm Res 2010;23:1019—28.

[44] Murphy DJ. The future of oil palm as a major global crop: opportunities and challenges. J Oil Palm Res 2014;26:1—24.

[45] Murphy DJ. From bioinformatics to Brazil: the future of oil palm as a major 21st century global crop. Malaysian Oil Science & Technology 2014;23:1—12.

[46] McCallum CM, Comai L, Greene EA, Henikoff S. Targeting Induced Local Lesions IN Genomes (TILLING) for plant functional genomics. Plant Physiol 2000;123:439—42.

[47] Lusser M. Workshop on public-private partnerships in plant breeding which was organised by JRC-IPTS 2014. Joint Research Centre—Institute for Prospective Technological Studies; 2014. Available from: http://dx.doi.org/10.2791/80891.

[48] Morris M, et al. The global need for plant breeding capacity: what roles for public and private sectors. Hort Science 2006;41:30—9.

[49] Ronald PC. Lab to farm: applying research on plant genetics and genomics to crop improvement. PLoS Biol 2014;12(6):e1001878. Available from: <http://dx.doi.org/10.1371/journal.pbio.1001878>.

[50] Carana Corporation. Trends in public—private partnerships and inclusive business models for improving food security and rural development through agriculture. Prepared for the Food Systems Innovation Initiative. <http://foodsystemsinnovation.org.au/sites/default/files/study_ppps_ibms_2-2015.pdf>; 2014.

CHAPTER 14

Genomics and Traditional Indian Ayurvedic Medicine

Mitali Mukerji and Bhavana Prasher
CSIR Ayurgenomics Unit-TRISUTRA, CSIR-Institute of Genomics and Integrative Biology, Sukhdev Vihar, New Delhi, India

Contents

INTRODUCTION

With rapid advances and new developments in genomics and related 'omics' fields, the medical and healthcare practices globally are under pressure to plan and implement effective changes in the current modern and traditional medical practices. This is not only to meet the challenges in diagnosis and treatment of rare and common

Genomics and Society
DOI: http://dx.doi.org/10.1016/B978-0-12-420195-8.00014-8

chronic and complex medical diseases but also to address the variability in therapeutic outcomes and develop effective and efficient public and population health strategies.

The advent of genomics has provided a tremendous impetus to bring much needed changes in the current medical and health care practices. However, there are a number of challenges before this could be implemented. A number of countries notably China (see Chapter 15) and India are leaders in this endeavour. Ayurveda, the ancient traditional Indian system of Medicine is probably best suited for this paradigm shift. Its concepts and approaches closely matches to the contemporary predictive and personalized medicine practice and fulfills the core principles (P4)- predictive, preventive, personalized, and participatory) medicine and probably also has a promotive component.

The Ayurveda has documented proven methods for maintenance of health and personalized management of chronic diseases. It is also widely practiced in most Indian communities despite sociocultural variations and many aspects for preventive health are also integrated into Indian traditional living. Despite this a large number of challenges exists in getting this system to mainstream and for its global acceptability. This chapter highlights some of these aspects in the genomics context and proposes a novel 'omics' field, the Ayurgenomics as a new paradigm for modernizing the current medical and health practices with global applications.

NEED FOR A PARADIGM SHIFT IN MODERN HEALTH CARE PRACTICE

The western medicine system, the allopathy, forms the mainstream of global health care system. Discovery of antibiotics and a number of vaccines in the twentieth century, have led to successful eradication of huge number of microbial diseases. In addition, technological advancements in imaging, surgical, and diagnostic as well as life support devices have led to a substantial reduction in morbidity and mortality. These have been the primary reasons for its wide global acceptance as it has hugely reduced perinatal. Infant, maternal and adult mortality as is evident from average increased life expectancy. Perhaps a good example would be the early detection of chromosomal anomalies and inborn errors of metabolism in genetic programs for carrier and prenatal screening reducing the burden of common inherited, for example Down syndrome, cystic fibrosis, sickle cell anemia, hemophilia, beta-thalassemia, Tay Sachs disease and other metabolic diseases. In addition, detection of late onset monogenic disorders and syndromes such as cardiomyopathies, muscular dystrophies, etc. is now possible and adopted in many public health programmes.

In the recent times, with the capability to sequence an individual's complete genome at a very affordable cost, the catalog of human genetic variations associated with Mendelian diseases has seen a steep rise [1–5]. The most exciting potential has been realized in the area of pharmacogenomics where now in many cases it has been become possible to predict responsiveness/nonresponsiveness to therapy using genetic markers [6,7].

This has been useful in predicting the outcome of different cancer therapies, use of warfarin, clopidogrel, etc. or managing the dosage of drugs that are used in chronic ailments and have frequent side effects [8–12]. However, management of common and chronic diseases still remain a challenge especially in the contemporary times as there is a threat of a global epidemic of lifestyle disorders. Compounded with an increase in number of aging population this now threatens the economics of health care management systems even in the developed countries. A major focus besides reducing the burden of disease is now on maintenance of quality of life both in health and disease through preventive interventions. There are also a number of other challenges that has highlighted the need for a paradigm shift in the current practice of modern medicine [13,14].

Challenges in Diagnosis and Treatment

In the allopathic system, classification and treatment of the disease is primarily on the basis of the principal organ or body system that exhibits major signs and symptoms of the disease. The disease is established by measuring gross anatomical pathology, histopathology, or a biochemical test. Diseases are given a nomenclature either based on the organ system or into syndromes if it encompasses a defined set of clinical features. Syndromic classification restricts the clinical phenotypes within a set boundary and also creates artificial boundaries between two diseases which might share the same origin and have overlapping sets of clinical features [13–17]. Also this is ascribed as a cause of ineffectiveness of drugs in a large spectrum of affected individuals since during clinical trials the recruitment of subjects are based on the primary set of features that are used to describe the syndrome. A patient on the basis of the overt symptoms most often approaches a doctor who specializes in the treatment of a specific organ system. Since the treatment and success of treatment is also causative and feature centric this approach not only limits treatment options but also does not heal the system holistically. It is being increasingly acknowledged by practitioners that most of the diseases exhibit clinical and population level heterogeneity and there is a need for methods that would enable delineating the phenotypic variation within a disease.

Variability in Therapeutic Outcome

Majority of the treatment procedures are targeted toward curbing the activity of the disease by a drug that either kills the pathogen/curbs the activity of the biomolecules that are elevated or by chemotherapy or radiotherapy which restrict the growth of proliferating cells. A last resort is surgery which removes the affected part is acknowledged that there is an enormous variability in success of treatment with respect to disease and individuals. As a result, only a finite set of diseases are entirely cured from the root cause. A majority of chronic diseases require medication throughout the lifetime of an individual and in diseases such as cancer and infectious diseases recurrence or resistance to drugs respectively are quite common. With the increase in life expectancy in most of the population, the cost of health care especially in chronic diseases is unaffordable for a large fraction of the population. Coupled with this, the quality of life is severely compromised due to the side effects of lifetime medications as well as other therapeutic and surgical interventions. The inability to predict the progression and prognosis of the disease in an individualized manner and also the management of the side effects of the therapeutic interventions are additional challenges.

Preventive Measures in Health and Disease

A third aspect is with respect to the preventive measures that are currently available. Prevention has been mostly successful by vaccination for common infectious diseases as there is increasing emergence and spread of new viruses and emergence of new multi-drug resistance pathogens. The list of vaccinations in children as well as the number of drugs that an individual has to take in a complex disease seems unending. Evolving lifestyle recommendations for disease such as diabetes mellitus and metabolic disorders is still in infancy as the genetic and epigenetic causes of these disorders are not fully established. Moreover the recommendations of lifestyle practices which are deemed important in different diseases are mostly generic and restrictive in nature. For instance excessive restriction of carbohydrate in a person with diabetes or fat and salt restriction in hypertensive/cardiac patients is mostly arbitrary as it is generally inferred that the amount of consumption would be proportional to the disease severity. A major caveat of this system is limited applications of individual aspects in terms of predisposition, susceptibility to disease as well as responsiveness to treatment and management.

ADVENT OF GENOMICS IN PERSONALIZED MEDICINE

The primary goal of the human genome project was to provide complete sequence of a reference human that would be a reference genetic blueprint for predictive and person-alized medicine [18,19]. It was envisaged that understanding of the human genome would allow discovery of predictive markers that could enable assessment of genetic risk to diseases and identification of actionable points for interventions, prognosis of disease, genetic modifiers, and drug targets as well as markers for assessing the responsiveness to drugs, diet, and lifestyle. This is the primary objective of predictive, preventive, person-alized, and participatory (P4) medicine [20,21]. Identification of predictive markers in the area of pharmacogenomics as well as in the discovery of specific gene variants and mutations for many monogenic disorders has helped in preventing and reducing the burden of disease. However, there are a number of challenges and we still have a long way to go. Recently, an ambitious genomics-based project was launched where nearly a million people are anticipated to participate in a prospective study [22]. The dynamics of health as well as deviations from individual participant's baseline would be monitored through sequencing an individual's genome one time as well as the transcriptome and biochemical markers a couple of times over a period of 2−3 years. It is anticipated that this would allow detection of individualized actionable points for interventions and herald a new era of preventive and precision medicine.

However, as more and more complete genome sequences become available from different populations across the world, the definition of human genome in terms of what would comprise as a reference is becoming elusive. There are now nearly 38 million single nucleotide polymorphisms (SNPs), 1.4 million in-dels, 14,000 deletions and 20,000 structural variations represented in the variation databases [23]. The frequency of these variations differs between populations and among individuals as a consequence of migration, admixture, natural selection, pathogen load, or cultural practices [24−32]. These give rise to enormous combinatorial possibilities whose effects impact the entire system [33−41]. There is a further impact of environmental and epigenetic changes and also the enormous human microbial diversity on the phenotype of an individual [24,27,42−48]. Each individual is thus an ecosystem harboring a unique subset of variations and the phenotype of an individual is the net outcome of the ecosystem. This unanticipated extent of human genome variations has nearly ruled out the possibility of defining or reconstructing a reference healthy human from mere reading of the genomic sequences [23,49,50].

A comprehensive assessment of individuality that encompasses different systems and connects it to outcome in health and disease and their relation with personalized therapeutics is not yet available. A primary challenge remains in first defining a healthy individual as we still use disease state as a reference for defining health and not health as a reference point since there are no comprehensive methods for stratifying healthy individuals. It is well acknowledged that 1−5% of the population is at risk of one or the other complex disorder and also the frequency of any monogenic disorder if we consider all the Mendelian disease genes is also 1%. The hope is that if we were able to identify these predisposed individuals at preclinical stage, prevention and management of disease could become much more tractable and so also the burden of the disease could be reduced substantially in Mendelian disease through early interventions. Most of the genetic markers discovered till date are not useful as predictive markers in preclinical stage.

AYURVEDA: ANCIENT INDIAN SYSTEM OF MEDICINE

Contemporariness of the Practice as P5 Medicine

Ayurveda is an Indian system of life sciences, documented, and practiced since 1500BC with personalized approach to predict, prevent, and promote the state of health in healthy and alleviation of disorder in the diseased [51]. Understanding of human individuality through assessment of his/her constitution type forms the fundamental basis for P5 medicine. According to Ayurveda an individual is born with a specific constitution *Prakriti* that not only determines his overall phenotype but also predicts the susceptibility to diseases and responsiveness to extrinsic and intrinsic [51–54,54a]. Ayurveda describes the subject matter through "TRISUTRA," meaning the three interconnected axes of causes (*Hetu*), features (*Linga*), and therapeutics (*Aushadha*) both for healthy and diseased person [51,54]. *Hetu*, the causes of diseases documented are from the lifestyle, dietary regimen, and thought process that affect the behavior of various metabolic pathways. These are noticed as signs and the alteration of disturbed metabolic pathways is done with natural interventions from required adjustments in lifestyle; dietary regimen, detoxification with *panchakarma* therapeutics, use of herbal compounds depending on the nature and state of disease and strength of the diseased compared to the level of baseline health state of the person. It also takes care of particular geo-climatic environmental variations during treatment.

It is an often asked question as to how a system of medicine which was practiced 5000 years back would be relevant even today since some of the diseases seem to be the effect of modern day living and unlikely to have been observed in the ancient times. It might be worthwhile to mention that population genomics methods that has substantiated the evidence for existence of humans for nearly 100,000 years has also allowed us to trace the origin and spread of diseases across the world and many of the infectious disease have been traced to be old as 8000–10000 years back [55]. So it is not inconceivable that 5000 years which would be substantially recent in evolutionary terms, these diseases would have been nonexistent or not prevalent in India. Some of the founder mutations linked to diseases reported in Indian populations are also shared with African and other world populations [56]. Surprising as it may sound the phenotypic features of most of the diseases that are described contemporarily have descriptions in the ancient texts. The similarity is more at the feature level than at the syndrome level. The practice covers all the aspects of P4 with an additional promotive component (P5).

- Prediction of disease susceptibility and responsiveness to diet, drug, and environment in an individual right from the time of birth
- Prevention through identification of actionable points for early interventions
- Personalized based on an individual's constitution in health and diseased condition
- Promotive for optimizing the homeostatic and rejuvenating potential of the system
- Participatory through engagement of an individual in his/her own health management through awareness and proactive reporting of the same.

Potential of Ayurveda in Chronic and Complex Diseases

Potential of the Ayurveda in the treatment of chronic and complex lifestyle originated disease has been realized world over. There are reports mentioning use of Ayurveda by ~60% of the world populations at some point of time or the other during the treatment of these diseases. Ayurveda describes the methods of deciding the line of treatment after assessing the prognosis of a disease. All diseases are classified into sub-types based on their being curative or palliative nature which is further decided on the basis of strength and severity of disease as well as that of an individual's strength. Most of the times for chronic and complex diseases a patient approaches an Ayurveda physician with an ongoing allopathic medical treatment. Ayurveda physician treats every individual based on the subtype, stage, and chronicity of the disease, patient's constitution with additional consideration of outcome of the ongoing treatment. In general, it is observed that when an individual adopts Ayurveda there is an improvement in the quality of life with a better control of the disease with ongoing allopathic medicines. Not only this, a fraction of the patients get cured and also a large majority can get better control of disease only with Ayurveda treatment. In a number of instances, for example in case of thyroid problems though the level of thyroid can be reduced by modern medicines sometimes the primary causes for which a thyroid test is conducted such as problems of dryness of skin, constipation, muscular pain, frequent attacks of cough and cold, anxiety are not alleviated. It has been observed that a combination therapy of Ayurveda along with modern medicine can alleviate or totally cure all the problems in 1−3 months. Ayurveda has also been useful in treating asthma, diabetes, diabetic nephropathy, chronic heart failure (CHF), high blood pressure, thyroid, hypercholesterolemia, fatty liver diseases, psoriasis, chronic sinusitis, migraine, rheumatoid arthritis, gout, multiple sclerosis, Parkinson's disease, Alzheimer's disease, prostate enlargement, paralysis, etc.

It might be worthwhile to mention that most of the discoveries of drugs such as levodopa, reserpine, chloroquine, aspirin, codeine, vincristine, vinblastine, bromhexine, digitalis, etc. have their origins in Ayurveda [57−61]. Besides, the same source herbs are used to treat these diseases in Ayurveda. Knowledge of usage of the herbs, their method of preparation, formulation, and routes of administration for different diseases is extensively documented in Ayurveda and is also available in Traditional Knowledge Digital Library developed (http://www.tkdl.res.in/) and maintained by the Indian government. In some cases where it is not possible to cure the disease, food regimen/therapeutic diet/daily usable herbal products can help maintain their healthy state along with the advised lifestyle changes. There are some diseases for which there is hardly any treatment available in modern system of medicine, like inflammatory bowel disease (IBD), ulcerative colitis, NAFLD (Nonalcoholic fatty liver disease), some viral diseases like Hepatitis B, Hepatitis C, etc. where people seek Ayurveda as an option.

Principles of Ayurveda in Practice
Human Individuality as a Primary Basis
Understanding of human individuality through assessment of his/her constitution type forms the fundamental basis for P5 medicine. According to Ayurveda an individual is born with a specific constitution *Prakriti* that not only determines his overall

phenotype but also predicts the susceptibility to diseases and responsiveness to extrinsic and intrinsic environment [53,54]. *Prakriti* is determined by the relative proportions of three physiological entities *Tridoshas* in an individual which forms the common organizing principle. The three entities *Vata*, *Pitta*, and *Kapha* govern and determine the kinetic, metabolic, and structural components of the system respectively that are established at the time of birth and remain invariant throughout the lifetime in an individual [51,53,54,54a]. The ethnicity, geography, heritability, age of the transmitting parents, and intrauterine conditions contribute to the relative proportion of *Tridoshas* in an individual and determines the constitution type of an individual. This can be assessed through a comprehensive analysis of anatomical, physiological, metabolic, and psychological attributes. Individuals of a population can be broadly stratified into seven broad constitution types *Vata*, *Pitta*, *Kapha*, *Vata-Pitta*, *Pitta-Kapha*, *Vata-Kapha*, and *Vata-Pitta-Kapha* based on the relative proportions of *doshas* [53,54]. There is a broad continuous spectrum of healthy states in which individuals fall and can be categorized into some groups with three of them described to be at the end of the spectrum. Any perturbation from the baseline state of *doshas* in a particular constitution type leads to diseases and the purpose of Ayurveda interventions is to bring back the *doshas* to the homeostatic threshold. Though the proportions of *doshas* are invariant in an individual, they fluctuate within an allowable range during different times of the day, season, and age of an individual. These are taken into account during *Prakriti* assessment. Their levels could also be modulated by geo-climatic conditions, food, and drugs as each of these are also described to impact levels of different *doshas* [54a].

Maintenance of Health

Dynamic state of health is maintained through personalized recommendations of diet, exercise, rest, sleep, and other lifestyle practices including yoga with respect to time and amount based on an individual's *Prakriti* considering his/her age, place, season, etc. This also includes special care of all body orifices including skin as well as periodic cleansing of entire body during the rhythmic peaks of VPK (*Kapha* during spring; *Pitta* during autumn, and *Vata* during monsoon season) to prevent accumulation of excess toxins and other excretory metabolites. This is carried out following specific protocols which includes a preparatory phase for the system to expel out the toxins, followed by a post procedural care that ensures the proper restoration and rejuvenation. This preemptive approach of Ayurveda toward maintenance of health is aimed at preventing the manifestation of diseases to which an individual is predisposed [51,54a,62].

Ayurveda ad vocates special types of therapy (*Rasayana and Vajikarana*) for enhancement of regenerative potential and reproductive health of an individual. These therapies are administered only after cleansing the body of toxins, and accumulated *Tridoshas*. By definition, this therapy is meant to enhance the strength and robustness of the systems by augmenting their cellular functions. It improves the higher functions of brain and mental faculties like cognition, memory, speech, intelligence, etc. It thus acts as a preventive therapy for aging and age related disorders, increasing the longevity in an individual. This at times is also administered in the advanced stages of diseases where the recovery from them is expected to come through tapping the regenerative potential of the system rather than through corrective mechanisms of drugs [51,54a,62,63].

Personalized Management of Diseases

The descriptions of the diseases are in terms of perturbations of *Vata*, *Pitta*, and *Kapha* and their manifestation in various systems. Classification of diseases is not organ based in Ayurveda, although all the organs are documented in the text and also in the respective diseases pathophysiologies. The aim of treatment is alleviation of disorder in the manner that does not provoke the pathogenesis of others or disturb the healthy tissues. Ayurveda describes clinical examination points pertaining to disease and diseased, by physician in order to analyze not only the nature and strength of disease, but also that of the diseased to select the line of treatment and decide if the therapy and the drug as well as dosage advocated for the treatment would at all be tolerated by him. Thus a triad of drug, disease, and diseased is analyzed to arrive at the right combination for an individual [54a].

- Examination of variables related to disease includes clinical presentation with subtypes, severity and stage, strength and multiplicity of triggers, etiological factors of disease—extrinsic and intrinsic
- Affected individual related-baseline-*Prakriti*, suitability toward therapy and drugs, and age
- Present status of health—physical, psychological, and individual's present status of metabolism and waste clearance organs, including external environmental factors like geo-climatic and time.

Indian Traditional Living

Traditional medicine is very much the part as a domestic remedy in Indian society where locally available herbs are used to treat various diseases and usually this knowledge is acquired with the use in any of the family or community. Some of the modalities of preventive treatment recommendations of Ayurveda have become a part of kitchen in every house of Indian sub-continent as herbs and spices.

These are used in traditional preparations to impart flavor and have led to an extensive diversity of food across the country. Usage of herbs and spices in food ensures maximization of the range of nutrients consumed on a regular basis as well as increased consumption of vegetables through a variety of preparations. Besides, the usages of these herbs have also been demonstrated in decreasing food poisoning as well as protection of the food from early decompositions. Similar to the medicines described earlier, their usage may differ with the sociocultural and geo-climatic conditions however their usage is universal. The most common herbs are turmeric, cinnamon, cardamom, cumin seeds, dried ginger, pepper, red chilies, curry leaves, and use of rock salt. These herbs and spices have been described for their medicinal properties in Ayurveda [51,62,63]. Most of them have been scientifically proven to be effective in prevention of various allergic, metabolic, cognitive, and degenerative disorders all of which are closely associated with increased oxidative processes [58,64,65]. Limited clinical trials in humans as well as extensive studies in animal models have revealed the antioxidant as well as anticarcinogenic properties of herbs and spices [65,66]. For instance, in cancer it has been shown that a pro-inflammatory stimulus leading to increased oxidative stress could result in DNA damage, breaks in chromosome, telomere shortening which ultimately leads to chromosomal instability and increased cancer risk [42]. These pro-inflammatory stimuli also activate mitogen activated protein kinase leading to activation of NFkappa B which in turn increases the expression of cyclooxygenase-2 resulting in cell proliferation and increase cancer risk. Herbs and spices have been shown to inhibit many of these steps of carcinogenesis. Other herbs such as garlic have been found to be useful in atherosclerotic conditions and blood clotting, ginger in alleviating arthritic knee pain and is considered as a nonsteroidal anti-inflammatory alternative to ibuprofen, etc. [58,67]. Thus herbs and spices which have been used since ancient times not only in India but in other ancient civilizations would be what is considered as a part of functional foods in the contemporary times.

Besides, as a part of routine practice, regular breathing exercise and yoga, meditation, social functions, fasting, long walks by way of annual pilgrimage and a routine in sleep and dietary practice during different times of the year have been advocated and practiced in traditional households as cultural practice since centuries.

Practice of Ayurveda in Most Indian Communities Despite Sociocultural Variation

Ayurveda is practiced across the Indian sub-continent and nearly 70—80% of rural India uses it as the preferred primary health care system. On the other hand, in urban settings where people have access to modern allopathic medicine avail Ayurveda for cases where there is little or no help from modern medicine is available, for example in chronic and complex diseases as well as terminally ill patients with no relief possible from allopathy. In contrast to modern medicine, the globalization of the Ayurvedic practice entails adjustment for geo-climatic and sociocultural variations for better suitability. Thus, so far as principles of understanding the disease, their diagnosis, and treatment are concerned there is not much difference across communities. Ancient classical texts, *Charaka* and *Shushruta Samhita* are followed by physicians across the country. However shorter/easier compiled versions like *Ashtanga Samgraha*, *Ashtangahridhyama*, *Sarangdhara*, *Madhava*, and other about 12 *Nighantus* and *Bhaishajya*

Ratnavali, Chakradutta, Sahasrayoga, etc. have been developed in different parts of the country to make it easier to understand and practice with the need of time. In clinical practice, in the southern part of India most of Ayurveda institutes including pharmacies use Sahasrayoga as reference text, in western part *Ashtanga Hridayam* is commonly followed whereas northern part of India *Charaka Samhita* and *Bhaishajya Ratnavali* remains main texts of reference. Thus when it comes to delivery of health care and recommendations for healthy life, different communities use Ayurveda with some variations to suit biological requirements of those populations based on geo-climatic conditions and sociocultural practices. The variations are also sometimes noted because the same medicinal preparations are called by different names in different communities. Also the variations are not always done in the choice of medicines rather there are different formulations/dosage forms developed from same combinations. For example, medicated oils and decoctions are more common in south India than in northern or western parts of India, whereas in northern India preferred formulations are easy to take dosage forms like tablets, capsules, etc. It would be worthwhile to mention that Ayurveda's claim of being a personalized medicine matches with the descriptions of diverse ways in which the same disease can be managed in order to suit the person in a particular place, age, and strength.

In an attempt to make Ayurveda treatments acceptable, modern diagnostic methods are also integrated by Ayurveda physicians in their practice as they provide clinical, biochemical, or radiological markers that are helpful in monitoring the disease as well as effect of treatment. Although this also has its disadvantages as the limits of diagnostics would be limited to the extent measurable or defined in the modern times. A very illustrative example is diabetes which is commonly measured by blood sugar levels. In Ayurveda, diabetes is classified as one of the subgroups of a class of diseases under *Prameha* and in this class, subtyping of the subject besides clinical evaluation is also carried out by looking at the properties of urine in terms of color, odor as well as the dispersion behavior (specific gravity) of a drop of oil in urine of the subject. In modern language it could well be the outcome of metabolites in the urine which had been used for diagnosis and measurable by metabolomics. In the prevalent practice of Ayurveda, in majority of the clinics the ancient practice of detection has been replaced by blood sugar estimation.

GETTING AYURVEDA TO MAINSTREAM

Perception of the Common Man

Understanding and awareness about the practice of Ayurveda is not uniform across the cross section of population including common people both from rural and urban background, modern medicine practitioners, academic and research faculty, public health scientists as well as policy makers. A clear demarcation does not exist in the mind of people of what is herbal, traditional or Ayurvedic and therefore all these are perceived similarly. This has created a collateral damage since the perception has propagated the belief that anyone can practice Ayurveda, any herbal medicine off the shelf is Ayurveda and a recommendation from a neighbor who had a success with some medicine can be taken without consulting a clinician. Since Ayurveda-based herbs are used as spices as well as medicines, when used as latter, the recommendation has to be made under clinical supervision is not very evident to a common man. The most common perception about Ayurveda in western world is that it is a kind of traditional knowledge maintained and passed on as a grandmother's recipe. This mind set prevails because of the ignorance about the existence of its extensive documentation and scientific literature. Although originally written in Sanskrit, translations in several languages are available today. At times practice of Ayurveda is also equated with the practice of Hindu religion since all ancient Indian literary work including religious texts referred in religious ceremonies of Hindu household are also written in Sanskrit, which was the prevailing language of that time. Over the last 100–150 years, this system of medicine which was once mainstream is now the 'alternative or complementary system' of medicine. There are historical, socioeconomic, and political reasons which can be ascribed to these gradual loss of connection of modern Indian community with this system of medicine. Today, most often exposure to Ayurveda in school is through a general knowledge question related to Charaka and Sushruta from ancient India. The prevailing perceptions shrouding Ayurveda also emanates from a lack awareness of its success stories. These are not available in mainstream in a way that a large section of society can understand and acknowledge the practice. At this point recommendations are more by word of mouth, by claims of practitioners or in some cases of prominent personalities who are not qualified Ayurveda practitioners but can influence a large section of society.

Acceptability of Ayurveda in Health Care
Attitude of the Patient
In most cases, a patient approaches an Ayurveda practitioner in an advanced stage of a disease when the cost of treatment become prohibitive or living with the disease affects the quality of life. The hope and anticipation from the system is for a safe, speedy, and permanent cure.

However, the approach to adopting this system is taken with mixed apprehensions stemming from awareness about the potentials on one hand and the belief created by studies which has demonstrated the presence of heavy metals in Ayurveda products. Another aspect which has also got into our mindset is our faith that a long prescription of drugs can only cure the disease or that an overcrowded medical hospital only demonstrates that people are getting healed. The importance of adopting changes in lifestyle practices, for example in diet and lifestyle routine as well as exercise is secondary. Since the practice of Ayurveda encompasses a combination of medicine and change in lifestyle practices which is very much closer to our traditional lifestyle, it does not sound contemporary and patients are willing to adopt only the prescription part of the practice. Now with the social media in place where it is possible to do a google search for anything, many are still not able to get a satisfactory answer as to how this medicine can claim to cure cold and cancer and that too by the same practitioner; why would one get treated by such an ancient system when already state-of-the-art systems exist in modern times; how could a system be expected to be successful when it has not worked in big hospitals; why the aid of modern diagnosis for Ayurveda practitioners and more importantly how it could treat diseases which might not have been described in Ayurveda since these are emerging diseases and unlikely to be present 5000 years back. There are no ready references available to connect the modern and Ayurveda sciences. Also even if none of the modern medicines is known to cure a chronic illness for instance diabetes, asthma, Parkinson's, rheumatoid arthritis and cancer, most patients often do not question the curability of the modern medicine. However, this is the first question that is posed to an Ayurveda physician and there is a need for reassurance for the same from the practitioner about a permanent cure before the patient avails the treatment.

Another aspect often encountered is when an individual patient takes both Ayurvedic and modern drugs, the clinician might not be aware of this since in most cases this information is not shared. Whilst the Ayurvedic practitioner might have some knowledge of the nature of modern drug, in most cases, the modern western style clinician would be unlikely well informed. It is also perceived in general that since Ayurveda is an herbal-based medicine the cost of treatment should be very low whereas sometimes it is observed that is not the case. There could be more than one reason for the same. One, some of the Ayurveda herbs used for treatment of chronic diseases are not readily available and also the methods of preparations of formulations are not trivial. In most chronic and complex disease, for personalized therapeutic interventions it may be required to prepare medicine for the given patient, which might make it more expensive. However if the overall cost of therapy is evaluated against its outcomes in terms of recovery or prevention of complications that might require costly medical care and/or hospitalization, improvement in quality of life, abstinence from work for the patient as well as family attending to the person, etc., it may be found that overall cost of treatment becomes much lesser. This would also be true in case of treatment that aims toward maintenance of health and prevention of disease to which an individual may be predisposed or for better quality of life so as to prevent future therapeutic interventions, frequent hospitalization and loss of work hours. Thus, one needs to consider the higher cost of treatment *vis-a-vis* the type of health conditions Ayurveda caters to along with the overall outcomes.

When diagnosed with a complex and lifestyle related disease, a substantial fraction of people who are aware of the outcomes of disease and treatment, opt for Ayurvedic treatment. They take this decision with the understanding that Ayurveda treatment will have no or less number of side effects even when taken for long time. Also many of them are ready to adopt lifestyle and dietary changes that could help contain the disease. In fact, some patients in their own words say that in spite of not having much difference in the levels of blood sugar they feel healthier. Some patients are also relieved of the fact that the generalized recommendations that follow after a diagnosis of a disease for instance no exercise on treadmill because of tiredness or avoiding sweets in chronic diabetes can be managed to a large extent by combining treatment with Ayurveda. In some cases after a long usage of Ayurveda drugs along with modern medicine, patients are prevented from progressing to complications of the disease for instance progression to fatty liver diseases due to poorly controlled cholesterol levels even after a long usage of cholesterol lowering drugs. In this category, many people have a family history of disease and therefore seek such ayurvedic medicines as a preventive measure/or at least help in delaying the process of occurrence of that particular disease by adopting a biological required lifestyle and dietary guidelines to minimize the chances of occurrence of diseases guided by Ayurveda physician.

Yet another fraction of individuals adopt this system as a last ray of hope. There are some diseases for which although treatment options are available in modern medicine like surgery or chemotherapy in cancer or switch from oral hypoglycemic drugs to insulin in case of poorly controlled diabetes, dialysis in chronic renal failure or steroid therapy in multiple sclerosis, patients are unwilling to go in for these options and resort to Ayurveda. Among all these conditions there are times when a stage comes, that there is very little survival time left and no hope available from modern medicine side, patient think of Ayurveda for ease of life. Though it may sound a little surprising, there are many instances when people have been living longer than was expected and that too with much more ease. A tag of adopting a Vedic lifestyle in modern era has become a status symbol/matter of boasting about the uses of natural and organic food products along with yoga for a substantial fraction of people all over the world including India especially in upper classes, and this trend is increasing very fast. Though this has gained mass appeal, a suitability of the usage of natural products as well as the amount of exercise as recommended in Ayurveda is not followed.

Education and Research in Ayurveda

It is a common perception even among clinical practitioners and researchers that the education and research in Ayurveda is not exhaustive or systematized. Whereas, the undergraduate degree in Ayurveda "BAMS" (Bachelor of Ayurveda Medicine and Surgery) is awarded after five and half years course including 1 year compulsory internship similar to modern medicine graduation "MBBS." Almost all subjects, anatomy, physiology, pharmacology, pathology, medicines, etc. that are a part of the undergraduate program of MBBS in India are part of syllabi of Ayurveda degree course in addition to the study of these subjects from Ayurveda point of view. Also the standard of education is regulated and governed by Central Council of Indian Medicine (CCIM), Department of AYUSH, Government of India. Postgraduate course in Ayurveda "MD" is a 3 years training and research program. In India there are more than 250 educational institutes offering Ayurveda graduation/postgraduate courses. There are a large number of MD and PhD research programs in Ayurveda that carry out research toward understanding the principles and practice described in Ayurveda. There are annual conferences and meetings where these are presented and discussed. However, this is mostly restricted to the Ayurveda fraternity and the language of these discussions is mostly based on Sanskrit textual references and also the publications are primarily restricted to University journals and books with few peer reviewed journals in the field. A large fraction of Ayurveda clinicians either get into public health, clinical trials, drug discovery programs as a resource person for providing herbs or in some cases also get into practicing modern medicine. There is a very limited scope for a crosstalk between the two medical fraternities and one can rarely spot either of the practitioners making case presentations in the other audience. Unavailability of a ready reference in modern language has been one of the major reasons for unawareness of the extent of documentation encompassing all areas of P5 medicine.

AYURGENOMICS: APPROACH FOR INTEGRATION OF AYURVEDA INTO CURRENT MEDICAL PRACTICE

As described and highlighted in the earlier sections, both the modern and the Ayurveda system of medicine have merits which can be complemented for mutual benefit and for the society at large. The science of Ayurveda needs to be understood in totality such as a "TRISUTRA" approach in genomics encompassing a broader spectrum for its usage. If this is achieved then there is a potential for development of innovative solutions for finding newer targets and medicines to reverse the chronic complex disorders. This aspect is claimed in Ayurveda Institutes and centers in India but for its globalization, applicability as well as for demonstrating the uniformity in results on the scientific basis as is described in the texts needs to be established by transdisciplinary approaches. Some attempts have been done in this direction in drug discovery programs. However these studies just use the drug enlisted for the treatment of some diseases and it mostly entails identification of the most active molecule in that herb on modern disease models.

Definition of baseline health state with a noninvasive phenotypic assessment as described in Ayurveda if correlated in the molecular terms could be one of the most affordable way of preventive and personalized management of health. The objective measures of anatomical, physiological assessments available in modern medicine could be integrated with the clinical methods of Ayurveda for giving objectivity to the more subjective assessments. There is not only an enormous data generation possibility in genomics but also methods are being evolved for pinpointing precise targets in pathogenesis of disease in modern medicines. The integration of these strength areas of modern day sciences with the documented comprehensive applied aspect of personalized approach in evaluating the state of health and disease and the relation of practical issues not only from diet and nutrition perspective but also, lifestyle, daily-seasonal routines, exposure to sense objects, etc. from Ayurveda can provide holistic health solutions without much untoward effects. However, this would need de-convolution of the intricate and exhaustive literature of language of Ayurveda.

In order to integrate the concepts of Ayurveda into modern health practices, the first step would be to use a shared vocabulary to denote the properties and interrelationships of these concepts (shared ontological descriptions) in the language of system biology or modern network medicine. A primary place to initiate this crosstalk could be integration of *Prakriti* concepts with Genomics for understanding human individuality. Evidence for this potential has already been provided in the first of its kind genomics study on extreme constitution types of Ayurveda [53]. This Ayurgenomics approach provided the molecular and genomic correlates of *Prakriti* [53]. It also highlighted that the normal range of biochemical parameter for different constitution types may be different in a population and so also their subclinical ranges. It might well be that what is subclinical might need to be redefined amongst healthy individuals in a population. However, this has to be validated across diverse ethnic populations from different geographical regions for the method to be globally applicable.

It was also hypothesized that an Ayurgenomics approach could help identify axes of variation that can help predict risk for diseases and enable predictive marker discovery. Using this method, we discovered a genetic variation in an oxygen sensor gene *EGLN1* that differed between *Prakriti* types and conferred differences with respect to high altitude adaptation and susceptibility to high altitude pulmonary edema (HAPE) [68]. Further, extension of this study to a cellular model also demonstrated how difference in this axes could be relevant in modulating asthma [69]. This gene is now being reported as a therapeutic target in diseases such as recovery from stroke, cartilage repair, outcome of visceral surgeries or in cancer progression where modulation of hypoxia has been shown to be important [70–77]. Therefore, it seems possible to explore the three axes of cause (*Hetu*), consequence (*Linga*), and therapy (*Aushadha*) in a common molecular language if we are able to integrate both the sciences through a common genomics framework [77a]. Thus this TRISUTRA framework of (i) generation of data by using the samples selected with *Prakriti* concept of Ayurveda in understanding the health markers and process, (ii) integration with the functional and disease genomics approach, and (iii) validation in cellular models can provide clinically actionable points.

Understanding of mechanism of therapeutic action of Ayurveda interventions as well as drugs is also essential for the acknowledgment of the scientific basis and its acceptability and adoption at the global scale. In parallel it could also result in the development of genomic signatures that would enable monitoring not only the quality but also the therapeutic outcome. Most often a test result and an assurance for the quality of the drug are the most critical aspects of a treatment. At present, the phenotypic outcome of a treatment of Ayurveda is monitored based on measurement of an outcome that is described in modern medicine. This has caveats. In Ayurveda, similar to modern medicine, diabetes is classified into different subtypes and the prognosis as well as treatment differs between the subtypes. Use of glycated Hb1ac which is a sensitive (not specific) measure for assessing chronic glycemia in monitoring a diabetic patient or a therapeutic outcome in Ayurveda system may impact its validation as well as acceptability. Noteworthy, it is now being acknowledged that glycated Hb1ac is present in a fraction of individuals who are not diabetic and it is felt that measurement of this would not be useful for assessing diabetes in specific endotypes.

The major emphasis of *panchkarma* therapy is on detoxification and rebuilding the homeostatic state of the system through dietary as well as therapeutic regimes that are likely to work through restoration of the microbiome. The emerging area of metagenomic studies has revealed altered microbiome dynamics in a large number of clinical conditions such as obesity, arthritis, psoriasis, multiple sclerosis, and inflammatory bowel disease (IBD) etc. [45,46,78]. This has opened up new opportunities in modulating the microbiome through dietary interventions [79−83]. Integration of these well-developed methods with Ayurveda interventions might also allow identification of metagenomic signatures that could be used to monitor the outcome/success of the ayurvedic treatment.

Another important aspect would be to develop a resource that would allow integration of already available large amount of human genomic variation with the practice of Ayurveda for preventive health maintenance. Lack of confidence and trust in routine and traditional dietary practices as a consequence of globalization is now being considered to be a cause of many emerging diseases, like obesity, depression, common cancers and probably some neuro-degenerative diseases. Some recent reports provide compelling evidence of the importance of maintenance of sleep timings, being in synchrony with circadian rhythms, benefits of exercise, and meditation in recovery from diseases as well as importance of fasting in detoxification process

[84–86]. Evidence for these has been provided in epidemiological as well as model system and molecular studies [87–90].

Whilst there is overwhelming evidence for the revival and reshaping the traditional Ayurvedic medicine, this would need developing medium to long term strategic projects centered around genomics and judicious use of the next generation sequencing methods in developing specific databases for lifestyle, common diseases and response to expensive and life-saving medications. This genomic information resource could then be put together as a package with the Ayurveda in the acceptable format for Ayurgenomics.

SUMMARY

To conclude, integration of genomics in Ayurveda could help in:
1. Determination of baselines of health in an individualized manner that would be important in management of health and disease
2. Identification of predictive (genomic biomarkers) markers and noninvasive measures that would be important in assessing health and disease in diverse populations
3. Development of personalized therapeutic (drugs, nutrition and lifestyle related) recommendations for good quality life in health and disease
4. Increase the awareness and participation of individual in his/her own health care.

There are many challenges that need to be overcome. A primary one is to create awareness which would include redefining it in a modern language to remove the perception and biases linked to the Ayurveda practice. This would enable us to develop a critical mass that is needed to embark on a diverse genomic research into this area to validate the scientific basis of Ayurveda and translate the concepts for affordable health care solutions. This would need research to reduce the gap between the two sciences, dissemination and sharing information of success stories in a language that sounds scientific and convincing by the current definition, appreciation of weakness, and strengths of both the modern and Ayurveda clinicians. It is important that genomic scientists, modern medical practitioners and the Ayurveda medical practitioners engage in an effective and positive manner and develop strategically important projects utilizing the next generation sequencing methods, bio-informatic tools and other skills. The future of Ayurgenomics would largely depend on these developments.

ACKNOWLEDGMENTS

We would like to acknowledge CSIR for funding the work (MLP901). We would also like to thank Dr Ram Niwas Prasher (MD Ayurveda) for extensive discussion on clinical practice of Ayurveda, valuable suggestions and critical comments, Dr Bharat Khuntia and Atish Gheware in help with manuscript editing.

REFERENCES

[1] Botstein D, Risch N. Discovering genotypes underlying human phenotypes: past successes for Mendelian disease, future approaches for complex disease. Nat Genet 2003;33(Suppl.):228−37.

[2] Chakravarti A. Genomic contributions to Mendelian disease. Genome Res 2011;21(5):643−4.

[3] Yang Y, Muzny DM, Reid JG, et al. Clinical whole-exome sequencing for the diagnosis of Mendelian disorders. N Engl J Med 2013;369(16):1502−11.

[4] Bamshad MJ, Ng SB, Bigham AW, et al. Exome sequencing as a tool for Mendelian disease gene discovery. Nat Rev Genet 2011;12(11):745−55.

[5] Gilissen C, Hoischen A, Brunner HG, Veltman JA. Disease gene identification strategies for exome sequencing. Eur J Hum Genet 2012;20(5):490−7.

[6] Eichelbaum M, Ingelman-Sundberg M, Evans WE. Pharmacogenomics and individualized drug therapy. Annu Rev Med 2006;57:119−37.

[7] Evans WE, Relling MV. Pharmacogenomics: translating functional genomics into rational therapeutics. Science (New York, NY) 1999;286(5439):487−91.

[8] Wheeler HE, Maitland ML, Dolan ME, Cox NJ, Ratain MJ. Cancer pharmacogenomics: strategies and challenges. Nat Rev Genet 2013;14(1):23−34.

[9] Wall AM, Rubnitz JE. Pharmacogenomic effects on therapy for acute lymphoblastic leukemia in children. Pharmacogenomics J 2003;3(3):128−35.

[10] Li J, Wang S, Barone J, Malone B. Warfarin pharmacogenomics. P & T 2009;34(8):422−7.

[11] Topol EJ, Schork NJ. Catapulting clopidogrel pharmacogenomics forward. Nat Med 2011;17(1):40−1.

[12] Yin T, Miyata T. Pharmacogenomics of clopidogrel: evidence and perspectives. Thromb Res 2011;128(4):307−16.

[13] Vanfleteren LE, Kocks JW, Stone IS, et al. Moving from the Oslerian paradigm to the postgenomic era: are asthma and COPD outdated terms? Thorax 2014;69(1):72−9.

[14] Feramisco JD, Sadreyev RI, Murray ML, Grishin NV, Tsao H. Phenotypic and genotypic analyses of genetic skin disease through the Online Mendelian Inheritance in Man (OMIM) database. J Invest Dermatol 2009;129(11):2628−36.

[15] Faner R, Tal-Singer R, Riley JH, et al. Lessons from ECLIPSE: a review of COPD biomarkers. Thorax 2014;69(7):666−72.

[16] Agusti A. Phenotypes and disease characterization in chronic obstructive pulmonary disease. Toward the extinction of phenotypes? Ann Am Thorac Soc 2013;10(Suppl.):S125−30.

[17] Barabasi AL, Gulbahce N, Loscalzo J. Network medicine: a network-based approach to human disease. Nat Rev Genet 2011;12(1):56−68.

[18] Lander ES, Linton LM, Birren B, et al. Initial sequencing and analysis of the human genome. Nature 2001;409(6822):860−921.

[19] Venter JC, Adams MD, Myers EW, et al. The sequence of the human genome. Science (New York, NY) 2001;291(5507):1304−51.

[20] Hood LaF SH. Predictive, personalized, preventive, participatory (P4) cancer medicine. Nat Rev Clin Oncol 2011;8:184−7.

[21] Agusti A, Sobradillo P, Celli B. Addressing the complexity of chronic obstructive pulmonary disease: from phenotypes and biomarkers to scale-free networks, systems biology, and P4 medicine. Am J Respir Crit Care Med 2011;183(9):1129−37.

[22] Gibbs WW. Medicine gets up close and personal. Nature 2014;506(7487):144−5.

[23] Consortium GP. An integrated map of genetic variation from 1,092 human genomes. Nature 2012;491(7422):56−65.

[24] Coop G, Pickrell JK, Novembre J, et al. The role of geography in human adaptation. PLoS Genet 2009;5(6):e1000500.

[25] Fu W, Akey JM. Selection and adaptation in the human genome. Annu Rev Genomics Hum Genet 2013;14:467−89.

[26] Fumagalli M, Sironi M, Pozzoli U, Ferrer-Admettla A, Pattini L, Nielsen R. Signatures of environmental genetic adaptation pinpoint pathogens as the main selective pressure through human evolution. PLoS Genet 2011;7(11):e1002355.

[27] Hancock AM, Witonsky DB, Alkorta-Aranburu G, et al. Adaptations to climate-mediated selective pressures in humans. PLoS Genet 2011;7(4):e1001375.

[28] Jablonski NG, Chaplin G. Human skin pigmentation as an adaptation to UV radiation. Proc Natl Acad Sci U S A 2010;107(Suppl. 2):8962–8.

[29] Tishkoff SA, Reed FA, Ranciaro A, et al. Convergent adaptation of human lactase persistence in Africa and Europe. Nat Genet 2007;39(1):31–40.

[30] Patterson N, Moorjani P, Luo Y, et al. Ancient admixture in human history. Genetics 2012;192 (3):1065–93.

[31] Consortium GP. A map of human genome variation from population-scale sequencing. Nature 2010;467(7319):1061–73.

[32] Tishkoff SA, Verrelli BC. Patterns of human genetic diversity: implications for human evolutionary history and disease. Annu Rev Genomics Hum Genet 2003;4(1):293–340.

[33] Hawkins RD, Hon GC, Ren B. Next-generation genomics: an integrative approach. Nat Rev Genet 2010;11(7):476–86.

[34] Baryshnikova A, Costanzo M, Myers CL, Andrews B, Boone C. Genetic interaction networks: toward an understanding of heritability. Annu Rev Genomics Hum Genet 2013;14(1):111–33.

[35] Jeong H, Tombor B, Albert R, Oltvai ZN, Barabási A-L. The large-scale organization of metabolic networks. Nature 2000;407(6804):651–4.

[36] Lusis AJ, Weiss JN. Cardiovascular networks systems-based approaches to cardiovascular disease. Circulation 2010;121(1):157–70.

[37] Park J, Lee DS, Christakis NA, Barabási AL. The impact of cellular networks on disease comorbidity. Mol Syst Biol 2009;5(1).

[38] Vidal M, Cusick ME, Barabasi A-L. Interactome networks and human disease. Cell 2011;144 (6):986–98.

[39] Rual J-F, Venkatesan K, Hao T, et al. Towards a proteome-scale map of the human protein–protein interaction network. Nature 2005;437(7062):1173–8.

[40] Stelzl U, Worm U, Lalowski M, et al. A human protein-protein interaction network: a resource for annotating the proteome. Cell 2005;122(6):957–68.

[41] Pan Q, Shai O, Lee LJ, Frey BJ, Blencowe BJ. Deep surveying of alternative splicing complexity in the human transcriptome by high-throughput sequencing. Nat Genet 2008;40(12):1413–15.

[42] Jackson SP, Bartek J. The DNA-damage response in human biology and disease. Nature 2009;461 (7267):1071–8.

[43] Hancock AM, Witonsky DB, Gordon AS, et al. Adaptations to climate in candidate genes for common metabolic disorders. PLoS Genet 2008;4(2):e32.

[44] Dolinoy DC, Jirtle RL. Environmental epigenomics in human health and disease. Environ Mol Mutagen 2008;49(1):4–8.

[45] Cho I, Blaser MJ. The human microbiome: at the interface of health and disease. Nat Rev Genet 2012;13(4):260–70.

[46] Pflughoeft KJ, Versalovic J. Human microbiome in health and disease. Ann Rev Pathol 2012;7:99–122.

[47] Consortium HMJRS. A catalog of reference genomes from the human microbiome. Science 2010;328(5981):994–9.

[48] Consortium HMP. Structure, function and diversity of the healthy human microbiome. Nature 2012;486(7402):207–14.

[49] Olson MV. Human genetic individuality. Annu Rev Genomics Hum Genet 2012;13(1):1–27.

[50] Olson MV. What does a "normal" human genome look like? Science 2011;331(6019):872.

[51] Sharma P. Charaka samhita: text with english translation. Varanasi, India: Chaukhambha Orientalia; 1981. p. 240

[52] Valiathan M. The legacy of caraka. Chennai, India: Orient Longman; 2003.

[53] Prasher B, Negi S, Aggarwal S, et al. Whole genome expression and biochemical correlates of extreme constitutional types defined in Ayurveda. J Transl Med 2008;6:48.

[54] Sethi TP, Prasher B, Mukerji M. Ayurgenomics: a new way of threading molecular variability for stratified medicine. ACS Chem Biol 2011;6(9):875–80.

[54a] Prasher B, Gibson G, Mukerji M. Genomic insights into Ayurvedic and Western approaches to personalized medicine. J Genet 2015; In press.

[55] Karlsson EK, Kwiatkowski DP, Sabeti PC. Natural selection and infectious disease in human populations. Nat Rev Genet 2014;15(6):379−93.

[56] Singh I, Faruq M, Mukherjee O, et al. North and South Indian populations share a common ancestral origin of Friedreich's ataxia but vary in age of GAA repeat expansion. Ann Hum Genet 2010;74(3):202−10.

[57] Holland BK. Prospecting for drugs in ancient texts. Nature 1994;369(6483):702.

[58] Heinrich M. Ethnopharmacology and drug discovery. 2010: 30.

[59] Patwardhan B. Ethnopharmacology and drug discovery. J Ethnopharmacol 2005;100(1−2):50−2.

[60] Jain S, Murthy P. The other bose: an account of missed opportunities in the history of neurobilogy in India. Curr Sci 2009;97.

[61] Patwardhan B, Mashelkar RA. Traditional medicine-inspired approaches to drug discovery: can Ayurveda show the way forward? Drug Discov Today 2009;14(15−16):804−11.

[62] Sharma P. Susruta-Samhita: with english translation of text and dalhana's commentary along with critical notes. Varanasi: Chaukhamba Vishwa Bharati; 1999.

[63] Krishna murthy S. Ashtanga sangraha with english translation text. Varanasi, India: Chaukhambha Orientalia; 2000.

[64] Gurib-Fakim A. Medicinal plants: traditions of yesterday and drugs of tomorrow. Mol Aspects Med 2006;27(1):1−93.

[65] Tapsell LC, Hemphill I, Cobiac L, et al. Health benefits of herbs and spices: the past, the present, the future. Med J Aust 2006;185(Suppl. 4): S4−24

[66] Diwanay S, Chitre D, Patwardhan B. Immunoprotection by botanical drugs in cancer chemotherapy. J Ethnopharmacol 2004;90(1):49−55.

[67] Chopra A, Lavin P, Patwardhan B, Chitre D. A 32-week randomized, placebo-controlled clinical evaluation of RA-11, an Ayurvedic drug, on osteoarthritis of the knees. J Clin Rheumatol 2004;10 (5):236−45.

[68] Aggarwal S, Negi S, Jha P, et al. EGLN1 involvement in high-altitude adaptation revealed through genetic analysis of extreme constitution types defined in Ayurveda. Proc Natl Acad Sci USA 2010;107(44):18961−6.

[69] Ahmad T, Kumar M, Mabalirajan U, et al. Hypoxia response in asthma: differential modulation on inflammation and epithelial injury. Am J Respir Cell Mol Biol 2012;47(1):1−10.

[70] Haase VH. Hypoxic regulation of erythropoiesis and iron metabolism. Am J Physiol Renal Physiol 2010;299(1):F1−13.

[71] Harten SK, Ashcroft M, Maxwell PH. Prolyl hydroxylase domain inhibitors: a route to HIF activation and neuroprotection. Antioxid Redox Signal 2010;12(4):459−80.

[72] Kalucka J, Ettinger A, Franke K, et al. Loss of epithelial hypoxia-inducible factor prolyl hydroxylase 2 accelerates skin wound healing in mice. Mol Cell Biol 2013;33(17):3426−38.

[73] Nagel S, Talbot NP, Mecinovic J, Smith TG, Buchan AM, Schofield CJ. Therapeutic manipulation of the HIF hydroxylases. Antioxid Redox Signal 2010;12(4):481−501.

[74] Selvaraju V, Parinandi NL, Adluri RS, et al. Molecular mechanisms of action and therapeutic uses of pharmacological inhibitors of HIF-prolyl 4-hydroxylases for treatment of ischemic diseases. Antioxid Redox Signal 2014;20(16):2631−65.

[75] Soni H. Prolyl hydroxylase domain-2 (PHD2) inhibition may be a better therapeutic strategy in renal anemia. Med Hypotheses 2014;82(5):547−50.

[76] Zhao S, Wu J. Hypoxia inducible factor stabilization as a novel strategy to treat anemia. Curr Med Chem 2013;20(21):2697−711.

[77] Ziello JE, Jovin IS, Huang Y. Hypoxia-Inducible Factor (HIF)-1 regulatory pathway and its potential for therapeutic intervention in malignancy and ischemia. Yale J Biol Med 2007;80 (2):51−60.

[77a] Aggarwal S, Gheware A, Agrawal A, et al. Combined genetic effects of EGLN1 and VWF modulate thrombotic outcome in hypoxia revealed by Ayurgenomics approach. J Transl Med 2015;13:184.

[78] Peterson J, Garges S, Giovanni M, et al. The NIH human microbiome project. Genome Res 2009;19(12):2317—23.

[79] Flint HJ. The impact of nutrition on the human microbiome. Nutr Rev 2012;70(Suppl. 1): S10—13.

[80] Conlon MA, Bird AR. The impact of diet and lifestyle on gut microbiota and human health. Nutrients 2015;7(1):17—44.

[81] David LA, Maurice CF, Carmody RN, et al. Diet rapidly and reproducibly alters the human gut microbiome. Nature 2014;505(7484):559—63.

[82] Kau AL, Ahern PP, Griffin NW, Goodman AL, Gordon JI. Human nutrition, the gut microbiome and the immune system. Nature 2011;474(7351):327—36.

[83] Graf D, Di Cagno R, Fak F, et al. Contribution of diet to the composition of the human gut microbiota. Microb Ecol Health Dis 2015;26:26164.

[84] Moller-Levet CS, Archer SN, Bucca G, et al. Effects of insufficient sleep on circadian rhythmicity and expression amplitude of the human blood transcriptome. Proc Natl Acad Sci USA 2013;110 (12):E1132—41.

[85] Mattson MP. Energy intake and exercise as determinants of brain health and vulnerability to injury and disease. Cell Metab 2012;16(6):706—22.

[86] Galluzzi L, Pietrocola F, Levine B, Kroemer G. Metabolic control of autophagy. Cell 2014;159 (6):1263—76.

[87] Archer SN, Laing EE, Moller-Levet CS, et al. Mistimed sleep disrupts circadian regulation of the human transcriptome. Proc Natl Acad Sci USA 2014;111(6):E682—91.

[88] Longo VD, Mattson MP. Fasting: molecular mechanisms and clinical applications. Cell Metab 2014;19(2):181—92.

[89] Chaix A, Zarrinpar A, Miu P, Panda S. Time-restricted feeding is a preventative and therapeutic intervention against diverse nutritional challenges. Cell Metab 2014;20(6):991—1005.

[90] Carnio S, LoVerso F, Baraibar MA, et al. Autophagy impairment in muscle induces neuromuscular junction degeneration and precocious aging. Cell Rep 2014;8(5):1509—21.

CHAPTER 15

Genomics and Traditional Chinese Medicine

Wei Wang[1,2]
[1]School of Medical Sciences, Edith Cowan University, Western Australia, Australia
[2]Beijing Municipal Key Laboratory of Clinical Epidemiology, Beijing, China

Contents

INTRODUCTION

Traditional Chinese medicine (TCM) is a broad range of medicine practices sharing common concepts and an integrated theory system that has evolved through over 3000 years of clinical and pharmacological trails as one of the longest medical practices in China and Asia-Pacific areas [1]. As the second most practiced medical system in the world, one-quarter of the world's population uses one or more of TCM therapies including acupuncture, moxibustion, Chinese herbal medicine, Tui-Na massage, dietary therapy and physical exercise (Tai Chi and Qi Gong) to maintain health and wellness. In China, TCM shares equal status and has been integrated with Western medicine in the healthcare system to diagnose, prevent, and treat many types of diseases. Meanwhile, compared to Western medicine, TCM approach has not been recognized internationally as a lack of systemic research and evidenced investigation [2]. Nevertheless, scientists have begun to provide novel insights into the essence and molecular basis of TCM, with globalization of life sciences (global health) and arrival of mega data mining (big data) and the progress of genomics research [3]. In this chapter, the fast emerging integration of the science of genomics and the practice of TCM are reviewed with the emphasis on the genomic medicine.

Genomics and Society
DOI: http://dx.doi.org/10.1016/B978-0-12-420195-8.00015-X

GENOMICS, GENOMIC MEDICINE, AND PERSONALIZED MEDICINE

Genomics is the study of total or part of genetic sequence information of organisms, and attempts to understand the structure and function of these sequences and downstream biological products, which differs from genetics, the study of genes, heredity, and variation in living organisms. *Genomics medicine* involves clinical care including diagnostic, therapeutic, and the other implications of diseases by using genomic information. As an application of rapidly expanding knowledge of the human genome to clinical practice, genomic medicine plays an important role in the modernization of TCM by combining TCM theory with modern biological and genomic/genetic concepts, for example, to elucidate the active components of TCM herbal medications and their pharmacodynamic mechanisms at molecular level. Personalized medicine, an approach to emphasize the customization of health care is used to help tailor interventions to maximize health outcomes, is rapidly becoming a reality for a variety of conditions. Study on genomics led to insights on gene regulation and the complex interplay of factors responsible for both Chinese medicine and personalized medicine [2].

Clinicians and life scientists, therefore, are currently at a critical junction to accelerate both TCM and its evidence base with the availability of genomics as well as postgenomics technologies such as functional genomics, proteomics, metabolomics, glycomics, and lipidomics at molecular and cellular levels. This chapter provides an outlook on the enormous promise anticipated from the integration of TCM with genomics/genetics as a new driver for novel molecular-targeted personalized medicine, and the future directions and challenges in this hitherto neglected dimension of postgenomics personalized medicine. Recently, this has mainly involved the systematic use of patients' genotype and clinical phenotype to optimize individual's preventive and therapeutic care [2,4,5] (Figure 15.1).

KEY FUNDAMENTALS OF TCM CONCEPTS

Yin—Yang

Yin—Yang theory comes from the philosophy rather than nature sciences. Chinese believed light and darkness, as the ideal opposites, when united, yielded creative energy. The two opposites were further conceived as matter and energy which became dual-natured but as one. The two opposites were *Yin—Yang* [6]. *Yin—Yang* is the core of TCM. The concept of *Yin—Yang* originates in ancient Chinese philosophy of Confucianism, Buddhism, and Taoism, dated back 3000 years ago. The ancient Chinese discovered that all things in the universe contains two opposite sides (dimensions), for example, heaven versus earth, sun versus moon, day versus night, cold versus heat, water versus fire, upward versus downward, and male versus female. TCM doctors classify different body parts as Yin and Yang, for example, the head (Yang) versus foot (Yin), back (Yang) versus abdomen (Yin), the body surface (Yang)

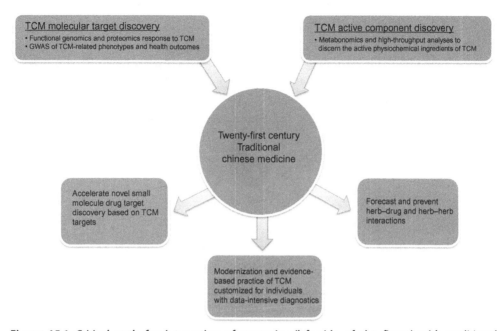

Figure 15.1 *Critical path for integration of genomics (left side of the figure) with traditional Chinese medicine (TCM) (right side of the figure) and anticipated outputs and its added value for a twenty-first century evidence-based and personalized TCM practice.* Data-intensive genomics and omics technologies can help identify both physiochemical active components of TCM and the molecular targets they interact. GWAS, genome-wide association study. *Modified from Ref. [2].*

versus interior of the body (Yin), to explain human physical structure (anatomy) and mental activities (physiology and psychology), and then to analyze the etiology of the disease: disharmony of the Yin versus Yang [7,8].

Five elements

Five elements of TCM indicate metal, wood, water, fire, and earth, which are the basic material units consisting of the world based on the ancient Chinese philosophy. As shown in Figure 15.2, it elucidates that the five elements represent five attributes of materials in the universe: water moistens all things, tending to flow downward; fire generates heat, tending to ascent; wood grows, tending to be flexible; metal shapes, tending to be transformed into any shapes, and earth generates all things, being the mother of all materials in universe. Although diversified in appearance, all things were classified into five categories: metal, wood, water, fire, and earth [7].

Of the interactions among the five elements, the most basic principles are the "mutual generation" and "inter-restriction." The "mutual generation" indicates to generate and support each other, that is, wood generates fire, fire generates earth, earth generates metal, metal generates water, and water generates wood, and thus

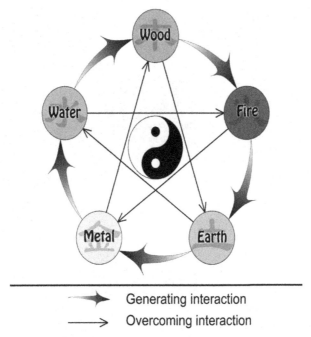

Figure 15.2 Yin and Yang and five elements.

constitutes an endless cycle. The "inter-restriction" means to inhibit and control, that is, wood restricts earth, earth restricts water, water restricts fire, fire restricts metal, and metal restricts wood. Therefore, changes in any one element may result in changes in the other elements. The cyclic generation and restriction maintain dynamic and equilibrium of the entire system. The role of the five elements theory in TCM manifests in the following two aspects: (i) the five elements correspond to physiological system via five anatomical organs as we can see in Table 15.1: heart (fire), liver (wood), spleen (earth), lung (metal), and kidney (water). Imaging that tree is growing in spring, the liver dominates growth, spreading and dispersing, and as a result, liver is classified as wood in the five elements theory; (ii) the mutual generation and inter-restriction among the five elements could be used to explain human physiology and pathology as well as the subsequent treatment. Normal mutual generation and inter-restriction indicates healthy status, whereas abnormal mutual generation and inter-restriction causes unhealthy or illness status, which requires treatments through TCM, such as Chinese herb medicine, and acupuncture [8].

Yin and Yang are constantly moving and so is the physiological function of the human being. Normal balance of Yin and Yang indicates a health status, whereas abnormal imbalance of Yin and Yang tells unhealthy or illness. But unfortunately, some of the TCM conceptions have not been recognized internationally due to the lack of systemic evidenced

Table 15.1 The role of the five elements theory in TCM

Phenomenon	Wood	Fire	Earth	Metal	Water
Direction [9]	East	South	Center	West	North
Color [10]	Green/blue	Red	Yellow	White	Black
Climate [11]	Wind	Heat	Damp	Dryness	Cold
Taste [9]	Sour	Bitter	Sweet	Acrid	Salty
Zang organ [12]	Liver	Heart	Spleen	Lung	Kidney
Fu organ [12]	Gallbladder	Small intestine	Stomach	Large intestine	Bladder
Sense organ [10]	Eye	Tongue	Mouth	Nose	Ears
Facial part [10]	Above bridge of nose	Between eyes, lower part	Bridge of nose	Between eyes, middle part	Cheeks (below cheekbone)
Eye part [10]	Iris	Inner/outer corner of the eye	Upper and lower lid	Sclera	Pupil

supports [13]. Suboptimal health status (SHS) is such an example. Since the ancient time, TCM has been identifying a physical status between health and disease coined as SHS [14]. SHS is a physical state between health and disease (the plasticity of Jing) and is characterized (i) by the perception of health complaints, general weakness, and low energy within a period of 3 months; and (ii) as a subclinical, reversible stage of chronic disease [14]. Effective intervention on SHS may be a cost-effective way for preventing chronic diseases. In addition, laboratory-based screening and development of disease-related biomarkers (genomics, proteomics, glycomics, and metabolomics) is also the key promise for the early diagnosis, prevention, and management in the practice of preventive, predictive, and personalized medicine (PPPM) [15−17], and for improving the ecology of medical care globally [14].

Jing and acupoints

Jing (meridian) are organized into 12 primary systems named after their respective organ systems. This is often confusing because Chinese medicine organ systems share the same names as anatomical organs, for example, heart, kidney, liver, lung, and spleen. Other names used in Chinese medicine include the San, Jiao, Chong, Ren, Du, and Dai channel systems. Each organ system described a specific set of physiological functions and symptoms as well as connected by *Jing*. It is important to note that this understanding is different from, and should not be confused with the functions of these organs as understood in Western anatomy and physiology. Moreover, there are 12 primary, 12 divergent, 8 extraordinary, and 15 luo-connecting channels belonging to this system; 361 *acupoints* are located on the pathways of these channels standardized by the World Health Organization [18]. Each of these points has a particular function when stimulated alone, but in clinical practice it is more effective to use a combination of points. The constituents in TCM preparations are influenced by three principal factors: heredity (genomic composition), ontogeny (stage of development), and environment (climate, associated flora, soil, method of cultivation). The concept of Jing (meridian), one of the foundational principles of TCM, plays a central role in the TCM theoretical frame (Figure 15.3). Presumably, *Jing* has

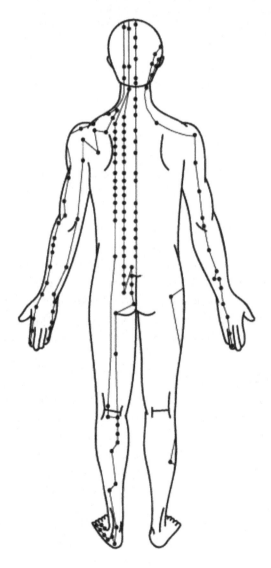

Figure 15.3 Meriden.

referred to the genetic information as well as its plasticity, as Jing is thought to be "the substance essential for development, growth, and maturation" and "conception is made possible by the power of Jing, growth to maturity is the blossoming of Jing, and the decline into old age reflects the weakening of the Jing" [2,13].

The Huangdi's *Internal Classic* and *Hippocratic Corpus* are ancient masterpieces in Eastern and Western medical practices. The former medical text is dated back to 475–221 BC (Chinese late warring states and early Han periods), which established the fundamental doctrinal source for TCM. The latter is the collection of Hippocrates and his students compiled in the 300 BC that represents the core teaching of medical school in Western medicine. Despite the difference between the Eastern and Western civilizations, the two books do share similarities in the aspects of how people thought and recognized the world. Body constitution is highlighted in both books, influenced by geography, demography, and heredity. Body constitutions were closely associated with disease occurrence, magnification, prevalence, and prognosis. Based on the Huangdi's *Internal Classic*, people across China had different constitutions owning to diversity in geography (the east is closer to the sea, west is sandier, south is more foggy, the north is cold, and central is more humid), dietary habits (hot, spicy vs. sweet, salty), and lifestyles (farmers vs. Normandy) [19]. While *Hippocratic Corpus* also mentioned that people who live in mountainous area, hollow zones, plateaus, and urban cities manifested different constitutions due to factors such as cold wind, hot weather, sunrise, and sunset.

Both books described the physical build, skin, and fat and thin. However two books did show the difference in understanding of the human constitution. In Huangdi's *Internal Classic*, the stria between the skin, muscles, and viscera was recognized as meridian "Jing," the pathway of circulation. While in the *Hippocratic Corpus*, people with different constitutions had different health curves, a balance between the best water and weakest fire, which could help people stay in healthy based on the "water–fire" theory of ancient Greek philosophy. Health preservation was also highlighted in both books. Health maintenance was related to geographic regions and seasons, influenced by "nonaction" concept, a leading ethical principle of Taoism, focusing on inner cultivation, man and environmental harmony, and body and mind balance. The *Hippocratic Corpus* believed that moisture of the purest water and fire could constitute most intelligent people with persistent concentration [20]. It was advisable for these people to avoid overheating and do more exercise. Ancient Greeks preferred logical thinking to objectify observations while ancient Chinese practiced concrete or imagery thinking to conclude observations through analogy. However both systems mentioned close interactions between man and nature, and maintained a holistic view of peoples' constructions in the clinical observations [8].

From the perspective of Western medicine, the innovative tools, pointing to an awareness of the totality of human health, have arisen as a direct outcome of the

Western medicine urge to penetrate phenotypes and to unravel the transcendent truth behind them. Western medicine has been nourished by the constant tension between the unknown and known, and imperfect and perfect [13,21]. In TCM, on the other hand, the Chinese doctors direct their attention to the complete individual, including psychological aspects that Western physicians often see as unrelated to a specific health and disease issue. All relevant information, including the symptoms as well as the patient's other general characteristics, is gathered and woven together until it forms what TCM called a "pattern of disharmony" [13,21]. This pattern of disharmony describes a situation of "distress" or "imbalance" in a patient's body. The oriental diagnostic technique was described as "It does not turn up a specific disease entity or precise cause, but renders an almost poetic, yet workable, description of a whole person. The therapy then attempts to bring the configuration into balance, to restore harmony to the individual" [4,21].

APPLICATION OF GENOMICS TECHNOLOGIES IN THE AUTHENTICATION OF TCM AND ACUPUNCTURE

Natural products are gaining increased applications in drug discovery and development. DNA mutation and polymorphism analysis lead the applications of DNA microarrays in pharmacodynamics, pharmacogenomics, toxicogenomics, and quality control of herbal drugs and extracts. Genomic analyses of Chinese herbs provide the botanical identity of TCM constituents [22]. One of the well-studied herbs is Ginseng that has long been used to maintain physical vitality in China and the Far East. Ginsenosides, a main element of Ginseng, can inhibit early antigen activation of Epstein—Barr virus, and also show anticarcinogenic effects in a two-stage mouse skin model with 9,10-dimethyl-1,2-benzanthracene and in lung carcinogenesis induced by 4-nitroquinolin-1-oxide [23,24]. DNA probe method for the identification of host-specific DNA fragments is employed in DNA fingerprinting analysis of Ginseng and generated a distinctive banding pattern with a homologous index of 0.55 between Chinese and American Ginseng [25]. In another investigation, random amplified polymorphic DNA (RAPD) technique has been used to identify the Panax species and their adulterant, and distinct RAPD fingerprints of American and Chinese Ginseng have been obtained, irrespective of sources and ages [26]. Restriction fragment length polymorphic DNA is also applied on Ginseng authentication based on ribosomal ITS1-5.8S-ITS2 region and 18S rRNA gene and showed promising results on authentication of American and Chinese Panax species [13,21]. Most Chinese herbal formulae consist of several individual herbal components, which is obviously more complicated than individual herbs. Genetic technologies have also been used to reveal the relationship between formula and components, for example, in a research of the herbal formula of San—Huang—Xie—Xin—Tang (SHXXT), microarray technology was used to analyze the putative mechanism of SHXXT and to define the relationship between SHXXT and its individual herbal components. Gene expression profiles of HepG2 cells treated with SHXXT's components were obtained by DNA microarray, indicating that SHXXT's components display a unique antiproliferation pattern via p53 protein through DNA damage signaling pathways in HepG2 cells. In addition, hierarchical clustering analysis showed that Rhizoma Coptis, the principle herb, shares a similar gene expression profile with SHXXT [27]. These findings may explain why Rhizoma Coptis exerts a major effect in the herbal formula of SHXXT. This is one of the good examples to reveal the relationship between formulae and their herbal [13].

Elucidation of putative biological mechanisms of TCM by genomic approaches

In recent years, genomic and molecular approaches have been extensively used to illustrate potential mechanisms and biological functions of TCM. As mentioned earlier, genomic approaches were used to authenticate Ginseng from different countries. In addition, based on DNA microarray analysis, Ginseng was reported to up-regulate the expression of a set of genes involved in adhesion, migration, and cytoskeleton [28].

As we can see from Table 15.2, research on royal jelly demonstrated that it has diverse nutritional and pharmacological functions in humans, including vasodilative and hypotensive, or antihypercholesterolemic and antitumor activity. Royal jelly competes with 17β-estradiol for binding to the human estrogen receptors α and β. Treatment of MCF-7 cells (a breast cancer cell line, acronym of **Michigan Cancer Foundation-7**)

Table 15.2 Application cases of genomic/molecular approaches in TCM*

TCM	Genomic/molecular approaches	Conclusion
Ginseng	Microarray analysis	Regulation of genes in cellular adhesion, migration, and cytoskeleton
Royal jelly	RT-PCR; ligand binding assay, reporter gene assay	Estrogenic activities through interaction with estrogen receptors, transcriptional activation via an estrogen responsive element
Berberine	RT-PCR, western blot, migration assay	Modulation of the expression and function of the cell cycle regulatory molecules
Artemisinin	Microarray analysis	Regulation of Wnt/β-catenin signaling
E. fischeriana Steud	Luciferase assay, RT-PCR, western blot, receptor binding assay	Inhibition of NF-κB signaling and inducing of cell apoptosis
Mushroom	Microarray analysis	Cell cycle regulatory genes expression inhibition

*RT-PCR, Reverse transcription Polymerase chain reaction.

with royal jelly enhances proliferation, and concomitant treatment with tamoxifen blocks this effect [29]. A reporter gene assay showed that royal jelly enhances transcription of the luciferase gene through the estrogen responsive element in MCF-7 cells. Subcutaneous injection of royal jelly could restore the expression of vascular endothelial growth factor gene in the uteri of ovariectomized rats [30].

Berberine, a well-known component of the Chinese herb medicine of Huanglian (*Coptis chinensis*), is capable of inhibiting growth and endogenous platelet-derived growth factor (PDGF) synthesis in vascular smooth muscle cells after *in vitro* mechanical injury. It also acts on suppressing PDGF-stimulated cyclin D1/D3 and cyclin-dependent kinase (Cdk) gene expression. Moreover, berberine increased the activity of AMP-activated protein kinase, which leads to phosphorylation of p53 and increased protein levels of the Cdk inhibitor p21Cip1. These observations offer a molecular explanation for the antiproliferative and antimigratory properties of berberine [31].

Numerous Chinese herbs have been suggested to have antitumor potential. Scientists indicated that Chinese herbs have shed light on possible mechanisms and provided biological clues for development of new modern drugs. Konkimalla et al.'s study showed that cytotoxicity of its derivative, artesunate, is associated with inhibition of inducible nitric oxide syntheses (iNOS). The fact that a number of genes are involved in nitric oxide (NO) signaling and are significantly up- or down-regulated by artesunate indicates that artesunate may not only inhibit iNOS, but also affect other NO-related genes. Microarray analysis also showed that Wnt/β-catenin signaling pathway, which plays an important role in colon cancer etiology, is regulated by artesunate, and colon cancer cell lines are the most sensitive ones toward artesunate among all solid tumor cell lines. These results collectively suggest that artesunate might attenuate the growth of human colorectal carcinoma by inhibition of the Wnt/β-catenin pathway [32].

Another important element nuclear factor-κB (NF-κB) is critically important for tumor cell survival, growth, angiogenesis, and metastasis. One of the key events in the NF-κB signaling is the activation of inhibitor of NF-κB kinase (IKK) in response to stimuli of various cytokines. The root of *Euphorbia fischeriana* Steud has been used as a traditional Chinese herb for more than 2000 years. The 17-Acetoxyjolkinolide B (17-AJB), one of the components of *E. fischeriana* Steud, is a novel small molecule inhibitor of IKK. Indeed, 17-AJB has been shown to effectively inhibit tumor necrosis factor-A-induced NF-κB activation and induce apoptosis of tumor cells. Detailed analysis revealed that 17-AJB keeps IKK in its phosphorylated form irreversibly to inactivate its kinase activity, leading to its failure to activate NF-κB. The effect of 17-AJB on IKK is specific and has no effect on other kinases such as p38, p44/42, and JNK. The effects of 17-AJB on apoptosis also correlate with inhibitions of expressions of the NF-κB regulated genes. It is suggested that 17-AJB is a novel type NF-κB pathway inhibitor and its unique interaction mechanism with IKK may render it a strong apoptosis inducer of tumor cells and a novel type anticancer drug candidate [33].

Certain types of mushrooms are also considered as Chinese herbs and have widely been used as dietary supplements in the United States. Recent studies demonstrated that mushroom intake could protect against cancer, which might be linked to the modulation of the immune system [34].

Using DNA microarray analysis, Jiang and Sliva [35] have found that mushroom could inhibit expression of genes involved in cell cycle regulation, thus inhibit invasiveness of breast cancer.

It is interesting to note that Zang—Fu theory is one of the foundational principles of TCM where the concept of Jing plays a central role. Presumably, the Jing concept has referred to the genetic information as well as its plasticity, as Jing is thought to be "the substance essential for development, growth, and maturation" and "conception is made possible by the power of Jing, growth to maturity is the blossoming of Jing, and the decline into old age reflects the weakening of the Jing" [36]. With the availability of functional genomics and proteomics, it is now possible to examine the precise molecular targets impacted by TCM preparations and related health interventions. Ultimately, this can lead to a more evidence-informed practice of TCM that can in the near future account for individual and population variability in these molecular TCM targets. Similarly, genome-wide association studies can offer molecular leads on the molecular genetic substrates of TCM mechanism of action. A plausible critical path for integration of genomics with TCM and the anticipated outputs from such a convergence are highlighted in Figure. 15.1.

Application of genomics theory to support acupuncture practice

Acupuncture utilizes fine needles to pierce through specific anatomical points (positioned "Jing"), has been extensively used, and has emerged as an important modality of complementary and alternative therapy to Western medicine. Systems biology has become practically available and resembles acupuncture in many aspects and is current key technology that serves as the major driving force for translation of acupuncture medicine revolution into practice, will advance acupuncture therapy into health care for individuals. High-throughput genomics, proteomics, and metabolomics in the context of systems biology have been able to identify potential candidates for the effects of acupuncture and provide valuable information toward understanding mechanisms of the therapy. To realize the full potential of TCM acupuncture, the current status of principles and practice of acupuncture integrated with systems biology platform in the postgenomic era. Some characteristic examples are presented to highlight the application of this platform in omics and systems biology approaches to acupuncture research and some of the necessary milestones for moving acupuncture into mainstream health care [24].

Application of genomic medicine to investigate Herb—Drug interactions

In view of the increasing use of herbal medicines not only in China and the Asia-Pacific but also in many other parts of the world, including Western countries, concerns have been raised about herb—drug interactions. Pharmacogenomic studies required for better understanding of the genomic components of kinetic and dynamic effects of TCM preparations and their physiochemical active ingredients. Similarly, more studies are needed for the role of genetics for herb—drug and TCM—drug interactions [36]. Accumulating evidence has demonstrated that concurrent administration of herbal remedies may alter the pharmacokinetic or pharmacodynamic behaviors of certain drugs and severe adverse effects may occur [37]. However, mechanisms underlying herb—drug interactions remain an understudied area of pharmacokinetic and pharmacotherapy. Systematic evaluation of herbal product—drug interaction liability, as is routine for new drugs under development, necessitates identifying individual constituents from herbal products and characterizing the interaction potential of such constituents [38]. Genomic approaches applied in herb—drug interaction research gradually illuminates the interaction may be influenced by multiple environmental and/or genetic factors.

Tian Xian is a traditional Chinese herbal anticancer remedy that activates human pregnane X receptor (PXR) in cell-based reporter gene assays. Tian Xian products are herbal dietary supplements manufactured in China that are distributed worldwide and aggressively marketed as anticancer herbal therapy. These products are also marketed as herbal therapies that alleviate the unpleasant side effects associated with Western medicine anticancer treatments. Activation of PXR in liver regulates the expression genes encoding proteins that are intimately involved in the hepatic uptake, metabolism, and elimination of toxic compounds from the bodies. PXR-mediated herb—drug interactions can have undesirable effects in patients receiving combination therapy. Tian Xian can alter the strength of interaction between the human PXR protein and transcriptional cofactor proteins. It can increase expression of Cyp3a11 in primary cultures of rodent hepatocytes and induce expression of CYP3a4 in primary cultures of human hepatocytes. These data indicate that coadministration of Tian Xian is probably contraindicated in patients undergoing anticancer therapy with conventional chemotherapeutic agents [39].

Another example is the recent report on the interaction of PHY906, a TCM for chemotherapy-associated side effects relief, with Irinotecan (CPT-11), a drug used for the treatment of cancer using a systems biology approach. A study demonstrated that PHY906 significantly amplifies the effects of CPT-11 in tumor tissues. Furthermore, administration of PHY906 together with CPT-11 could trigger unique changes that are not triggered by either alone, suggesting the enhanced antitumor activity of the combination [40]. Thus, genomic medicine may provide a crucial link to help understand the complicated interactive relationships between TCM and Western drugs.

CONCLUSIONS AND OUTLOOK

Our understanding of TCM has advanced greatly with the data-intensive genomics and omics biotechnologies. However, a number of critical challenges remain ahead to fully articulate the vision and roadmap (Figure 15.1). First, it is difficult to combine multidimensional omics data emerging from epigenomics, proteomics, metabolomics, and clinical phenotype data attendant to TCM health outcomes. Second, there are complicated exogenous components of TCM [41]. Potential personalized medicine could be found if we can utilize genomics and metabolomics techniques for full spectrum of metabolites of TCM. *Genomics*, together with metabolomics, transcriptomics, and proteomics, jointly inform the "systems biology" approaches that are crucial for integration of postgenomics knowledgebase with TCM [42,43].

Up to date, China has launched several projects related to the human genome. The earliest project started in 1991 when the Human Genome Diversity Project was launched. The Chinese Human Genome Diversity Project has focused on the analyzed polymorphisms of the 56 ethnic groups resident in China. In 1999, China began to participate in the international Human Genome Project and undertook the sequencing of 1% of the human genome working draft. China is also a contributor to about 10% of the International HapMap Project launched in 2002 jointly by America, Canada, China, Japan, and United Kingdom. In January 2008, the "Yanhuang" (Emperor Yan) Project was started by the Beijing Genomics Institute to sequence the entire genome of 100 Chinese individuals [44]. The sheer demographic weight of China's population undergoing rapid and profound health transitions is also of enormous importance in regard to the global health and the global disease burden. China is also a major contributor in the control and spread of global health risks, as this is an inevitable aspect of China's growing international participation in the global trade of goods and services [15]. Genomic medicine in China has thus played an important and long-standing role in the shaping of health care and research policy in the country while also participating collaboratively with other countries in the Asia-Pacific and beyond. However, integrating TCM with modern medicine at omics level to contribute to more targeted personalized medicine is a long and arduous process that requires accumulation and synthesis of knowledge in many fields, including genomics/genetics, molecular biology, pharmacology, epidemiologic studies on gene—disease associations, gene—environment—behavior interactions, and genomics-informed clinical trials of TCM health interventions. Advances in omics may also provide new opportunities for TCM in public health that focus on disease prevention because genomic studies can provide genetic information at individual level as well as in population scale, thus increasing the interaction and the interdependence between the traditional healthcare delivery system, which focuses on treatment of individuals, and the public health system, which focuses on prevention, suboptimal health management, and control in populations [16,17,45,46,47]. Notably, this enhanced interaction is creating a shared population health information focus on using genomic advances appropriately and effectively to promote health and to prevent disease, which can be applied to TCM for its prevention use. Ultimately, with increasing integration, the modern TCM will continue to progress in the postgenomics era for a more integrated, evidence-based, and personalized health care. Hence, TCM has the potential to be a new rapidly emerging driver for novel molecular-targeted personalized medicine in the coming decade.

REFERENCES

[1] Yu F, Takahashi T, Moriya J, Kawaura K, Yamakawa J, Kusaka K, et al. Traditional Chinese medicine and Kampo: a review from the distant past for the future. J Int Med Res 2006; 34(3):231—9.

[2] Yun H, Hou L, Song M, Wang Y, Zakus D, Wu L, et al. Genomics and traditional Chinese medicine: a new driver for novel molecular-targeted personalized medicine? Curr Pharmacogenomics Pers Med 2012;10:101—5.

[3] Wang P, Chen Z. Traditional Chinese medicine ZHENG and omics convergence: a systems approach to post-genomics medicine in a global world. OMICS 2013;17(9):451—9.

[4] Golubnitschaja O, Watson ID, Topic E, Sandberg S, Ferrari M, Costigliola V. Position paper of the EPMA and EFLM: a global vision of the consolidated promotion of an integrative medical approach to advance health care. EPMA J 2013;4(1):12.

[5] Zheng S, Song M, Wu L, Yang S, Shen J, Lu X, et al. China: public health genomics. Public Health Genomics 2010;13(5):269—75.

[6] Mahdihassan S. Comparing Yin—Yang, the Chinese symbol of creation, with Ouroboros of Greek alchemy. Am J Chin Med 1989;17(3—4):95—8.

[7] Zhang Y. Introduction to Yin—Yang and five elements. Chinese Medical Culture 2014;1:1—5.

[8] Li H. Introduction to Yin—Yang and five elements. Chinese Med Culture 2014;1:6—8.

[9] Ergil MC, Ergil KV. Pocket atlas of Chinese medicine. Stuttgart: Thieme; 2009.

[10] Aung S, Chen W. Clinical introduction to medical acupuncture. London: Thieme Medical Publishers; 2007.

[11] Kaptchuck TJ. The web that has no weaver. 2nd ed. Chicago, III: Contemporary Books; 2000.

[12] Deng T. Practical diagnosis in traditional Chinese medicine. 5th reprint, 2005 ed.: Elsevier; 1999.

[13] Ngan F, Shaw P, But P, Wang J. Molecular authentication of Panax species. Phytochemistry 1999;50(5):787—91.

[14] Wang W, Russell A, Yan Y. Traditional Chinese medicine and new concepts of predictive, preventive and personalized medicine in diagnosis and treatment of suboptimal health. EPMA J 2014;5(1):4.

[15] Han Q, Chen L, Evans T, Horton R. China and global health. Lancet 2008;372(9648):1439—41.

[16] Khoury MJ, Gwinn M, Burke W, Bowen S, Zimmern R. Will genomics widen or help heal the schism between medicine and public health? Am J Prev Med 2007;33(4):310—17.

[17] Ling RE, Liu F, Lu XQ, Wang W. Emerging issues in public health: a perspective on China's healthcare system. Public Health 2011;125(1):9—14.

[18] Zhang YX. Standardizalion of acupunclure nomenclature, WHO. Hua Xia Med 2008;5:344.

[19] Zhang WB. [Analysis on the concepts of qi, blood and meridians in Huangdi Neijing (Yellow Emperor's Canon of Internal Classic)]. Zhongguo Zhen Jiu 2013;33(8):708—16.

[20] Touwaide A. [Recent research on Hippocrates and the Hippocratic corpus]. Nuncius 1994;9 (1):305—19.

[21] Fushimi H, Komatsu K, Isobe M, Namba T. Application of PCR-RFLP and MASA analyses on 18S ribosomal RNA gene sequence for the identification of three Ginseng drugs. Biol Pharm Bull 1997;20(7):765—9.

[22] Hon CC, Chow YC, Zeng FY, Leung FC. Genetic authentication of ginseng and other traditional Chinese medicine. Acta Pharmacol Sin 2003;24(9):841—6.

[23] Konoshima T, Takasaki M, Tokuda H, Masuda K, Arai Y, Shiojima K, et al. Anti-tumor-promoting activities of triterpenoids from ferns. I. Biol Pharm Bull 1996;19(7):962—5.

[24] Zhang A, Sun H, Yan G, Cheng W, Wang X. Systems biology approach opens door to essence of acupuncture. Complement Ther Med 2013;21(3):253—9.

[25] Ho IS, Leung FC. Isolation and characterization of repetitive DNA sequences from *Panax ginseng*. Mol Genet Genomics 2002;266(6):951—61.

[26] Shaw PC, But PP. Authentication of Panax species and their adulterants by random-primed polymerase chain reaction. Planta Med 1995;61(5):466—9.

[27] Cheng WY, Wu SL, Hsiang CY, Li CC, Lai TY, Lo HY, et al. Relationship Between San—Huang—Xie—Xin—Tang and its herbal components on the gene expression profiles in HepG2 cells. Am J Chin Med 2008;36(4):783—97.

[28] Yue PY, Mak NK, Cheng YK, Leung KW, Ng TB, Fan DT, et al. Pharmacogenomics and the Yin/Yang actions of ginseng: anti-tumor, angiomodulating and steroid-like activities of ginsenosides. Chin Med 2007;2:6.

[29] Mishima S, Suzuki KM, Isohama Y, Kuratsu N, Araki Y, Inoue M, et al. Royal jelly has estrogenic effects *in vitro* and *in vivo*. J Ethnopharmacol 2005;101(1—3):215—20.

[30] Suzuki KM, Isohama Y, Maruyama H, Yamada Y, Narita Y, Ohta S, et al. Estrogenic activities of fatty acids and a sterol isolated from royal jelly. Evid Based Complement Alternat Med 2008;5(3):295—302.

[31] Liang KW, Yin SC, Ting CT, Lin SJ, Hsueh CM, Chen CY, et al. Berberine inhibits platelet-derived growth factor-induced growth and migration partly through an AMPK-dependent pathway in vascular smooth muscle cells. Eur J Pharmacol 2008;590(1—3):343—54.

[32] Konkimalla VB, Blunder M, Korn B, Soomro SA, Jansen H, Chang W, et al. Effect of artemisinins and other endoperoxides on nitric oxide-related signaling pathway in RAW 264.7 mouse macrophage cells. Nitric Oxide 2008;19(2):184—91.

[33] Yan SS, Li Y, Wang Y, Shen SS, Gu Y, Wang HB, et al. 17-Acetoxyjolkinolide B irreversibly inhibits IkappaB kinase and induces apoptosis of tumor cells. Mol Cancer Ther 2008;7(6):1523—32.

[34] Wasser SP. Medicinal mushrooms as a source of antitumor and immunomodulating polysaccharides. Appl Microbiol Biotechnol 2002;60(3):258—74.

[35] Jiang J, Sliva D. Novel medicinal mushroom blend suppresses growth and invasiveness of human breast cancer cells. Int J Oncol 2010;37(6):1529—36.

[36] Hu M, Wang DQ, Xiao YJ, Mak VW, Tomlinson B. Herb—drug interactions: methods to identify potential influence of genetic variations in genes encoding drug metabolizing enzymes and drug transporters. Curr Pharm Biotechnol 2012;13(9):1718—30.

[37] Fugh-Berman A. Herb—drug interactions. Lancet 2000;355(9198):134—8.

[38] Brantley SJ, Argikar AA, Lin YS, Nagar S, Paine MF. Herb—drug interactions: challenges and opportunities for improved predictions. Drug Metab Dispos 2014;42(3):301—17.

[39] Lichti-Kaiser K, Staudinger JL. The traditional Chinese herbal remedy tian xian activates pregnane X receptor and induces CYP3A gene expression in hepatocytes. Drug Metab Dispos 2008;36(8):1538—45.

[40] Wang E, Bussom S, Chen J, Quinn C, Bedognetti D, Lam W, et al. Interaction of a traditional Chinese medicine (PHY906) and CPT-11 on the inflammatory process in the tumor microenvironment. BMC Med Genomics 2011;4:38.

[41] Tang HWY. Metabonomics: a revolution in progress. Prog Biochem Biophys 2006;33(5):401—17.

[42] Joshi K, Ghodke Y, Shintre P. Traditional medicine and genomics. J Ayurveda Integr Med 2010;1 (1):26—32.

[43] Zhang A, Sun H, Wang P, Han Y, Wang X. Future perspectives of personalized medicine in traditional Chinese medicine: a systems biology approach. Complement Ther Med 2012;20(1—2):93—9.

[44] Salter B, Cooper M, Dickins A. China and the global stem cell bioeconomy: an emerging political strategy? Regen Med 2006;1(5):671—83.

[45] Yan YX, Liu YQ, Li M, Hu PF, Guo AM, Yang XH, et al. Development and evaluation of a questionnaire for measuring suboptimal health status in urban Chinese. J Epidemiol 2009;19(6):333—41.

[46] Yan YX, Dong J, Liu YQ, Yang XH, Li M, Shia G, et al. Association of suboptimal health status and cardiovascular risk factors in urban Chinese workers. J Urban Health 2012;89(2):329—38.

[47] Yan YX, Dong J, Liu YQ, Zhang J, Song MS, He Y, et al. Association of suboptimal health status with psychosocial stress, plasma cortisol and mRNA expression of glucocorticoid receptor in lymphocyte. Stress 2015;18(1):29—34.

CHAPTER 16

Human Genetics and Genomics and Sociocultural Beliefs and Practices in South Africa

Himla Soodyall and Jennifer G. R Kromberg

Division of Human Genetics, School of Pathology, Faculty of Health Sciences, University of the Witwatersrand and National Health Laboratory Service, Johannesburg, South Africa

Contents

INTRODUCTION

The finalizing of the Human Genome Project in 2003 with the generation of the first complete human genome sequence heralded a new era for understanding the role genetic factors play in human health and in solving problems related to disease and disability. However, more than a decade later, many of the anticipated outcomes of whole genome-sequencing projects and genome-wide association studies with respect to complex diseases have still to be realized. While genomic science and molecular studies continues to grow in developing countries involving indigenous populations, the promises and perils of these advances need to be seen against a backdrop of global health disparity, political vulnerability, and cultural diversity. Cultural and ethnic identity and folk beliefs play a decisive role in shaping peoples' perceptions, attitudes, and practices regarding health care and illness.

In South Africa, indigenous health systems have provided care to people for many years prior to the introduction and integration of Western health systems into traditional cultures. With a population of about 52 million people [1] distributed throughout the nine provinces, and a complex history, the present-day population is an amalgam of diverse beliefs, religions, ethnic, and cultural backgrounds. Southern Africa was occupied exclusively by the San who were hunter gatherers, and the Khoe who practiced pastoralism, from about 2000 years ago, prior to the arrival, in the past

Genomics and Society
DOI: http://dx.doi.org/10.1016/B978-0-12-420195-8.00016-1

1200 years, of people who spoke Bantu languages, from the northern parts of the continent. These groups engaged with each other on various levels with varying degrees of admixture as evident in the linguistic, cultural, and genetic diversity. The colonial period that developed, following the arrival of the Dutch in 1652, resulted in sea-borne immigrants from various parts of Europe settling in South Africa. Later, slaves from Madagascar and various parts of Indonesia, indentured laborers from India, traders, as well as survivors of shipwrecks, also integrated with the local populations thereby contributing to the rich genetic and cultural diversity in South Africa.

GENETIC SERVICES IN SOUTH AFRICA

The country has a health care system, based on national, provincial, and private health programs, which is in transition. The genetic services, which operate within this system, are small, but comprehensive and well-organized. They are based mainly in the academic centers, provincial health departments, and the National Health Laboratory Service (NHLS), in the major cities [2]. A limited number of trained medical geneticists, genetic counsellors, and medical scientists are available to provide the service, but they can meet only 10% of the country's needs. Other urgent health issues, such as the HIV/AIDS epidemic and the high number of tuberculosis cases, as well as the general shortage of health professionals at all levels, tend to take priority. Nevertheless, genetic services offer genetic counselling clinics (7300 patients were counselled in 2008) and genetic testing facilities (16,000 tests were carried out in 2008), and research is conducted on many genetic conditions of concern to the country. Patients, the majority of whom are from the black population groups, from all over Southern and Central Africa make use of the service [3].

Altogether, five out of six people in South Africa are either black or Asian and many do not have the language or the words to describe the complex concepts used in genetics. There are various genetic terms and technologies, thousands of different genetic disorders, including chromosome and biochemical disorders, as well as new techniques such as prenatal diagnosis. The counsellors need to be aware that these concepts are often difficult to understand, and difficult to explain, especially to those who have health-related beliefs in direct conflict with Western medicine. Misunderstandings and misconceptions can easily occur and so counsellors need to be empathic and knowledgeable about local customs and beliefs, and learn to tolerate ambiguity, in order to provide quality care and ensure comprehensive understanding.

UNDERSTANDING CULTURE IN THE CONTEXT OF HEALTH

Culture is all those things people have learned to do, to believe, to value, and to enjoy in their history. It includes the ideals, beliefs, skills, tools, customs, and institutions into which each person is born. It is a set of assumptions widely shared by a group of people [4]. It shapes the lives of, and behavior of, the group from birth to death. It is derived from the family and the community and it is passed on from one generation to the next. Culture determines what we believe, think, and do; how we relate socially to other people; and who we are. Jegede [5] states that culture is the "totality of socially transmitted behavior patterns, arts, beliefs, institutions, and all other products of human work and thought characteristic of a community or population" which differentiates human populations in characteristics relevant to health.

In South Africa there are two dominant forms of culture, the so-called "Anglo" culture and the other cultures, mainly black "African" cultures. In order to understand the possible cultural issues and conflicts that may arise, within the provision of genetic services, between these two cultures (although each consists of the amalgamation of the cultural beliefs of their many different subgroups) the beliefs and values of each need to be understood (see Table 16.1). In Anglo-based cultures there is emphasis on individualization, and health care is based on scientific discoveries, human/animal studies, empirical data, and experience; while values are based on patient autonomy, open communication, truth telling, honesty, and are patient-focused. In other cultures there is emphasis on the group, and health care is based on superstition, myths, and experience; while values are based on respect for authority, relevant communication, cultural appropriateness, and are family-focused.

Table 16.1 Beliefs and values [4]

Anglo culture	Other cultures
Materialism	Spiritualism
Individualism	Group identity
Directness	Indirectness
Handshake	No personal touching
Change	Tradition
Personal control	Fate
Competition	Cooperativeness

Decision-making, which is required in many genetic counselling scenarios, in the Anglo culture, is generally individual-/patient-based, risks and benefits need to be explained, the right to information is important, and informed consent is upheld. In many other cultures decision-making is family-based and, for example, some women may not be able to make a decision in favor of prenatal diagnosis until they have discussed the matter with key members of their family. In the Anglo culture, religion is meaningful and philosophical, and helps with coping and spiritual harmony; while in other cultures religion is involved with spiritual and physical health and superstition.

Many sociocultural factors are relevant in the provision of genetic services. For example, the family in Anglo cultures is generally nuclear, the role of women is important, the care of children may be shared, some families are patriarchal, there is freedom of choice regarding marriage partner, and there are rituals around birth and death. In other cultures, the extended family may be emphasized, the role of women may be limited, care of children may be allocated to females, matriarchal families may occur, arranged marriages may be preferred, and consanguinity may be common.

The views on causes of illness may differ according to the cultural background. In Anglo cultures the cause may be understandable and associated with the views on the body, mind, and spirit (but still the question might arise: "A genetic disorder! Why me?"). Others may view the causes of illness as natural/unnatural, or as punishment for wrongdoing, or the result of imbalance between body, mind, and spirit [6].

The barriers to health care, then, may be involved with language and medical terminology, verbal and nonverbal communication, eye contact as well as silences, in Anglo cultures; while in others the barriers may involve language too, but also poor verbal skills (nonverbal interaction becoming more important), difficulty with eye contact, proximity, physical contact, and silences, and in addition poverty and ignorance may become issues.

Views of health and illness also differ according to the prevailing cultural beliefs. In Anglo cultures questions like what, why, how, and why some and not others succumb, arise, and physical causes, prevention, and cure are discussed. Also the interrelationship between body, mind, and spirit is considered. Regarding the body the physical attributes, such as gender, age, nutritional status, metabolism, and genetic background of the individual, are taken into account; while the cognitive processes of the mind, thoughts, memories, and feelings, as well as learned spiritual practices, intuition, dreams, and metaphysical forces associated with the spirit, may also be discussed. The body, mind, and spirit must be in balance, since ill health may be believed to be due to an imbalance.

The use of alternative medicines varies and people of other cultures show increased use compared with patients with an Anglo background. Also, one study showed that about 26% of people in the black population, for example, made use of the services offered by Western medicine, as well as traditional healers, and 45% had consulted a healer at some point [7]. The use of traditional services, however, may depend partly on the nature of the condition, and partly on a belief that Western medicine cannot cure nor explain the cause of that particular condition. In a local study on epilepsy, for example, findings showed that 42% of parents consulted traditional healers only for treatment, while a further 22% consulted both these healers and Western medical services [8].

Clinical interactions, such as in a genetic counselling clinic, may be affected by the personalized/familial culture of the counsellor, the culture of the client/patient, the culture of the primary medical system (especially bearing in mind South Africa's apartheid history and the poor access, for disadvantaged people, to medical services in the past), as well as the traditional medical culture [9]. The way in which these cultures interact and overlap can vary considerably, however, the greater the shared knowledge the less likely that misunderstandings will occur. The cultural and ethnic background of the patient and family, facing a genetic disorder, needs to be considered by the health professional who is offering counselling. The background differences between the counsellor and counsellees can affect communication, the perception of the disorder and support structures, as well as the acceptance within the family and the community, the willingness to seek help and receptivity to medical interventions [10]. This knowledge, however, should only be used to provide general guidelines in identifying and dealing with the problems caused by the genetic condition in the family, and generalizations and stereotyping should be avoided.

During genetic counselling sessions the behavior of the clients might be partly determined by their cultural background [11]. Rapport might develop quickly with those with an Anglo culture, while in cross-cultural counselling (where the culture of the counsellor and counsellee differ), with those of other cultures, rapport might develop more slowly. Most counselling sessions in South Africa are of this type, due to the fact that the majority of counsellors are of European extraction and the majority of clients (about 70%) are from the black population [3]. The counsellors are, however, trained with respect to the various cultures found in South Africa, during their academic educational and counselling courses.

SOCIOCULTURAL ISSUES, GENETIC DISORDERS, AND OTHER HEALTH ISSUES

There are several examples from South Africa that exemplify the complexity between sociocultural beliefs and practices when dealing with genetic disorders and other health issues.

In Johannesburg, staff of the Department of Human Genetics have conducted research spanning a period of four decades on more than 400 families, with at least one member with albinism, living in Soweto. Oculocutaneous albinism, a recessively inherited condition, is common (1 in 4000 is affected) in the local black population [12]. Interviews with these families gave insight into the sociocultural issues surrounding the condition. The first and most prominent of these issues was the widely believed myth that people with albinism do not die naturally [13]. This myth had an impact on most of the genetic counselling sessions conducted for this condition, especially those carried out with mothers with a new affected baby [14]. Both mothers and fathers were unhappy about the birth and mothers feared ridicule, for themselves and their infants, as well as the negative attitudes of the community. As a result maternal—infant bonding was delayed and, due also to their reduced visual acuity, the children were slow to reach their milestones.

A second issue uncovered in the counselling sessions with families with an affected child was the lack of understanding of the inherited nature of albinism [15]. When the recessive inheritance was explained the parents often refused to believe they were both carriers (unless there was another affected person in the family when the parent on whose side this person was might accept carrier status). They stated that the condition was due to punishment from God, for laughing at, chasing, or being afraid of albinos, to maternal impression, eating the wrong foods in pregnancy, using bleaching creams, or was evidence of infidelity. However, some believed that an affected baby was a gift from God. During the research data collection one participant said, "being black by birth but not by colour may be one of their major problems, this deprives them of the companionship of other black children," an astute observation that indicates one of the social side-effects of the condition [15].

Further discussion revolved around attitudes in the community and the stigma attached to albinism. It was acknowledged that the child would have eye and skin problems and these should be cared for and prevented as much as possible, so that teachers at school should place the child at the front of the class, sun exposure should be limited and anti-actinic cream used. However, the child might be stigmatized by the peer group who might believe the condition to be infectious [13]. Also research showed that affected people were not accepted as marriage partners, but this did not stop them from producing children (which were normally pigmented in most cases).

However, some myths become life-threatening, like the recent spate of killings of persons with albinism in East Africa, especially in Tanzania, and southern Africa [16]. Reports from various sources indicate that about 100 persons with albinism may have been killed in Tanzania and Burundi in the past few years. The killings are carried out in order to harvest body parts that are used in the making of charms by traditional witch doctors. It is believed that charms made with body parts of persons with albinism, especially hair, genitals, limbs, breasts, fingers, the tongue, and blood, make strong magic portions and can be sold at high prices. In Tanzania, for example, a leg or an arm can fetch between US$1000 and US$3000. Similar cases have been reported in Mali, West Africa. Also, people in Benin believe that the blood of persons with albinism has magical properties and that it brings prosperity and luck.

Another example of how cultural beliefs can interfere with the efficient delivery of genetic services is a case of *Xeroderma pigmentosa* and prenatal diagnosis. A woman from Soweto had three sons with the condition, one had already died because of this and the other two had numerous cancers particularly on the face. She was referred for genetic counselling early in the following pregnancy and after careful counselling on several occasions she decided she wanted prenatal diagnosis. This technique was duly performed and the result showed that the fetus in utero was also affected. The women said she did not want to bring another affected child into the world to suffer as her other children did. However she needed to discuss a termination decision with her family. When she did not return for a follow-up a nurse visited her at home and learned that she had been to a traditional healer who told her that this fetus would not be affected. She therefore decided to continue the pregnancy. Soon after birth the child was confirmed as affected, as indicated by the prenatal testing.

Sociocultural beliefs and practices vary according to population group. Indians have been in the country since the 1860s. One of the authors (Himla Soodyall) is of Indian origin and her cultural upbringing was shaped by Hinduism and traditional practices of her forebears in Durban. In her family, and that of many Hindi-speaking Hindus in the area where she was raised, a particular practice was followed when a child was infected with measles. Even though measles has been linked with a virus and a vaccine has been available since 1963, the popular belief in this culture was that a child suffering with measles was possessed by one of the demi-gods and deserved to be respected in a particular way to appease the Gods, or else the child would be scarred with the pox. The child was isolated from other family members and people from outside the immediate family and visitors were usually not encouraged to be in contact with the child. Furthermore, the affected child was not allowed to be associated with leather (e.g., wearing of a leather belt or shoes as is the practice when conducting prayer), all leather items were removed from the room occupied by the child; all members of the family resorted to eating boiled vegetarian food (no spicy food or cooking involving braising of onions with sizzling sounds), and at the entrance of the house there was a brass jar filled with water that had turmeric powder added to it and a few stems with leaves from the Syringa berry tree. When people entered the house, they would usually sprinkle this water over themselves to cleanse or purify themselves before entering the house. At the end of the child's infectious period, the mother would wear simple traditional clothes (sari) and go to seven homes like a beggar and ask for raw ingredients (e.g., sugar, flour, and milk) that she would then use to make food to offer to the Gods, during prayer, to steer them onwards on their journey. Since the immediate neighbors were familiar with the practice, they all participated knowingly with this ritual.

This ritual is not practiced by all Hindu religious groups in South Africa. It must have had a regional origin in a few practicing families who eventually grew in numbers, then the ritual became intertwined with a religious component. The association with respect for the Goddess in the religious context, and fear of scarring with pox if this ritual was not practiced, has contributed to the perpetuation of this tradition despite the scientific knowledge on measles.

ORAL HISTORIES AND BELIEF SYSTEMS ASSOCIATED WITH ORIGINS

There are a few examples of interesting oral narratives concerning the origins of some black population groups in South Africa. The Lemba, for example, have a strong oral history of non-African origins, and a culture similar to that practiced by Jewish/Arabic people, for example, abstinence from eating pork, certain rituals associated with the moon, male circumcision, etc. It should also be noted that some of these cultural attributes are not unique to the Lemba and are also practiced to varying degrees among other black groups. The Lemba live among other larger groups of people in southern Africa, mainly the Venda (Limpopo Province, South Africa) and Northern Sotho or Pedi in Sekhukhuneland (Mpumalanga, South Africa) and among the Shona (Kalanga) in the southern parts of Zimbabwe. While some Lemba, particularly those from South Africa, claim Jewish origin, the Jewish link is not universally accepted, and there have been several studies suggesting Islamic connections with Arabs. Moreover, some of the Lemba who live in Zimbabwe referred to here as Remba (their name for themselves, since there is no "L" sound in their spoken language, Shona), identify with Arabic ancestry, and several clan names in use are Arabic in origin [17].

A few genetic studies have attempted to elucidate the ancestry of the Lemba. When "classical" serogenetic markers (unpublished) and mitochondrial DNA studies were used, no differences between the Lemba and other southern African populations could be detected [18]. However, initial Y chromosome DNA molecular studies conducted by Spurdle and Jenkins [19] provided the first definitive evidence that the male gene pool of South African Lemba was derived, in part, from non-African sources. Although they could not distinguish between Jewish and non-Jewish Y chromosomes at this level of resolution, they concluded that the non-African Y chromosomes in the Lemba were of Semitic origin.

Using a higher level of resolution on the Y chromosome, Thomas and colleagues [20] were able to further resolve the Y chromosomes in the Lemba. They found that a particular Y chromosome haplotype—referred to as the Cohen modal haplotype (CMH)—was present in the Lemba at a frequency of 8.8% (12/136). The CMH had previously been reported at frequencies of 44.9% in Ashkenazi and 56.1% in Sephardic Cohanim (descendants of Jewish priests) and among Ashkenazi and Sephardic Israelites at 13.2% and 9.8%, respectively [21]. A few years later Hammer's group used additional Y chromosome markers to better resolve and define what they referred to as the *extended*-CMH [22]. When the same set of markers were screened for in the Lemba, none of the Lemba males examined was found to have the *extended*-CMH [23].

As can be seen from this example, cultural and group identities can be derived from oral histories, but genetic ancestries are determined by molecular studies and the genetic markers being used. When blood groups and serum protein markers were used, the Lemba were indistinguishable from the neighbors among whom they lived; the same was true for mitochondrial DNA which represented the input of females in their gene pool. However, the Y chromosomes, which represented their history through male contributions, showed the link to non-African ancestors. When trying to elucidate the most likely geographic region of origin of the non-African Y chromosomes in the Lemba, the best that could be done was to narrow it to the Middle Eastern region. While no evidence of the CMH was found in the higher resolution study, no inferences can be made about their claims about being Jewish—all that can be said is the lineage commonly associated with the Cohanim is not found in the Lemba.

Another example is represented by the origins of the abeLungu. In stories handed down from generation to generation, the abeLungu regard themselves as "white" and can trace their origins to "white" men. The Xhosa word for Europeans, abeLungu, means "foam from the sea," and their oral history traces some of their male forebears to specific people like Jekwa, Bhayi, and Pita from whom clan affiliation is claimed. These men are thought to have been survivors of ships that were wrecked on the treacherous strip of the Wild Coast coastline. They eventually integrated into the local Xhosa groups, introducing their genes into the population. Using Y chromosome DNA it has been shown that the Y lineages associated with four clan names are in fact derived from Eurasian sources (David de Veredicis, personal communication).

CONCLUSIONS

Scientific interest in genomes in Africa is on the rise with a number of funding initiatives aimed at supporting research in this area [24]. In 2011 the Department of Science and Technology in South Africa funded the Southern African Human Genome Programme (SAHGP) which resulted in whole genome sequencing of 24 individuals genomes. One of the major objectives of the SAHGP is to develop capacity in bioinformatics in southern Africa and to better understand the evolutionary history of Africans. Through this program a workshop was convened in Pretoria in November 2014 to advance knowledge on ethical and legal issues around genomic research, and it also highlighted the need for educational initiatives and public engagement programs to build on the knowledge base of the general population on genomic research.

Moreover, it was emphasized that there was an urgent need for scientists to acknowledge the role of traditional healers and to encourage dialogue between scientists, policy makers, and traditional healers in genomic research and the integration of genomic medicine with sociocultural beliefs and practices. As mentioned earlier, in South Africa traditional medicine plays an important role in primary health care. There were approximately 200,000 traditional healers in South Africa in 1995 compared to 25,000 western medical doctors. In fact, in sub-Saharan Africa, the ratio of traditional healers to the general population is approximately 1:500, while doctors trained in "Western" medicine have a 1:40,000 ratio to the rest of the population. Also, it was estimated that between 60% and 80% of people in South Africa consult a traditional healer before going to a primary health care practitioner [25].

Given the scarce resources in the public health sector, people feel that traditional healers are more accessible and they can visit them for health-related problems as well as for other reasons. Since traditional healers are more familiar with culture-bound health issues and traditions, their relationship with patients and their families places them in a position to serve as an alternative to mainstream health providers. Breaking this tradition of the role of traditional healers in society is going to be difficult since they serve the role of physician, counsellor, psychiatrist, and priest [25]. However, if they are consulted and included in health-related research and in policy-making decisions, then they can play a significant role in closing the gap and building the bridges between genetic and genomic science and sociocultural beliefs and practices.

ACRONYMS

CMH Cohen modal haplotype
HIV/AIDS Human immunodeficiency virus/autoimmune deficiency syndrome
NHLS National Health Laboratory Service
SA South Africa
SAHGP Southern African Human Genome Programme
US United States

REFERENCES

[1] Census 2011 - Statistics South Africa. Available from: <www.statssa.gov.za/publications/p03014/p030142011.pdf>; [accessed 20.11.14].
[2] Kromberg JGR, Sizer E, Christianson ALC. Genetic services and testing in South Africa. J Community Genet 2013;4(3):413–23.
[3] Kromberg JGR, Wessels T-M, Krause A. Roles of genetic counsellors in South Africa. J Genet Couns 2013;22(6):753–61.
[4] Fisher NL. Ethnocultural approaches to genetics. Pediatr Clin North Am 1992;39:55–64.
[5] Jegede AS. Culture and genetic screening in Africa. Dev World Bioeth 2009;9(3):128–37.
[6] Kromberg JGR, Jenkins T. Cultural influences on the perceptions of genetic disorders in the Black population of Southern Africa. In: Clark A, Parsons E, editors. Culture, kinshop and genes. London: MacMillan; 1997. p. 147–57.
[7] Freeman M, Lee T, Vivian W. Evaluation of mental health services in the Orange Free State of South Africa. Report for the Centre for Health Policy. University of the Witwatersrand, Johannesburg; 1994.
[8] Christianson AL, Zwane ME, Manga P, Rosen E, Venter A, Kromberg JGR. Epilepsy in rural South African children—prevalence, associated disability and management. S Afr Med J 2000;90:262–6.
[9] Fitzgerald MH. Multicultural clinical interactions. J Rehabil 1992;58:38–42.

[10] Murray RF. Cultural and ethnic influences on the genetic counselling process. In: Weiss JO, Bernhardt BA, Paul NW, editors. Genetic disorders and birth defects in families and society: towards interdisciplinary understanding. Birth defects: original article series 20. New York, NY: March of Dimes; 1984. p. 71−4.

[11] Morris M, Glass M, Wessels T-M, Kromberg JG. Mothers' experiences of genetic counselling in Johannesburg, South Africa. J Genet Couns 2014. <http://dx.doi.org/10.1007/510897-014-9748-X>.

[12] Kromberg JGR, Jenkins T. Prevalence of albinism in the South African negro. S Afr Med J 1982;61:383−6.

[13] Kromberg JGR. Albinism in the South African negro. IV. Attitudes and the death myth. Birth Defects Orig Artic Ser 1992;28(1):159−66.

[14] Kromberg JGR, Zwane ME, Jenkins T. The response of black mothers to the birth of an Albino infant. Am J Dis Child 1987;141:911−16.

[15] Kromberg JGR, Jenkins T. Albinism in the South African negro. III. Genetic counselling issues. J Biosoc Sci 1984;16:99−108.

[16] Thuku M. Myths, discrimination, and the call for special rights for persons with Albinism in Sub-Saharan Africa. Available from: <http://www.underthesamesun.com/sites/default/files/MYTHS. Final_.pdf>; 2011 [accessed 20.11.14].

[17] Van Warmelo NJ. The classification of cultural groups. In: Hammond-Tooke WD, editor. The Bantu-speaking peoples of Southern Africa. London: Routledge and Kegan; 1974. p. 56−84.

[18] Soodyall H. Mitochondrial DNA polymorphisms in southern African populations [Ph.D. thesis]. Johannesburg: University of the Witwatersrand; 1993.

[19] Spurdle A, Jenkins T. The origins of the Lemba "Black Jews" of southern Africa: evidence from p12F2 and other Y-chromosome markers. Am J Hum Genet 1996;59(5):1126−33.

[20] Thomas MG, Parfitt T, Weiss DA, Skorecki K, Wilson JF, le Roux M, et al. Y chromosomes travel-ing south: the Cohen modal haplotype and the origins of the Lemba—the "Black Jews of Southern Africa". Am J Hum Genet 2000;66(2):674−86.

[21] Thomas MG, Skorecki K, Ben-Ami H, Parfitt T, Bradman N, Goldstein DB. Origins of old testa-ment priests. Nature 1998;394(6689):138−40.

[22] Hammer MF, Behar DM, Karafet TM, et al. Extended Y chromosome haplotypes resolve multiple and unique lineages of the Jewish priesthood. Hum Genet 2009;126(5):707−17.

[23] Soodyall H. Lemba origins revisited: tracing the ancestry of Y chromosomes in South African and Zimbabwean Lemba. S Afr Med J 2013;103(12 Suppl. 1):1009−13.

[24] deVries J, Pepper M. Genomic sovereignty and the African promise: mining the African genome for the benefit of Africa. J Med Ethics 2012;38(8):474−8.

[25] Truter I. African traditional healers: cultural and religious beliefs intertwined in a holistic way. S Afr Pharm J 2007;56−60.

CHAPTER 17

Genomics and Spirituality

Michael Ruse
Florida State University, Tallahassee, FL, USA

Contents

INTRODUCTION

What does one say about genomics and spirituality? I feel rather like I am revisiting the once-a-year event at my school debating club, where there was a hat with many suggestions written on slips of paper (we all contributed one) and you drew one out and were expected to lecture on the topic for 2 min. "The Prince of Wales on modern architecture," "The Chalcedonian Fathers' thinking about the relationship between Father and Son," and "Genomics and Spirituality." Well, here goes. Here is my two-minutes' worth sharing personal thoughts and perception of genomics in the context of Life and God.

GENOMICS

Let's play it straight. "Genomics" I take it is the understanding that we get of individual genomes—the collection of genes in some organism—using today's sophisticated techniques like recombinant DNA and next-generation genome sequencing, and (I am very much thinking of computers here) modern equipment to extract the information and to process it [1]. The triumph, at least in the public mind, is the Human Genome Project, its completion and its results. I take it that this is something that has grown up in the past 50 or so years. Is it revolutionary? As always, in these kinds of cases, it very much depends on what you mean by the words. Of course it is in one

Genomics and Society
DOI: http://dx.doi.org/10.1016/B978-0-12-420195-8.00017-3

sense, and we shall be looking at some of the results that make it so. In other senses, perhaps less so, in the way in which it has all seemed to be very sequential.

One doesn't have the feeling that there was a modern-day Copernicus, worrying about why the facts and figures didn't seem to work, until suddenly he or she had an insight and from then on the Earth moved and the Sun stood still [2]. Even if you go back to Watson and Crick and the discovery of the double helix, wonderful achievement though it may have been, it was in a way the natural culmination of a group effort, working out functions and structures at the molecular level [3,4]. In another way, it was less culmination and more start of more work, leading to the genetic code and onwards. A lot of very clever people were doing a lot of very clever things. But it wasn't like the French Revolution. King one day; headless corpse the next.[1]

But as I say, it doesn't mean that the coming of genomics has not been a fantastic achievement and surely something with major implications far from the necessarily narrow and restricted view of the laboratory. In this spirit, let me try out a number of personal reflections that I don't think are so entirely off the wall that others would think them quite irrelevant.

THE LIFE—THE PAGAN PERSPECTIVE

The one thing that strikes absolutely and overwhelmingly is the unity of life. Of course, we have known for a century or more that all organisms are driven by the same units, the genes, governed by the same laws, more or less [6]. Mendel and successors like Thomas Hunt Morgan didn't have to say: "Different processes for plants and animals and different processes for the macro and the micro and so forth." There are differences between, say, prokaryotic cells and eukaryotic cells, but in the end the differences wash out and we are all part of the same process. But it was still possible to think that the mammals were, as one might say, the top dogs and among the mammals the humans were the very top dogs—the Crufts Prize winners of the organic world [7]. My sense is that genomics has really torn all of this down. I can still remember about 50 years ago the late Ernst Mayr—one of the truly great figures of twentieth-century, evolutionary biology—lecturing me—a beginner in the history and philosophy of science—about homology, that is the isomorphisms between organisms of

---

[1] Always, philosophical discussions of scientific revolutions occur against the background of Thomas Kuhn's great book, *The Structure of Scientific Revolutions* [5]. After years of debate and explication, we now realize that Kuhn was using the term "revolution" in a very distinctive manner, to refer to an event that was a complete break with the past. That there are such revolutions is obvious—the French revolution in politics and probably the Copernican revolution in science—but many revolutions are more cumulative—the Industrial revolution in society and the molecular revolution in biology.

different species, like the parallels between the bones of the forelimbs of humans, horses, bats, and porpoises. He pointed out how this was the greatest evidence that one could have for evolution. Forelimbs with completely different functions— grasping, running, flying, and swimming—and yet sharing the same structure. Descent from a common ancestor! Then Mayr added, almost as an afterthought—"Of course, no one would be as foolish as to expect homologies between humans and fruit flies. They are just too far apart."

But that today is just what we do know! The functional genes that control the development of humans and fruit flies are identical [8,9]. Humans are not fruit flies, but we are all part of the same great schema. It is all a matter of the Lego principle. You start with the same building blocks and you do things one way and you get a model of the Eiffel Tower, and you do things another way and you get a model of Godzilla. Do things one way and you get Charles Darwin. Do things another way, and you get a little fellow with bright red eyes and an urge to eat bananas. I don't know where Mayr is today, but whether he is singing hymns and looking down or shoveling coal and looking up, I am sure he is taking a moment to reflect on the new findings. And I am sure that he, and Charles Darwin also, to their credits, are absolutely delighted. There are times when being wrong is better than being right. The evidence for evolution is better even than they thought. Moreover, the unity of life is stronger even than they thought.

Now I am not saying that you cannot now go on being a subscriber to a religion—and obviously I thinking above all of Christianity here—that privileges human beings [10]. A religion that says we humans uniquely are the beings made in the image of God. We are the organisms for whom God died on the cross. He didn't do it for warthogs, for all that He is aware of the fall of every sparrow. However things are now turned around a bit. This isn't to say that we cannot and must not value humans more than warthogs or fruit flies, although as it happens warthogs are so exquisitely ugly that when I was on safari in Zimbabwe last year, I enjoyed seeing them more than some of the nobler animals. But I am saying that somehow humans don't seem so very special. Somehow we should be venerating life as a whole and not just ourselves.

The people who do this above all are the Pagans [11,12]. I am not saying that we should now ourselves become Pagans, going sky clad (stark naked) and dancing around bonfires, while calling down the moon. For a start, I am not arguing for a full-blooded view of life, throughout the universe animate and inanimate. I am not into the Gaia hypothesis, seeing Earth as an organism or anything like that. Nor am I particularly driven to be a vegetarian and certainly not a vegan. Apart from anything else, cabbages are organisms as much as cows and we are part of the web of life with our needs and desires and so forth. But I do see something that the Pagans sense and make central, the essential oneness of life.

In this context, I am reminded of Thomas Hardy's great poem, "Drummer Hodge," about a lad from Wessex in the Boer War and his sad fate.

They throw in Drummer Hodge, to rest
Uncoffined — just as found:
His landmark is a kopje-crest
That breaks the veldt around:
And foreign constellations west
Each night above his mound.

Young Hodge the drummer never knew —
Fresh from his Wessex home —
The meaning of the broad Karoo,
The Bush, the dusty loam,
And why uprose to nightly view
Strange stars amid the gloam.

Yet portion of that unknown plain
Will Hodge for ever be;
His homely Northern breast and brain
Grow to some Southern tree,
And strange-eyed constellations reign
His stars eternally.

Here is a kid, not really a soldier—what he does is beat two sticks against an animal skin—who is killed in a place that he knows not for a cause that he knows not. And yet, somehow, meaning is given because Hodge is part of the great web or net of life and now he will go on to be part of a tree or some other living thing. He is gone and yet he thrives.

Pagans often think in terms of some life force—what Bergson [13] called the *élan vital*. I am not sure though that we need to go that far. I just want to say that for me the deepest spiritual aspect of genomics is the way that it underlines our oneness with all living things. I find that strangely comforting and inspiring.

VARIATION AND EVOLUTION

Let me talk now about variation. One incredibly important thing that Darwin [14] taught us is that populations of organisms are not uniform. They hold masses of variation. This of course is the corner stone of modern evolutionary biology. Without variation natural selection has nothing to work on and evolution simply grinds to a halt. The nature of this variation was something of an unknown to Darwin, except he was always stone-cold certain that this variation never occurs to need but is random in this sense—not random in the sense of uncaused, or even random in the sense of unquantifiable (Darwin knew nothing of this), but random in the sense of without direction. It was one of the triumphs of twentieth-century genetics to fill in a lot of the gaps in our knowledge of such biological variation. To pick out just one instance of major advance, the use of molecular techniques—gel electrophoresis—by evolutionists like Richard Lewontin in the 1960s to find the extent to which there was massive molecular variation within all naturally occurring populations [15].

In one sense one could say that these are just facts and add nothing to spiritual matters at all. In another sense, however, one can surely say that whether strictly spiritual or not, the variation does add to the experience of being human and to the richness of living with one's fellows. Not everything of course. No one wants to say that monozygotic twins are lesser beings than dizygotic twins. But something. We do not have a uniform blandness among humans but rich variation. Obviously in certain respects we have been selected to appreciate this. It is true that some people prefer certain types as potential sexual partners than others, but overall a great deal of the magic of falling in love is a function of not loving someone exactly like yourself—and one can think of good biological reasons why this might be so—as one can also think of reasons why going too far off the norm might equally not be a good thing.

THE HUMAN RACE(S) AND POPULATION VARIATION

More contentious is the matter of race [16,17]. I doubt that any alien biologist coming to Earth would doubt that *Homo sapiens* is divided into subspecies. The denizens of the Congo are not the denizens of Finland. Early evolutionists like Charles Darwin had no doubts on this score.

> There is, however, no doubt that the various races, when carefully compared and measured, differ much from each other,—as in the texture of the hair, the relative proportions of all parts of the body, the capacity of the lungs, the form and capacity of the skull, and even in the convolutions of the brain. But it would be an endless task to specify the numerous points of structural difference. The races differ also in constitution, in acclimatization, and in distinct; mainly as it would appear in their emotional, but partly in their intellectual, faculties. ([18], p. 1216)

They also had no doubts that some races are superior to others and that eventually they will win out in the struggle for existence. No prizes for guessing who will be the winners. "At some future period, not very distant as measured by centuries, the civilized races of man will almost certainly exterminate and replace throughout the world the savage races" (p. 1201). Although Darwin did suggest that Western diseases will do the job at least, as if not more efficiently than anything else.

For well-known reasons, by the middle of the last century, notions of race had fallen much out of favor. After what the Nazis did to the Jews, no one was very keen on seeing biological differences between groups of humans. And this kind of hostility seemed to be backed by more recent work in genetics. Since Lewontin has already been mentioned, it is appropriate to make further reference to him. He has argued strongly against race considered as a biological phenomenon. As much as 85% of human variation is to be found within populations, meaning that at most 15% exists between populations. This led him to conclude that: "Human racial classification is of no social value and is positively destructive of social and human relations. Since such racial classification is now seen to be of virtually no genetic or taxonomic significance either, no justification can be offered for its continuance" [19].

Others are not so sure. It has been argued that although most variation is within populations rather than between, it does not follow that the between-population variation precludes a kind of clustering and that this would lead to populations that can be delimited genetically. Indeed this seems so. For instance, one study of 1056 individuals from 52 groups found in Lewontin fashion that: "Within-population differences among individuals account for 93 to 95% of genetic variation; differences among major groups constitute on 3 to 5%" ([20], p. 2381). Yet it was still possible to pick out groups. For instance, this analysis picks out as anomalous a group in Northern Pakistan. These are the somewhat isolated Kailash. Tradition has it that they are not of the same ethnic background of the rest of their countrymen, but rather of European or Middle-Eastern origin. A finding that genetics backs. In short: "Genetic clusters often corresponded closely to predefined regional or population groups or to collections of geographically and linguistically similar populations" (p. 2384).

A very recent analysis has come up with very similar sorts of findings in Great Britain. The study was of 2039 Britons whose grandparents lived in rural areas and who had been born within a short distance (80 miles) of each other. The results were astounding. People differ genetically: "a statistical model lumped participants into 17 groups based only on their DNA, and these groupings matched geography. People across central and southern England fell into the largest group, but many groupings were more isolated, such as the split between Devonians and Cornish in Britain's southwest. People who trace their ancestry to the Orkney Islands, off the northeast coast of Scotland, fell into three distinct categories. They are likely so differentiated because the islands made it hard for different populations to mingle" [21].

What all of this means is obviously another matter. Even ardent evangelical Christians have moved beyond seeing different races as a result of God's punishment of Canaan for his father (Ham) having seen Noah naked, when drunk after the Flood was over. Interestingly some of the most ardent Creationists in America today are to be found in the Black Churches, and they are hardly likely to buy into that sort of argument. Moreover, despite the rivalries one gets at the level of county cricket, no one is saying that people from Devon and people from Cornwall belong to different subspecies.

However, I speak now as someone who lives in the American South in a city (Tallahassee, Florida) with a population that is 40% African American, it is simply false to say that differences however regarded do not count. When the results of the annual state-wide testing of children is announced, one can tell infallibly which are the predominantly white schools and which are black, except—as happened to one of my sons—where high-achieving white children are shipped into a black school to raise the overall average scores. One may deplore these findings—although African American leaders are the first to acknowledge them and give strong cultural reasons for their existence, starting with self-expectations—but they are true.

So also is the fact that black—white differences matter at the social level. We have now, thank goodness, got over the lunch counter issue. You can eat wherever you like in my town, whether you be black, white, sky-blue pink with yellow polka dots. But we still keep apart at a social level. I doubt I have seen 10 interracial couples since I moved down from Canada 15 years ago. (Black—white that is. No one notices white-Hispanic couples or—although there are few Asians here—white-Asian couples.) We also have two universities—mine (Florida State University), with less than 10% black students (many of whom are in sports affiliated programs) and a so-called historically black university (Florida Agricultural and Mechanical University) with almost exclusively black students (there are one or two white students in professional programs). The flip side is that sometimes identification of groups helps those most in need. Again in my city, the leading group of urologists are constantly putting out calls for African American men to get tested for prostate cancer. The fact is that African American men are 1.6 times more likely to be diagnosed with such cancer and 2.4 times more likely to die from it.

What does any of this mean, ethically or spiritually? Clearly as much or as little as we want to make of it. We obviously have to move on from—we surely have moved on from—the crude racial classifications and inferences of the nineteenth century. The sorts of views that even decent liberals like Charles Darwin—whose whole life was marked by a detestation of slavery—held as virtually *a priori* true. It is also the case that now we must no longer be so consumed by the guilt of the Holocaust, not to mention the many other abuses made on the basis of claims about race that we simply deny that there are differences however caused. Apart from anything else, to deny that there are differences would mean the immediate end of affirmative-action policies which, however contentious, have clearly meant a great difference to many people's lives.[2] Perhaps in this context the best we can say—and it is to say much—is that genomics is leading to ever greater understanding of genetic diversity and with this comes hope, at the medical level, at the social level, and much more.[3]

[2] One pertinent close-to-home item is that in 2000, Jeb Bush—then governor of the State of Florida—stopped affirmative-action policies aimed at increasing African-American enrollment in the state's universities. Since then, my own institution of Florida State University has seen black enrollment drop from about 9% to 6%. It goes without saying that many of not most of these still-enrolling students are in sports-related programs.

[3] But as the Florida affirmative-action example shows, it is never going to be a matter of pure science, but always of science in conjunction with social and moral policies.

SCOPE OF GENOMICS—REDUCTIONISM AND EMERGENTISM

One of the most striking findings of genomics is how much can be done by so little. There used to be the assumption that to get more and more complex organisms and sophisticated functioning, you needed more and more genes. Now we realize fully that that is not so. So much is a matter of timing and of interactions and so forth. Just before he died, Stephen Jay Gould [22] made an argument from this finding that resonates with many. First the facts.

> The fruit fly Drosophila, the staple of laboratory genetics, possesses between 13,000 and 14,000 genes. The roundworm C. elegans, the staple of laboratory studies in development, contains only 959 cells, looks like a tiny formless squib with virtually no complex anatomy beyond its genitalia, and possesses just over 19,000 genes.
>
> The general estimate for Homo sapiens — sufficiently large to account for the vastly greater complexity of humans under conventional views — had stood at well over 100,000, with a more precise figure of 142,634 widely advertised and considered well within the range of reasonable expectation. Despite the earlier estimates, Homo sapiens possesses around 25,000 genes, with the final tally almost sure to lie probably the lower than this figure. In other words, our bodies develop under the directing influence of only half again as many genes as the tiny roundworm needs to manufacture its utter, if elegant, outward simplicity.

Then the inference; the implications of this finding cascade across several realms. But the deepest ramifications will be scientific or philosophical in the largest sense. From its late-seventeenth-century inception in modern form, science has strongly privileged the reductionist mode of thought that breaks overt complexity into constituent parts and then tries to explain the totality by the properties of these parts and simple interactions fully predictable from the parts. ("Analysis" literally means to dissolve into basic parts.) The reductionist method works triumphantly for simple systems—predicting eclipses or the motion of planets (but not the histories of their complex surfaces), for example. But once again—and when will we ever learn?—we fell victim to hubris, as we imagined that, in discovering how to unlock some systems, we had found the key for the conquest of all natural phenomena. Will Parsifal ever learn that only humility (and a plurality of strategies for explanation) can locate the Holy Grail?

THE FUTURE—CONTINUING GENOMICS

The collapse of the doctrine of one gene for one protein, and one direction of causal flow from basic codes to elaborate totality, marks the failure of reductionism for the complex system that we call biology. The deflation of hubris is blessedly positive, not cynically disabling. The failure of reductionism doesn't mark the failure of science, but only the replacement of an ultimately unworkable set of assumptions by more appropriate styles of explanation that study complexity at its own level and respect the influences of unique histories.

What are we to make of this and what is the connection with spirituality? Start at the beginning with "reductionism." This is a term with many meanings, but in this context it means explaining the large in terms of the small [23,24]. A classic example of this kind of reductionism occurs in gas theory where we explain macrophenomena, like temperature and pressure, in terms of microphenomena, little balls whizzing around in a container. The speed at which they go relates to temperature and so forth. Now you might with reason say that this represents the best kind of science and surely is the trend since the Scientific Revolution in the sixteenth and seventeenth centuries. In biology for instance, one of the greatest triumphs of reductionistic science was when James Watson and Francis Crick discovered the double helix. Up to this point, the Mendelian gene had been considered an entity in its own right. Now it was seen to be a macromolecule, the DNA molecule, and this in turn was seen to be a number of smaller molecules strung together. By looking at the order of these smaller molecules, one could crack their "code" and thus find how information was carried and transmitted from one generation to the next. "Small is beautiful."

The thing that many find upsetting about this scenario—and clearly *prima facie* genomics is totally committed to this form of reductionism—is that it all seems cold and harsh, materialistic. All organisms, including humans, are nothing but molecules buzzing around in space—no further meaning or purpose. Moreover, it often seems connected to mechanistic philosophies entwined with socio-economic policies that many find uncongenial. Richard Dawkins' metaphor of the "selfish gene" seems to be the epitome of reductive thought and at the same time the epitome of a harsh, laissez faire view of society.

> We are survival machines, but "we" does not mean just people. It embraces all animals, plants, bacteria, and viruses. The total number of survival machines on earth is very difficult to count and even the total number of species is unknown. Taking just insects alone, the number of living species has been estimated at around three million, and the number of living insects may be a million, million, million.
>
> Different sorts of survival machines appear very varied on the outside and in their internal organs. An octopus is nothing like a mouse, and both are quite different from an oak tree. Yet in their fundamental chemistry they are rather uniform, and, in particular, the replicators which they bear, the genes, are basically the same kind of molecule in all of us—from bacteria to elephants. We are all survival machines for the same kind of replicator—molecules called DNA—but there are many different ways of making a living in the world, and the replicators have built a vast range of machines to exploit them. A monkey is a machine which preserves genes up trees, a fish is a machine which preserves genes in the water; there is even a small worm which preserves genes in German beer mats. DNA works in mysterious ways. ([25], p. 22)

We have no more status or ultimate meaning than an automobile or a refrigerator and we exist today only because our ancestral machines did a better job of pushing rival machines out of the way. Life seems to be nothing but a nightmare video game. Morality, feeling, love, compassion—they are merely epiphenomena of the genes, put in place simply because those genes that gave rise to them reproduced more efficiently than those that did not.

Not very spiritual! But in genomics, in the working of the human genome, people like Gould see a ray of hope. In this modern world, nobody working as a conventional scientist—and this includes religious believers—wants to make appeal to souls or occult forces or whatever, including the already-mentioned vitalistic forces of people like the French philosopher Henri Bergson, who supposed that there were such forces (*élans vitaux*) that were motivating and driving living things. But perhaps we can say that over and above the individual molecules there is a sense of order of some kind, an "organized complexity" as it has been called, and that this in a sense gives a wholeness to nature, to humans in particular [26]. We have therefore "emergent" properties that could not have been predicted simply by looking at the molecules in isolation. It is all rather like a triangle. You have three straight lines and that is what they are—three straight lines. But put them together in a particular order and you have a triangle with all sorts of properties—the internal angles adding up to two right angles for instance—that you could not have deduced or expected just from three straight lines. Something has emerged. In the words of the great Greek philosopher Aristotle, who was one of the first to seize on this notion: "the whole is more than the sum of the parts" ([27], *Metaphysics*, Book H 1045a 8–10).

Why is this a ray of hope? Why is this in some sense a more spiritual approach, where it is understood that this is not necessarily understood in any religious sense but more in a sense that gives greater dignity to life and to humans in particular? In part, of course, because almost by definition we are getting away from the molecules-in-isolation position. We are arguing that there is more to things—certainly more to life—than this. This "more" is something new and in a sense creative, rather as a picture by van Gogh is creative and something more than the paints and canvas that were there before he picked up a brush. A cornfield as painted by the great Dutch artist clearly has a meaning and a vitality that the paint and canvas molecules taken alone do not.

In part also one suspects that for people like Gould there is something morally better and more wholesome than the harsh reductive position allows. Things—genes, organisms, and populations—can be seen to come together and from them something integrated emerges. If we think about the human body, for instance, the heart and the kidneys are not independent entities doing their own thing. They are part of the whole working for the benefit of the whole. Gould sees this further down the chain. The molecules of DNA are not just doing their own thing in isolation. They are in some very real sense working together, influencing and being influenced, in pursuit of the whole, the finished organism. Thus talk of "selfish" genes—with the all-too-ready inference that selfishness exists further up the chain and that we humans are selfish and destined thus to remain—is simply bad science. Nothing is selfish in that sense, whether we are talking about genes, or bodily parts, or full-blown organisms, be they fruit flies or humans. Thus in this very real sense genomics does point to a fresh perspective, one where words like "spirituality" are not out of place.

What does one say in response to all of this? Much depends really one how much one is trying to get at a point like this. If one is simply saying—although perhaps "simply saying" is not quite right because one is trying to get quite a bit—that one is going to be in awe of the world of nature generally and the world of nature specifically as revealed by genomics, then few would differ. In fact, that arch-reductionist Richard Dawkins [28] makes this very point at the end of one of his books where he has taken us on a journey—on a "pilgrimage"—through the history of life.

> *"Pilgrimage" implies piety and reverence. I have not had occasion here to mention my impatience with traditional piety, and my disdain for reverence where the object is anything supernatural. But I make no secret of them. It is not because I wish to limit or circumscribe reverence; not because I want to reduce or downgrade the true reverence with which we are moved to celebrate the universe, once we understand it properly. "On the contrary" would be an understatement. My objection to supernatural beliefs is precisely that they miserably fail to do justice to the sublime grandeur of the real world. They represent a narrowing down from reality, an impoverishment of what the real world has to offer. (p. 506)*

"The sublime grandeur of the real world." Although Dawkins would probably shudder at the thought of being labeled "spiritual," in the secular sense that we are using it here, this is surely the sentiment he is expressing. So in this sense, we can all agree that there is something spiritual about genomics and its results.

The trouble and arguments come if more is being claimed—and the impression certainly is (and not just from Gould) that more is indeed being claimed. Perhaps, if you don't want to put things in an explicitly religious context—and there are those that do—we are veering again towards something akin to Paganism, seeing real meaning in the world, especially the world of nature. Here people will differ. People like Gould will think that science generally, genomics specifically, reveals an ever-emergent reality, not magical in the physical sense but certainly magical in the psychological sense. Probably, into the mix they will throw the phenomenon of sentience, of consciousness in some way. They will point out that molecules in motion are not to be identified with thinking, and yet they give rise to thinking [29]. This is the ultimate emergent, of which the glories of genomics are part of the foundation.

People like Dawkins will think that with the psychology you have said much if not all. Certainly we see new things as matter gets more complex and organized. Whoever thought otherwise? But let us not go overboard and read into existence real magic—the powers of the Wizard of Id and that sort of thing. Probably, although not necessarily, they will argue that consciousness raises no new problems in theory and that sometime (if not already in the opinion of some) it will be explained in terms of molecules in motion [30]. At least it will be seen to be no stranger than the explanation of Boyle's Law by gas theory. Heat and pressure are not balls in motion, but we see them fully and adequately explained in terms of balls in motion. Consciousness is not molecules in motion, but we shall see it fully and adequately explained in terms of molecules in motion.

Either way you get spirituality. The question is how far you want to take the notion of spirituality.

CONCLUSIONS

"I have answered three questions, and that is enough,"
Said his father; "don't give yourself airs!
Do you think I can listen all day to such stuff?
Be off, or I'll kick you down stairs!"

Old Father William in Lewis Carroll's glorious parody (of a dreadful, eighteenth-century poem about a nauseatingly pious old man) had the measure of things. One suspects that there are many more questions that could be asked about genomics and spirituality and other dimensions explored. One thinks most obviously of speculations about what one might do with genomics in the future and how this might affect the nature of humankind. Is there a new vista opened for eugenics, when we shall all be disease-free, have IQs that will get us into Mensa, have the physical beauty of Cary Grant or Grace Kelly, and play soccer in the World Cup? Or is that a future too awful to contemplate? It will be a bit of a waste if we all have the qualities to be President or Prime Minister, for those posts are occupied by one person at a time. But that is another paper, not by me!

REFERENCES

[1] Lesk AM. Introduction to genomics. Oxford: Oxford University Press; 2012.
[2] Kuhn T. The copernican revolution. Cambridge, MA: Harvard University Press; 1957.
[3] Watson J. Molecular biology of the gene. New York, NY: Benjamin; 1965.
[4] Watson J. The double helix. New York, NY: Signet Books; 1968.
[5] Kuhn T. The structure of scientific revolutions. Chicago, IL: University of Chicago Press; 1962.
[6] Bowler P. The mendelian revolution: the emergence of hereditarian concepts in modern science and society. London: The Athlone Press; 1989.
[7] Ruse M. Monad to man: the concept of progress in evolutionary biology. Cambridge, MA: Harvard University Press; 1996.
[8] Carroll SB. Endless forms most beautiful: the new science of Evo Devo. New York, NY: Norton; 2005.
[9] Carroll SB, Grenier JK, Weatherbee SD. From DNA to diversity: molecular genetics and the evolution of animal design. Oxford: Blackwell; 2001.
[10] Ruse M. Can a Darwinian be a Christian? The relationship between science and religion. Cambridge: Cambridge University Press; 2001.
[11] Adler M. Drawing down the moon: witches, druids, goddess-worshippers, and other pagans in America. New York, NY: Penguin; 2006.

[12] Ruse M. The Gaia hypothesis: science on a Pagan planet. Chicago, IL: University of Chicago Press; 2013.

[13] Bergson H. L'évolution créatrice. Paris: Alcan; 1907.

[14] Darwin C. On the origin of species by means of natural selection, or the preservation of favoured races in the struggle for life. London: John Murray; 1859.

[15] Lewontin RC. The genetic basis of evolutionary change. New York, NY: Columbia University Press; 1974.

[16] Ruse M. Darwinism and its discontents. Cambridge: Cambridge University Press; 2006.

[17] Ruse M. The philosophy of human evolution. Cambridge: Cambridge University Press; 2012.

[18] Darwin C. The descent of man, and selection in relation to sex. London: John Murray; 1871.

[19] Lewontin RC. Human diversity. New York, NY: Scientific American Library; 1982.

[20] Rosenberg NA, Pritchard JK, Weber JL, Cann HM, Kidd KK, Zhivotovsky LA, et al. Genetic structure of human populations. Science 2002;298:2381−5.

[21] Callaway E. UK mapped out by genetic ancestry, finest-scale DNA survey of any country reveals historical migrations. Nature News, March 18, 2015.

[22] Gould SJ. Humbled by the genome's mysteries. New York Times February 19, 2001.

[23] Nagel E. The structure of science, problems in the logic of scientific explanation. New York, NY: Harcourt, Brace and World; 1961.

[24] Ruse M. Science and spirituality: making room for faith in the age of science. Cambridge: Cambridge University Press; 2010.

[25] Dawkins C. The selfish gene. Oxford: Oxford University Press; 1976.

[26] Ruse M. Darwin and design: does evolution have a purpose? Cambridge, MA: Harvard University Press; 2003.

[27] Barnes J. The complete works of aristotle. Princeton, NJ: Princeton University Press; 1984.

[28] Dawkins C. The ancestor's tale. London: Weidenfeld and Nicolson; 2004.

[29] Chalmers DJ. Strong and weak emergence. In: Clayton P, Davies P, editors. The re-emergence of emergence: the emergentist hypothesis from science to religion. Oxford: Oxford University Press; 2006. p. 244−55.

[30] Dennett DC. Consciousness explained. New York, NY: Pantheon; 1992.

CHAPTER 18

The Sociodemographic and Economic Correlates of Consanguineous Marriages in Highly Consanguineous Populations

Hanan Hamamy[1] and Sura Alwan[2]
[1]Department of Genetic Medicine and Development, Geneva University, Geneva, Switzerland
[2]Department of Medical Genetics, University of British Columbia, Vancouver, BC, Canada

Contents

INTRODUCTION

Healthcare providers and genetics specialists have usually judged the overall impact of consanguineous marriage as being negative when assessed in terms of increased genetic risks to the offspring, as opposed to the potential social and economic benefits [1]. Consanguineous marriage is traditional and respected in most communities of North Africa, the Middle East, and West Asia, an area that stretches from Pakistan, Afghanistan, and South India in the East to Morocco in the West, with intrafamilial unions collectively accounting for 20–50% of all marriages [2–4] (Figure 18.1).

Genomics and Society
DOI: http://dx.doi.org/10.1016/B978-0-12-420195-8.00018-5
335

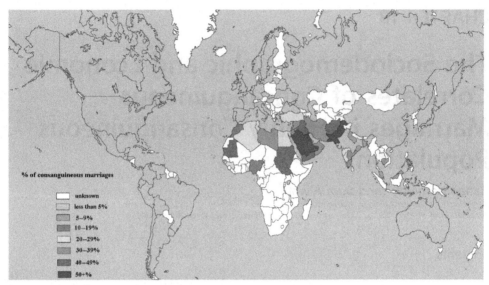

Figure 18.1 Global prevalence of consanguineous marriages.

Consanguineous marriages are also prevalent among emigrant communities from those highly consanguineous countries and regions, now resident in Europe, North America, and Australia, as well as among geographical isolates in these regions [5]. First-cousin unions are especially popular, comprising 20–30% of all marriages in some populations. On average, first cousins share one eighth of their genes inherited from their common ancestors (grandparents), so their progeny is autozygous at 1 of 16 of all loci, which is expressed as an inbreeding coefficient (F) of 0.0625 [4]. The prevalence of consanguinity and rates of first-cousin marriages can vary widely within and between populations and communities, depending on ethnicity, religion, culture, and geography [1]. The possible economic and social advantages offered by consanguineous marriage may contribute to the maintenance of this primarily cultural tradition in highly consanguineous societies. With such high prevalence of consanguinity among more than one billion of the world population, it is important to understand the underlying socioeconomic and other possible drivers for this practice.

At the Geneva International Consanguinity Workshop in 2010, a group of experts and international researchers discussed the known and presumptive risks and benefits of consanguineous marriages, and future prospects for research on consanguinity. The group highlighted the importance of counseling recommendations for consanguineous marriages and for undertaking genomic and evidence-based social research in defining the various influences and outcomes of consanguinity to better understand the balance between the risks and the benefits of consanguinity [1]. Marriage choice and decision-making is a complex interaction of various social and cultural patterns of

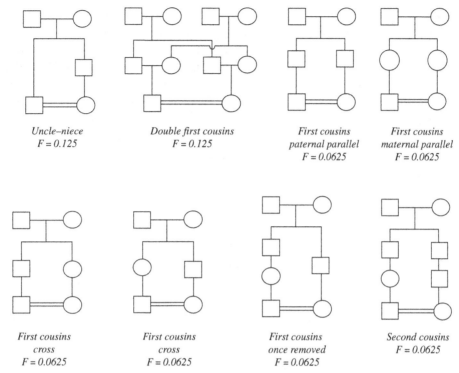

Figure 18.2 Categories of consanguineous marriages.

behavior and norms [6], and a custom that is so popular, would obviously have important social functions related to kinship pattern [7].

Although consanguinity rates are high in Arab countries, Turkey, Iran, Afghanistan, Pakistan, and South India, it seems that the socioeconomic and cultural drivers that favor consanguineous marriages vary in different countries and regions. The main subtypes of consanguineous marriages may also be different in highly consanguineous populations. For example, contrary to the custom in Arab countries, where a large majority of first-cousin marriages are preferentially contracted between a man and his father's brother's daughter (paternal parallel) (Figure 18.2), the sociocultural influences in Iran equally favor all four types of parallel and cross first-cousin marriages [8].

In Malakand District, Khyber Pakhtunkhwa Province (KPK), Pakistan, of the 1192 couples surveyed, 66.4% were related as second cousins or closer (F of 0.0338). First-cousin marriages were the most popular form of consanguineous union in Malakand, accounting for 40.3% of all marriages, with the maternal parallel first-cousin marriages between a man and his mother's sister's daughter (15.1%) and the cross-cousin marriages between a man and his father's sister daughter (16.4%) being the most common [9]. Conversely, in South India and according to Confucian tradition in China, although marriage between a man and his mother's brother's daughter is permitted, paternal parallel cousin unions are viewed as incestuous [10].

Sociocultural factors, such as the maintenance of family structure and property, ease of marital arrangements, better relationships with in-laws, and financial advantages relating to dowry, seem to be strong contributory factors in the preference for consanguineous unions [11]. In addition, there is a general belief that marrying within the family reduces the possibilities of hidden uncertainties in health and financial issues [10]. Contrary to common opinion, consanguinity is not confined to Muslim communities, as many other religious groups also practice consanguinity [12–15]. Close-kin marriage can be a strategy of conservation, with cousin marriage providing excellent opportunities for the transmission of cultural values and cultural continuity [16]. For these reasons, consanguineous unions are generally thought to be more stable than marriages between nonrelatives, although the data so far available on marital discord and divorce are small in number. In most Arab societies, paternal parallel cousin marriages are regarded as important in uniting members of the same descent group and keeping the education of offspring within the family line. These considerations may be particularly significant under conditions of social change and political or socioeconomic insecurity [17]. Higher rates of close-kin marriage have been observed among certain minority ethnic groups, especially during the initial phases of settlement of emigrant communities and refugees [18].

In the Pakistani group in the Born in Bradford research in the United Kingdom, 59.3% of the 5127 surveyed women were blood relatives of their baby's father [19]. Compared with nonconsanguineous counterparts, mothers in consanguineous relationships were socially and economically disadvantaged (e.g., never employed, less likely to have higher education). However, the Pakistani consanguineous group's social, economic, and health lifestyle circumstances were equivalent to, in some cases better than, women in nonconsanguineous relationships (e.g., up-to-date in paying bills, or in disagreeing that they wished for more warmth in their marital relationship). The consanguineous relationship group had less separation/divorce. Rates of cigarette smoking during pregnancy were lower in mothers in consanguineous relationships [19].

Marriage choices are not determined by cultural rules or preferences alone, and so they must be understood in terms of the wider political, economic, and social frameworks governing the highly consanguineous communities [20].

SECULAR TRENDS IN CONSANGUINEOUS MARRIAGES

The probability of consanguineous marriage is thought to be determined by such factors as the availability of consanguineous kin of comparable age, the similarity of socioeconomic conditions and physical traits among relatives, and traditions for or against specific types of consanguineous marriages [21].

Consanguinity rates are decreasing, increasing, or remaining stable among the highly consanguineous populations depending on factors such as mobility from rural to urban areas, trends in fertility rates, proportion of women attaining higher educational levels, and the presence of political or civil unrest. Significant secular changes in consanguinity rates have been reported in recent decades. In Jordan [22], Lebanon [23], among Palestinians [17,24], among the Israeli Arab population [25], and in South India [26], a decrease in the frequency of consanguineous marriage was seen and could be attributed to a number of factors. These could be the increasing higher levels of female education, declining fecundity with lower numbers of marriageable relatives, increased rural to urban mobility, and the improved economic status of families [1].

On the other hand, consanguinity seems to be increasing in some Arab countries including Qatar [27] and Yemen [28], possibly because of a belief that the social advantages of consanguineous marriages could outweigh the genetic risks and also because of misconceptions surrounding the nature of genetic risks among some members of the general public. In Yemen, another reason for the rising trend in consanguinity can be attributed to the increase in the availability of cousins due to high fertility coupled with the low socioeconomic conditions [28]. Social, religious, cultural, political, and economic factors still play important roles in favoring consanguineous marriages among the new generations. This is particularly the case in rural areas and among educated males but not among highly educated females [12]. In Iran, in spite of the rapid economic and social changes in the Iranian society from the 1950s to the 1970s, the overall incidence of consanguinity appears to have increased over this time period using data on 4667 women [29]. Rapid modernization and increased mobility from rural to urban settings may result in fewer opportunities to marry someone outside of one's family who is compatible with one's culture. Advances in technology and health care may cause families to perceive consanguinity as less risky. Reduction in infant and childhood mortality could lead to an increase in family size, with people having more cousins to marry from [30].

In other countries, such as Oman, consanguinity rates have remained stable with no appreciable change over the last four decades despite massive socioeconomic development and modernization [31].

The prevalence of consanguinity markedly declined in Europe, North America, and Japan in the last century [32,33], with a more recent reduction among some emigrant populations in Europe. For example, in the Norwegian Pakistani community, the proportion of women consanguineously related to their partner decreased from 45.5% in 1995–1997 to 27.3% in 2002–2005 for those born in Pakistan, and from 48.3% to 18.8% among women of Pakistani origin born in Norway [34]. Acculturation of the immigrant community in any country, with the gradual transition from their traditional consanguineous marriage preferences to those favored by the dominant group in their adopted country, could explain the decreasing trend in the consanguinity rates [35]. This, however, is not always the case. For example, the rates of first-cousin marriage in an Oxford sample and in the West Yorkshire sample of Pakistanis in the United Kingdom showed substantial increases in the younger generation in comparison with reported rates for the pioneer generation [20]. This is not simply a question of following a cultural preference for close-kin marriage; for many parents, it makes good sense to meet the obligation to consider your siblings' children as spouses for your own children [20].

Within the same country, marriage patterns are preserved when moving from a less developed to a more developed region, but could vary depending on the period spent in the new settlement. For example, in an urban slum of a metropolitan area in Izmir, Turkey, consanguineous marriages were 4.7 times more frequent for women who have moved to their new settlement within less than 10 years as compared to those who had settled for more than 10 years [36].

Among Palestinians, the decrease of marriages to distant relatives was occurring more rapidly than the decrease of marriages to first cousins, possibly due to the increased urbanization and movement restrictions. Furthermore, if fertility continues to decrease, this can have an effect on decreasing consanguinity rates and particularly first-cousin marriages through a decline in the number of cousins to choose from [17]. In the Palestinian areas, the marriage market is known to be responsive to the larger political and socioeconomic environment. It is possible that with the recent increase in restrictions on mobility and in political strife, consanguineous marriages could actually increase again due to lack of exposure to men outside the family caused by restricted mobility and also for reasons of strengthening the family unit [37].

Consanguinity rates are higher among couples whose parents are also consanguineous as documented in studies in Saudi Arabia [38], Jordan [22], among the Israeli Arab population [25], and in Mangalore, India [39]. In Tabriz city in the North-West of Iran, among 166 consanguineous couples, parents' consanguinity was associated with 84% more chance of consanguineous marriage in men but no association was seen in women [40]. In Izmir, Turkey, consanguinity was 3.8 times higher if the woman or her husband had consanguineously married ancestors [36].

WHAT WILL BE THE FUTURE TREND OF CONSANGUINITY IN HIGHLY CONSANGUINEOUS POPULATIONS?

Most studies on consanguinity rates in highly consanguineous populations have detected minor changes, whether a rise or fall in the rate in recent generations, with the conclusion that this cultural trend is likely to remain stable in the future, or possibly decline at a slow rate.

The following discussion is based on the analysis of results of a number of studies done in Iran that focus on changes in consanguinity rates [41,42]. Consanguineous marriages were still prevalent among the new (46% of all marriages), as compared with 42% for their parents' generation [42]. The apparent stability of consanguineous marriage may be attributed to the multiethnic nature of the society, as there are several ethnic groups in Iran with their own family structure, dynamics, and values. Increased rates were seen in the 1980s with a high increase in marriages between relatives more distantly related than first cousins, which may suggest increased community endogamy at the time of social change. This increase in relative marriage could be a response to traditional and religious values propagated during, and soon after the Islamic revolution [41]. The past high fertility rates also contributed to stable consanguinity rates despite the rise in numbers of females with higher levels of education [41]. It is possible that females moving up the educational scale are more conservative than those educated to the higher level in the previous cohort [41].

Conversely, a possible factor that could contribute to the decline of consanguinity in Iran is the decrease in the rate of the arranged marriage custom (53% in the older generation of women surveyed and 32% in the new generation), since marriage to a biological relative was more common among marriages arranged by parents. Moreover, within a decade or so, the much smaller birth cohorts of the 1990s onward in Iran will be reaching the ages of marriage and it may be increasingly difficult to find a cousin of the right age to marry [41,42].

A different view from Oman was seen in a cross-sectional study conducted using a self-administered questionnaire to 400 Omani adults aged 20−35 who attended primary healthcare institutions at the South Batinah Governorate in Oman. The majority (72.4%) of the participants who were single did not prefer to get married from among their relatives whereas 27.6% preferred consanguineous unions [43].

It seems that the trend for marrying within the family in highly consanguineous populations will continue, probably with some increase or decrease depending on the factors that influence people's decisions to marry cousins.

HEALTH IMPACT ON OFFSPRING OF CONSANGUINEOUS PARENTS

Consanguinity does not seem to be associated with elevated rates of miscarriages, as a large majority of studies have failed to detect any significant increase in fetal loss rates among consanguineous couples [3]. A meta-analysis of stillbirths showed a mean excess of 1.5% deaths among first-cousin progeny [44]. Multinational studies among first-cousin offspring indicated a mean 1.1% excess in infant deaths [44] with an equivalent excess of 3.5% in overall prereproductive mortality [2]. The prevalence of congenital anomalies in the offspring of first-cousin marriages has been estimated to be 1.7−2.8% higher than the background population risk, mostly attributable to autosomal recessive diseases [1,3,45]. In a recent study among Pakistani population in Bradford, consanguinity was associated with a doubling of risk for congenital anomaly (multivariate RR 2.19, 95% CI 1.67—2.85) with no association with increasing deprivation [46].

WHY ARE CONSANGUINEOUS MARRIAGES PREFERRED AND RESPECTED IN SOME POPULATIONS?

Consanguinity is a highly compound and versatile topic because it has both negative and positive impacts on society and health. The positive impacts on society include easier marriage preparations, improvement of female autonomy, more stable marital associations, lower domestic violence, and the easier economic settlement for dowry [2,17,47]. The negative impact on health is the increased risk of mortality and morbidity in offspring of consanguineous parents [1,44,48].

Consanguinity is a practice that has enjoyed much support historically in certain populations and continues to be popular among these communities to the present day. A first explanation for consanguinity is that it is the outcome of personal preference that is influenced by cultural or religious aspects [49]. The question here is why should this practice continue even with changes in religious and cultural practices concerning marriage, or why this practice is unaffected by recent scientific evidence that suggests that undertaking such a marriage can increase the likelihood of congenital anomalies in the children of such unions. Factors other than culture may have an appreciable impact in sustaining this trend in present day high consanguinity societies [49].

Consanguineous marriages are thought to strengthen the extended family ties, allow better knowledge of the spouse economic and health issues, and simplify the cultural premarital negotiations. Economic reasons for favoring consanguineous marriages could be the reduced requirement for dowry in low economic settings, and keeping the land and wealth within the family in higher economic settings [1,50]. Another reason for consanguinity is that it may be a favored form of marriage simply because it can significantly reduce the costs of searching for a suitable partner, and both partners and their parents know each other which reduces the uncertainty about the compatibility of spouses and families [49].

Among a sample of 400 Muslim women in Nepal, the main reasons that were given for supporting consanguineous marriage were culture/tradition (40.8%), no financial burden (26%), family pressure (11.5%), ease of identifying partner before marriage (10%), and illiteracy [47]. Women in Bangladesh reported that cousin marriage improved their relationships with their in-laws and reduced dowry payments; men reported that cousin marriage improved the spousal relationship, allowed the wife to stay closer to her family, and avoided the splitting of inherited property [51]. The vast majority of respondents who married a cousin did so based on their parents' wishes, which suggested that the benefits of consanguinity also accrue to parents of the couple [51].

In Karachi, the largest city of Pakistan, consanguineous marriages were preferred across all ethnic and religious groups to a varying degree. Parents continue to be the prime decision-makers for marriages of both sons and daughters. The major reasons for preferring consanguineous marriages were sociocultural rather than any perceived economic benefits, either in the form of consolidation of family property or smaller and less expensive dowries. Among Muslims, following religious traditions is the least commonly cited reason for such marriages [52]. Remaining within the inner bounds of clan endogamy was a matter of pride as it implied that the family was "sought after" and hence had a better social standing than one that opts for exogamous marriages. Also an exogamous marriage may imply that the parents were unable to solicit a match within the family because of some undesirable traits in their daughter. Across all social strata in Pakistan, consanguineous marriages attached far less importance to the physical attributes of the bride-to-be [52]. The economic rationale of consanguineous marriages in terms of consolidating property was mentioned by only 1.4% of the women as a major consideration in their own marriage [52].

Among the Pakistani community in the United Kingdom, the advantages to marrying within the family included: strengthening family ties, providing economic opportunities for family members, ease of finding a suitable match, and increased sustainability of the marriage. Respondents commented: "When you marry outside you don't know what you are getting. It's about security, about knowing your ancestry also." In many cases, going ahead with a cousin marriage is seen to be respecting the wishes of older relatives [53].

In an urban slum of a metropolitan area in Izmir, Turkey, the decision-maker for the consanguineous marriage was the family in 41.2% of cases. The major perceived advantage of marrying within the family was that "families knew each other" (14.2%), easier for the couple to understand each other (5.6%), and the strengthening of family ties (3.4%). The leading disadvantage perceived was "the children may be disabled" (35.2%), and that it may cause family conflict (12.5%) [36].

Among 552 households in Bekaa region in Lebanon, the most cited reason for preference of consanguineous marriages was reported as the social protection of the woman and the second being the promotion of stronger family ties and tradition [54].

Reasons given for the traditional popularity of consanguineous marriage in highly consanguineous populations usually focus on their perceived social and economic advantages, including simplified premarital negotiations, the assurance of marrying within the family, strengthening of family bonds, and minimal dowry requirements or demands resulting in the maintenance of family. It also has been proposed that female status is protected in a consanguineous union, because of the preexisting family relationship between a bride and her in-laws. There is evidence of the difficulties faced by families in arranging marriages with nonrelatives, when multiple family members are affected by a serious inherited health disorder. For such families the only realistic option may be to resort to even higher rates of intrafamilial marriage [55].

DOES CONSANGUINITY MINIMIZE INTIMATE PARTNER VIOLENCE?

The World Health Organization defines intimate partner violence as "behaviour within an intimate relationship that causes physical, sexual or psychological harm, including acts of physical aggression, sexual coercion, psychological abuse and controlling behaviours" [56]. Domestic violence is globally endemic and adversely impacts the health and economic well-being of women and society.

In a study on 400 women aged 15–49 years in Nepal, nearly two-thirds of the women had experience of marital violence. Marital violence was less common among consanguineous unions than nonconsanguineous unions [47].

Another study was undertaken to determine if there is any association between domestic violence and being married to one's cousin in the city of Rawalpindi in Pakistan, using the standardized and validated assessment instrument "Woman Abuse Screening Tool". Cumulatively, 1010 married women were interviewed. Emotional abuse was the most commonly reported abuse, reported by 71.4% of women as either often or sometimes, followed by sexual abuse and physical abuse, reported by 52.2% and 50.6% of women, respectively. Being married to one's cousin did not protect married women from being abused either emotionally or physically by their husbands, while sexual abuse was more likely to be reported by women not married to cousins. Respondents not married to cousins were more likely to report feeling frightened by what their husbands say or do, compared with those married to their cousins; knowing the man as one's cousin prior to marriage perhaps makes one feel less or not frightened, although abuse is still being experienced and endured [57].

The consanguinity rate among 5606 ever married women in Egypt was 24.2%. Among those married to a paternal cousin, 31% have ever experienced any form of domestic violence as compared to 34.9% of those married to a maternal cousin and 36.2% of those not married to a relative. Marital violence was somewhat less likely if the woman's husband was a first or second cousin, particularly a paternal cousin, than if the woman was married to a more distant relative or to a nonrelative [58].

Evidence from around the world suggests that potentially large numbers of women are at risk of domestic violence during pregnancy. In Jordan, a survey on a sample of 517 literate, ever-married women was conducted to estimate the lifetime prevalence of physical violence during pregnancy and examine the risk and protective factors associated with its occurrence. More than half of the respondents (54.1%) were not related to their spouses, 33.3% were married to their first or second cousin, and 12.6% were married to more distant relatives. Being more distantly related to their spouse conferred an increased risk of violence compared to being married to a first- or second degree. Among nonconsanguineous couples, 15.6% of women were abused during pregnancy, compared to 11.5% of women married to their first or second cousins, and among 24.5% of those married to more distant relatives. Thus being married to a close relative was found in this study to be protective against violence, but only to an extent [59]. In another study from Al-Mafraq in Jordan, spousal violence during pregnancy was reported among 40.9% of 303 women with no significant difference among consanguineous and nonconsanguineous couples [60].

In a study from Pakistan, a total of 300 women in the postnatal wards of a public tertiary hospital were administered a structured questionnaire; 44% of women reported lifetime marital physical abuse and 23% during the index pregnancy. Consanguineous marriages were common (49%) with most marriages being among first cousins (22% maternal; 20% paternal). Premarital family relationship appears to play a significant protective role in an abusive relationship. Women whose spouses were distantly related were nearly four times as likely to be in an abusive relationship as compared to women married to their first cousins. Maintaining spousal and family harmony is perhaps another reason for the continued high prevalence of first-cousin marriages in Pakistan in addition to maintaining wealth within the family [61].

It seems that women in first-cousin marriages are somewhat protected against intimate partner violence more than women married to distant relatives or in nonconsanguineous relationship. However, other reports indicated that there was no significant advantage or disadvantage reported for consanguineous marriage in association with domestic violence among low-income women in Syria [62] or Palestinian refugees in Lebanon [63].

A study compared forgiveness in secure and insecure Iranian married couples taking consanguinity into consideration [64]. Participants completed the Family Forgiveness Scale and the Adult Attachment Questionnaire. When one is hurt, a salient reaction is to be offended. However, unhealed psychological wounds, unexpressed feelings, and unforgiven and unresolved issues which remain within the hurt person cause the most pain and psychological damage. Results showed that the difference in reported forgiveness between individuals in consanguineous marriages versus those in nonconsanguineous marriages was only significant in one aspect of forgiveness within the present marital relationship, namely, recognition. Consanguineous couples reported significantly more recognition of the painful event requiring forgiveness than did those in nonconsanguine marriages, which means that they evaluated the offense committed as less painful and were able to forgive more readily. In light of the present study's finding, it seems that consanguinity may be associated with forgiveness, although this role needs to be clarified by further research. Higher ratings on the recognition subscale of the forgiveness scale by individuals in consanguine marriages reflect the possibility that blood ties with the spouse's family lead to underestimation of the offense experienced by the individual [64].

CONSANGUINITY AND DIVORCE

Although comparative data on the stability of consanguineous versus nonconsanguineous marriages are limited, all available evidence suggests that marriage dissolution and divorce is lower among cousin couples, in part because of the highly disruptive effect marriage failure could have on other relationships within the extended family [50,65].

A study was performed in Iran to investigate the association between consanguineous marriages and divorce risk among 496 couples at divorce time and 800 couples from general population who have no plan for divorce (as control group). Compared to unrelated marriages, first-cousin (OR = 0.39, 95% CI: 0.27–0.56, $P < 0.001$), first cousin once removed (OR = 0.18, 95% CI: 0.05–0.62, $P = 0.006$), and second-cousin marriages (OR = 0.37, 95% CI: 0.17–0.78, $P = 0.009$) decreased the risk of divorce. Survival of marriage was lower significantly for unrelated marriages than for first-cousin marriages, after adjustment for educational level. These findings indicate that consanguinity has some protective role against divorce [66]. In a study on a large sample size from the Born in Bradford cohort study, the authors reported that divorce was extremely rare being 0.8% in Pakistanis in consanguineous relationships and 1.7% in Pakistanis not in consanguineous relationships [19]. An early report from Sudan indicated greater marital stability in consanguineous unions, irrespective of the type of cousin relationship, with divorce in 3.6% of first-cousin marriages compared with 14.6% in other types of marriage [67].

From the selected households in the 2000 Oman National Health Survey, 2037 eligible women were interviewed. Consanguineous marriages were more likely to be stable than nonconsanguineous marriages, as indicated by the lower proportion of polygamous marriages and divorce among consanguineous couples than among nonconsanguineous couples. However, the difference was not statistically significant [31].

It seems that consanguineous unions are generally thought to be more stable than marriages between nonrelatives, although the data so far available on marital discord and divorce in highly consanguineous populations are very small in number.

CONSANGUINITY AND CIVIL UNREST

It has been hypothesized that in times of civil strife, intrafamilial marriage is an advantageous strategy to optimize the security of couples, families, and their communities [5].

In the Arab culture, social life and identity focus on the family and a small-intermarried group would tend to have tighter familial bonds. Thus, the family is a refuge of safety that will influence options for marriage partners [68], and at the local level, the clan traditions provide more support and stability than Western institutions [69]. The civil unrest and the hostility between tribal and religious factions that has developed in Iraq in recent years could have reinforced the cultural trend of marrying within the family. It is only logical that civil strife between social groups in the same society would tend to reduce the apparent number of marriage partners outside the family group because close family members can be trusted more than nonrelatives, and only family members will be deemed suitable marital choices [68]. Common saying include: "The traditional Iraqis who marry their cousins are very suspicious of outsiders," and "It is safer to marry a cousin than a stranger" [69].

Among the Bedouin community in Lebanon, consanguineous marriages were seen as reinforcing the unity and authority of the lineage. Rather than creating new ties, this form of marriage consolidated old ones among this community which is marginalized and discriminated against [70].

In Tehran, Iran, there has been a significant increase in the consanguinity rate in marriages contracted after 1979, as compared to the two previous generations. The Islamic Revolution in 1979 was a significant political event involving the change from a kingdom to a democratic system based on Islamic law. This may have had an effect on cultural practices such as consanguineous marriages by highlighting the importance of family as a core unit of the society [30]. In Iran, in parts of the country which continue to experience civil unrest, such as Sistan-Baluchistan, a decline in consanguinity is improbable, since close-kin marriage is regarded as a beneficial means of safeguarding personal and family security [8].

In Malakand District, KPK, Pakistan, among 1192 couples surveyed, data suggest that the prevalence of consanguineous unions has been increasing during the last decade, in response to the high levels of violence across KPK [9].

In the Palestinian Territories, the possible effect of political strife on consanguinity was reflected in the increase of first-cousin marriages between 1989 and 2000, a time that partly corresponds to the onset of the first Intifada in December 1987. The increase in consanguineous marriages was mostly among first cousins because this is seen as a closer and stronger relation, and first cousins would more likely be living in the same area than more distant relatives [17].

Close-kin marriage can be a strategy of conservation, with cousin marriage providing excellent opportunities for the transmission of cultural values and cultural continuity. These considerations may be particularly significant under conditions of social change and political or socioeconomic insecurity [1].

CONSANGUINITY AND SPOUSES EDUCATIONAL LEVELS

A number of studies have indicated that the rate of consanguineous marriages in highly consanguineous populations is affected by the educational levels of both women and men.

In Nepal, for example, consanguineous marriages constituted 36.7% of all marriages and higher consanguinity rates were associated with higher educational levels of husbands of 400 women interviewed in 2011 and 2012. The association of being in a consanguineous marriage among women whose husband's education level were secondary or higher was 3.35 times greater than among those whose husbands were unable to read and write. In this study, no significant differences were observed between consanguinity and women's educational level [47].

Among 1290 Israeli Bedouin women, where the total prevalence of consanguinity was 44.8%, there was a significant association between consanguinity and fewer years of schooling of the wife and higher years of schooling of the husband. It is possible that educated men are considered an asset to the family because they may earn more and can support large families and accumulate assets and property, and hence, their commitment to marry a relative is greater [71]. Similarly information on 6437 couples from the Israeli Arab population showed that among those who married in 2005—2009, there was a significant association between consanguinity and more years of schooling of the husband [25].

Using data on 4667 women from the 1976—1977 Iran Fertility Survey, the incidence of consanguinity is roughly the same for all three levels of husband's education which is in sharp contrast to wife's education, for which higher status leads to lower consanguinity [29]. In another study from Iran among 5515 women, consanguinity was higher among women with lower levels of formal education [8]. Again from Iran, data on the consanguinity status for a total of 1789 marriages in three generations in Tehran showed that a higher educational level for females was associated with a lower tendency to have a consanguineous marriage in the most recent generation [30]. Among 6550 ever-married women in four provinces in Iran, the modernization variable, education, had little significant effect upon behavior, the effect being only for those with tertiary education. There was no significant difference in the levels of consanguineous marriage between women who were illiterate and those who had primary and secondary education. Only those with tertiary education were significantly different, being 40% less likely than illiterate women to marry a relative [41].

In South India, the data from the National Family Health Survey, 1992—1993, among 16,969 ever-married women indicated that the chances of marrying a nonconsanguineous person increased with the higher level of education among women. [26]. Similarly, in Mangalore, India, consanguineous marriage was seen significantly more among illiterates [39].

In Tunisia, men with the highest education level and professionals had the lowest prevalence of consanguinity, but the differences were not statistically significant. For women in Tunisia, however, a strong trend of increasing prevalence of consanguineous marriages with decreasing education level was noted [72].

In Yemen, among 9726 women, it was shown that women with only primary schooling are more likely to be in consanguineous unions than women with more advanced education, 45.5% versus 38.5%, respectively. This reflected the still prevailing practice among Arab parents to drop their daughters from school when the prospective groom becomes available. As was seen in other studies in the region, consanguineous unions were more prevalent among highly educated men (43.2%) than among their illiterate counterparts (36.2%) [28].

In the Palestinian Territories, in 1995 roughly 44% of women with at least secondary education entered consanguineous marriages compared with 54% of women with less than elementary education; in 2004 it was 42% compared with 49%. This difference in the education level of the women as related to consanguinity was not significant [17].

The results seen in a study from Turkey indicated that consanguinity rates were higher among both males and females with lower levels of education than those with higher levels [73].

Among 2037 women interviewed in 2000 in Oman, the women's education showed little or no effect on first-cousin marriage rates, while more remote consanguineous marriages were associated with higher education levels. Overall, the differences in the rates of all consanguineous marriages across the different levels of education were very modest [31].

An interesting and converse finding to most other studies was seen in a study among the Bedouin in the Bekaa Valley of Lebanon, where an exploration of the relation between consanguinity and literacy rates revealed an inverse relationship between literacy and instances of consanguineous marriages. A high proportion (57.78%) of females who were unrelated to their husbands lacked literacy skills, in contrast with 42.86% of those married to their first cousins. Findings also indicated that 28.57% of those in first-cousin marriages held elementary education, in comparison to 17.78% of those unrelated to their husbands [70].

In Japan, a survey on 9225 couples showed that among the educational groups, the F value was highest in graduates of junior high school for husbands (0.0025) and wives (0.0024), whereas the F value was lowest in graduates of a college, university or graduate course for husbands (0.0006) and wives (0.0005) [74].

Most studies have indicated that higher education levels in females lowered the option of marrying a cousin, in contrast to higher education levels in males which increased or did not affect the option of marrying within the family. Few studies showed controversial results.

CONSANGUINITY IN RELATION TO SPOUSE EMPLOYMENT STATUS

Most studies researching the association of consanguinity with the work status of spouses in highly consanguineous populations showed that women in the work force are less likely to marry their cousins.

Among 9726 women in Yemen, it has been found that female occupational status prior to marriage is a significant determinant of whether unions are consanguineous or not, with women working for cash being less likely to be in consanguineous marriages. A similar statistically significant relation has been encountered between consanguinity and current occupational status, with self-employed women in the agricultural sector and white-collar workers being less likely to be in cousin marriages than their blue-collar or nonworking counterparts [28]. On the other hand, an unexpected finding was the lack of inverse association between husband's occupation and consanguinity. It was found that cousin unions are more prevalent among women whose husbands are engaged in the modern sector of the economy [28].

A study on 4667 women from the 1976—1977 Iran Fertility Survey revealed that it is not formal employment, but the nature of work that makes a difference in the secular trend of consanguinity rates. Women who worked in agriculture or in production jobs were just as likely as women who did not work to marry their cousins. The most common nonagricultural employment for Iranian women is carpet weaving where women typically work in their own homes in rural areas in quite traditional settings. Among the latter women, consanguinity rates were increasing. It is only for the very small number of women working in service occupations ("modern sector" employment) before marriage that the incidence of consanguinity decreased over time [29].

In Turkey, among 8075 women, examination of the working status revealed that the practice of consanguineous unions was significantly less prevalent among women working with social security (social and health insurance supported by the employer) (9%), which is mostly linked with higher education, compared with women not working (23%) or working without social security (24%) [73].

A study in Monastir, Tunisia, indicated that the proportion of housewives was greater in the consanguineous group (79.9%) than that in the nonconsanguineous group (64.3%), whereas the proportion of skilled women was higher in nonconsanguineous group (21.7%) than in the consanguineous group (14.7%) [72].

Among 2037 women interviewed in 2000 in Oman, a study indicated that women who were engaged in gainful employment and earning cash were less likely to have a consanguineous marriage than those who were housewives (39.8% vs. 53.4%) [31].

In Japan, in 1981, the F value was the highest in agriculture, forestry, and fishery for husbands occupation (0.0035) and wives (0.0032), whereas the rate was the lowest among salesmen, for husbands (0.0006), and in professional occupations and researchers, for wives (0.0004) [74].

Conversely, among 16,197 women in 1995 and 4971 women in 2004 in the Palestinian Territories, women's labor force status was found to be a nonsignificant predictor of consanguinity after controlling for the survey year [17].

Engagement of women in the more professional workforce could be a factor in lowering the consanguinity rate in highly consanguineous populations, but only to a certain extent. For men, this association of consanguinity rate with employment does not seem to hold true.

AGE AT MARRIAGE AND FERTILITY RATE IN CONSANGUINEOUS MARRIAGES

In most studies on the association of consanguinity with age at marriage, it was found that women married to their first cousins married at an earlier age than women not married to a relative [1,3]. The reasons for this association are many, among which is the fact that marriage arrangements for cousins are easier and less costly for the families. Moreover, the couple would have known each other for a long time before marriage.

Another suggested reason for the early female age of marriage in consanguineous unions is the economic driver. Unfortunately, the marriage of a very young daughter creates a secondary gain for her primary family if it relieves the low economy household of a member who cannot generate income. When an educated woman can contribute to the income of the family, there is no financial advantage to the household for the woman to marry earlier. Further, women in the workforce are more visible socially and typically in closer proximity to unrelated eligible males [68].

Among 1290 Israeli Bedouin women, there was a significant association between consanguinity and women younger age at marriage (20.1 ± 2.9 vs. 21.1 ± 3.4) [71]. A similar finding was reported from Yemen [28], and from Oman [31]. Information on 6437 couples from the Israeli Arab population showed that women in consanguineous marriages tended to marry at a younger age than did those who were in nonconsanguineous marriages [25]. Similarly, among 141,000 couples in the Gaza Strip of Palestine, the average age of those with first-cousin marriages is significantly lower (22.4 ± 4.4 years) than those with second-cousin marriages (24.3 ± 6.1 years) and the nonconsanguineous (26.5 ± 8.2 years) [75]. In another study in the Palestinian Territories, data from both the 1995 and 2004 surveys showed that over 70% of the women who marry over the age of 25 entered nonconsanguineous marriages [17]. Similarly, in South India, data indicated that a woman marrying at an older age has a relatively higher probability of marrying a person unrelated by blood than a woman marrying early [26]. In Mangalore, India, the mean age at marriage was found to be significantly lower among women married consanguineously [39].

In the latest demographic survey (TDHS-2003) in Turkey, it was seen that women in consanguineous unions were more likely both to be married and to start childbearing at an earlier age [73].

This trend of females marrying a cousin at an earlier age was also detected among immigrant populations. In a retrospective analysis of 1964 primary consultations at the Prenatal Genetic Counseling Outpatient Clinic at the Medical University of Vienna General Hospital in Austria, 8.9% of all patients lived in a consanguineous union. Consanguineous patients were significantly younger and had significantly more often a non-Austrian background than nonrelated control [76].

Studies have reported that consanguineous couples tend to produce more children than unrelated couples [3]. Reasons given for this finding could include that first-cousin couples may experience a higher risk for infant and child mortality than non-related couples, and the higher fertility could be regarded as a compensation mechanism. Another reason could be the longer reproductive span for consanguineous couples due to the earlier age at marriage. Some studies have documented a lower use of contraceptive methods among consanguineous as compared to nonconsanguineous couples [8]. In a meta-analysis of the fertility of first cousin and nonconsanguineous couples, first cousins had a higher mean number of live births in 33 of the 40 studies, which translated into a mean 0.08 additional births per family [44]. Currently, it is unclear whether the apparent greater fertility of first-cousin couples represents compensation for their increased risk of postnatal losses or may primarily be due to their younger mean age at marriage, earlier first pregnancy, and longer reproductive span with lower use of contraception methods [3].

Among 5515 women living in areas in Iran known to enjoy a high socioeconomic status with low fertility rates, 37.4% of the marriages were consanguineous. In general, consanguineous couples had higher mean numbers of pregnancies, live births, and surviving children. The lower mean age at marriage of consanguineous parents offers an enhanced opportunity to initiate childbearing at younger ages [8]. There is also evidence that, on average, women in consanguineous relationships continue child-bearing to comparatively later ages, thus extending their reproductive span. Interestingly, In Sardinia, the mean ages of women in consanguineous marriages at first and at last child are significantly higher than in the rest of Sardinia [77].

Among 400 Muslim women in Nepal, 36.7% were in a consanguineous marriage. Women in consanguineous marriages were less likely to have used contraceptive methods than those married to nonconsanguineous marriage, and consanguineous unions had more children compared to nonconsanguineous unions [47].

Analysis of the Karachi and the 1990–1991 Pakistan Demographic and Health Survey data sets of fertility differentials indicated that the higher fertility among consanguineous couples could be a reflection of reproductive compensation and greater exposure time (lower age at marriage and low use of contraception), rather than representing an underlying biological difference in fecundity which could be attributed to consanguinity [78]. However, in a study among 160,811 Icelandic couples from the deCODE Genetics genealogical database born between 1800 and 1965, results showed a significant positive association between kinship and fertility, with the greatest reproductive success observed for couples related at the level of third and fourth cousins. Owing to the relative socioeconomic homogeneity of Icelanders, and the observation of highly significant differences in the fertility of couples separated by very fine intervals of kinship, the authors concluded that this association is likely to have a biological basis [79].

ECONOMIC DRIVERS FOR CONSANGUINEOUS MARRIAGES

Economic factors that contribute to the preference for consanguineous marriages in some populations could be the maintenance of family property among the rich and the reduction of the dowry value among low-economy families.

Dowry, *jehez or mahr*, is the major economic transaction that underpins marriage decision-making. In Arab countries, dowry in the form of money or jewelry is presented to the bride from the bridegroom. In South East Asia, dowry (land, vehicle, gold, money) is given from the bride side to bridegroom side.

Negotiations on the dowry are more simple when the marriage is within the family, and the value of the dowry is generally less in consanguineous than in nonconsanguineous marriages. In consanguineous settings in Pakistan, the demands for dowry of the potential in-laws (who are also close relatives) are likely to be lower and more realistic than in nonconsanguineous marriages. The bride and her family are less likely to be "penalized" for any perceived shortcomings in the expected dowry [52]. In Bangladesh, marrying consanguineously reduces the need for dowry payments for the bride's parents who have neither cash on hand nor access to credit to make an up-front dowry payment and instead it becomes possible to promise *ex-post* payments [51]. It was suggested that dowry payments could be 25% lower in consanguineous than in the nonconsanguineous marriage [51].

More than 50% of the women interviewed in Nepal said that their family agreed on dowry and that the dowry practice was more common among nonconsanguineous union compared with consanguineous unions [47].

In Bangladesh, unlike the custom in Arab countries, the dowry is given by the bride family to the bridegroom's and the choice to marry a cousin is affected by the wealth of the family of the bride more than that of the bridegroom. Among 52,000 marriage decisions between 1982 and 1996, it was seen that changes in marital prospects for wealthier households relative to the less wealthy households were seen [51]. The brides from households protected from floods on one side of the river commanded larger dowries, married wealthier households, and became less likely to marry biological relatives than households on the other side of the river with no embankment construction (not protected from flooding). Because only brides have to pay the dowry, the embankment relaxed a relevant liquidity constraint only for the bride's family. Thus poor households engage in consanguinity partly in response to their inability to pay dowries up front at the time of marriage [51]. This study showed that indirect social equilibrium changes can be quite substantial when environmental and economic changes affect a population [51].

In an empirical analysis using data from Bangladesh, Do et al. [49] concluded that the practice of consanguinity is closely related to the practice of dowry, but that the relationship is nonlinear. At very high levels of wealth for example, households preferred to contract marriages within the kinship network and resisted the transfer of wealth to unrelated individuals. In low-economy settings when marriage takes place at an early age, parents might face steeper cash constraints as they have had less time to accumulate assets suggesting that a consanguineous union might be chosen instead [49].

Among the Bedouin community in Lebanon, a pattern of brother–sister exchanges—often to cousins of some degree—has become more common as a way of managing the rocketing cost of the "mahr" or bride price given to the bride and her family prior to marriage by the groom and his family. These exchanges are not usually noted in conventional surveys and thus fall to be detected as consanguineous marriages [70].

In the Palestinian Territories, the deteriorating economic situation caused by the political unrest could increase consanguinity rates as marrying within the family might imply lower dowry and would keep family wealth and land within the family. The findings of the study showed that the wealth index was significantly associated with the odds of consanguineous marriage [17].

Among, the Fulani of northern Burkina Faso, consanguineous marriages are frequent and arranged first marriages are accompanied by the payment of bridewealth. A study testing the hypothesis that inbreeding may be more frequent when there is a scarcity of cattle available, since bridewealth demands are thought to be reduced with close-kin marriage, found that no increase was found in population levels of inbreeding estimated from marriages contracted after the droughts of 1973 and 1984, which drastically reduced the Fulani's cattle stocks. However, a significantly higher rate of consanguineous marriage was found in families owning the fewest cattle. There was a strong relationship between poverty and consanguineous marriage. It seems that, while close-kin marriage is favored by wealthy and poor alike, the poorest and most destitute groups are more restricted to marrying very close kin while wealthier Fulani are able to contract marriages among a wider range of kin [80].

It could be concluded that with the generally accepted cultural and social drivers for consanguineous marriages, economic drivers markedly influence marriage decisions in highly consanguineous populations.

SOCIOECONOMIC STATUS AND CONSANGUINITY

Studies focusing on the social aspects of consanguinity have largely indicated that consanguineous marriages are more common among the lower socioeconomic settings.

In the Palestinian Territories, consanguinity was seen to be affected by women's standard of living as measured by the wealth index. The highest consanguinity rates were found in women with the lowest standard of living (or the first quartile of the wealth index) at 53.4% in 1995 and 50.2% in 2004 [17]. Similarly, among 1290 Israeli Bedouin women there was a significant association between consanguinity and lower monthly income of household [71]. Also, in India, among 16,969 ever-married women, the lowest socioeconomic cultural group had the highest level of inbreeding [26]. In Turkey, consanguinity decreased significantly with household welfare showing that consanguinity among the poorest women was 28% as opposed to only 13% among the richest women [73].

In Tabriz, Iran, a study indicated that the lower grade of father's job in both women and men was associated with higher occurrence of consanguineous marriage and remained significant after adjustment for other determinants in men but not in women. Men whose fathers were not working were five times more likely to marry consanguineously compared to men whose fathers were employees [40]. In Tehran, Iran, socioeconomic level of families was not significantly related to having a consanguineous marriage [30]. Another study from Tehran, Iran, however, could not detect an association between the socioeconomic level of families and consanguinity [30]. If modernization theory holds, then it would be expected that in Iran provinces with a higher level of development, a lower rate of consanguinity is expected. However, Yazd province with its high level of development had a high level of consanguinity that could be related to its conservative nature, as it is also known as "the home of prayer" with the women expressing more conservative attitudes toward family and childbearing aspects of life [41].

These generalizations are by no means uniform. For example, upper socioeconomic status males in Kuwait favored consanguineous marriage [81]. In all cases, it is not easy to compare reports on socioeconomic associations with consanguinity, because prevalence rates of consanguineous marriages could show major regional and ethnic differences within the same population.

CONSANGUINITY AND RELIGION

Contrary to common opinion, consanguinity is not confined to Muslim communities. Many other religious groups, including the Lebanese, Jordanian, and Palestinian Christian populations, also practice consanguineous marriage, although to a lesser extent than among co-resident Muslims [3], whereas in the Hindu population of

South India, more than 30% of marriages are consanguineous, with 20% between uncles and their nieces [1,15].

It has been argued that consanguinity must be seen as a cultural rather than an Islamic or religious trait [6]. It is commonly and somewhat erroneously believed that Islam favors marriage between cousins. However, no passage in the Koran can be interpreted as encouraging consanguineous marriages. Moreover, according to one of the *hadith* (a record of the pronouncements of Prophet Mohammad), the Prophet discouraged marriages to relatives, the underlying rationale perhaps was to ensure wider marriage alliances which would facilitate the spread of Islam through Arabia [52]. In in-depth interviews with consanguineous couples in Karachi, Pakistan, religion was never cited as a major reason for contracting cousin marriages [52].

Among 6550 ever-married women in four provinces in Iran, Sunni women were more likely to experience consanguinity. Being of the Shiite sect was a strong predictor of opposition to consanguineous marriage among women who themselves had married a relative [41]. The Shiite sect is a branch of Islam where, after death of the Prophet Mohammad, the customs of the 12 Imams are followed by the school of thought. Of the 11 marriages resulting in the birth of an Imam, there were only two familial marriages: Imam Ali with Fatima (first cousins once removed) and Imam Sajjad with his first paternal cousin Om Abdullah (first-cousin marriage). All the others (82%) were unrelated. Indeed, it is interesting to note that the majority of these marriages were with women from distinct ethnicities [82].

Information on 6437 couples from the Israeli Arab population showed that the highest rates of consanguineous marriages were tabulated for the Druze (33.1%), followed by Muslims (28.3%) and Christians (20.2%) [25].

The total number of couples surveyed in Japan in 1983 was 9225, chosen from six widely different areas. Among six religious groups, the mean inbreeding coefficient (F) was the highest in Buddhists (0.0019) and lowest in "no religion" (0.0008) and Catholics (0.0009). In the five areas, the F value is higher in Buddhists than in the "no religion" group in each marriage year [74].

Consanguinity rates remain to be highest in countries with mostly Muslim populations and among Muslim emigrants in Europe, Australia, and North America.

CONSANGUINITY AS RELATED TO ETHNICITY AND URBAN/RURAL SETTINGS

Ethnicity and place of residence are considered as important determinants of the frequency of consanguineous marriages as seen in many studies conducted in highly consanguineous populations. Urban to rural first-cousin rates in Algeria were 10% and 15% [83], in Egypt, 8.3% and 17.2% [84], and in Jordan, 29.8% and 37.9% [12], respectively. Likewise the mean inbreeding coefficient was lower in urban as compared to rural settings in Syria (0.0203 vs. 0.0265) [85].

Among 6550 ever-married women in four provinces in Iran, each ethnic group had its specific level of consanguineous marriage that was maintained across the 30 years of marriages. Of the four ethnic groups, Baluch had by far the highest level of relative marriage, followed by Kurd, Azari, Fars, and Gilak [41]. Another study among 5515 women in Iran indicated that consanguinity rate was higher among rural couples [8].

In Bekaa region in Lebanon, a significant association was found between region of residence and consanguinity [54], while in Beirut, a study among 1556 women found that Beirut suburb dwelling, presented the highest rates of consanguinity [23].

In the Palestinian Territories, surveys on 16,197 women in 1995 and 4971 women in 2004 indicated that consanguinity rates were significantly higher for women in the Gaza Strip compared with those in the West Bank, as well as for women living in rural areas compared to those in urban areas in both surveys [17].

In India, among 16,969 ever-married women, the lowest socioeconomic cultural group had the highest level of inbreeding. Girls brought up in urban areas seemed to have slightly less chance of marrying close blood relatives than those brought up in rural areas [26].

In the latest demographic survey (TDHS-2003) in Turkey, a nationally representative sample of 8075 ever-married women, consanguineous marriage rates were related to the ethnicity of the spouses. When both spouses were Kurdish they had somewhat higher levels of consanguinity (45%) compared with couples where either both spouses were Turkish (18%). Also, regional differences in consanguinity rates were reported, where, consanguineous marriages were 10% in West Marmara and of 42% in southeastern Anatolia, the latter being the least developed part of the country [73].

In Oman, overall consanguineous marriage rate shows little difference between urban and rural areas (51.9% vs. 51%), and urban/rural place of residence is not a significant correlate of consanguinity [31], probably because of the high rate of rural to urban migration in recent times. A substantial proportion of urban dwellers would likely have migrated from rural areas following the contraction of their marriages. Moreover, rural-to-urban migrants might maintain the tradition of consanguineous marriage after migration in urban areas due to their low socioeconomic status. Consanguineous marriages were significantly higher among women whose childhood place of residence was in rural (48%) areas than among women whose childhood place of residence was in urban areas (44%) [31].

CONCLUSIONS

Debate about consanguinity should balance the potential protective effect of consanguineous relationships with established genetic risk of congenital anomaly in children

[19]. Studies undertaken on the socioeconomic correlates of consanguinity in highly consanguineous populations have listed a number of advantages that may encourage the new generations to continue this trend of marrying within the family. Data indicate that there are unified reasons for the preference of cousin marriages in most highly consanguineous populations, but that there are also more specific factors exerting their influence on one population and not the other. Variation according to region of origin, socioeconomic status, civil unrest, education and employment levels, cultural beliefs, and upbringing must be considered in order to understand and research the factors influencing marriage decisions in highly consanguineous settings. It could be concluded that consanguineous marriages are still favored and respected in many parts of North Africa, the Middle East, West and South Asia, as well as among emigrants from those areas to Europe and North America. Evidence-based social research in defining the various influences and outcomes of consanguinity is recommended.

REFERENCES

[1] Hamamy H, Antonarakis SE, Cavalli-Sforza LL, et al. Consanguineous marriages, pearls and perils: Geneva International Consanguinity Workshop Report. Genet Med 2011;13:841−7.

[2] Bittles AH, Black ML. Evolution in health and medicine Sackler colloquium: consanguinity, human evolution, and complex diseases. Proc Natl Acad Sci USA 2010;107(Suppl 1):1779−86.

[3] Tadmouri GO, Nair P, Obeid T, et al. Consanguinity and reproductive health among Arabs. Reprod Health 2009;6:17.

[4] Bittles A. The global prevalence of consanguinity. Available from: <www.consang.net>; 2015.

[5] Bittles A. Consanguinity in context. New York, NY: Cambridge University Press; 2012.

[6] Anwar WA, Khyatti M, Hemminki K. Consanguinity and genetic diseases in North Africa and immigrants to Europe. Eur J Public Health 2014;24(Suppl. 1):57−63.

[7] Modell B. Social and genetic implications of customary consanguineous marriage among British Pakistanis. Report of a meeting held at the Ciba Foundation on 15 January 1991: Conference Report. J Med Genet 1991;28:720−3.

[8] Hosseini-Chavoshi M, Abbasi-Shavazi MJ, Bittles AH. Consanguineous marriage, reproductive behaviour and postnatal mortality in contemporary Iran. Hum Hered 2014;77:16−25.

[9] Sthanadar A, Bittles A, Zahid M. Civil unrest and the current profile of conanguineous marriage in Khyber Paktunkhwa Province, Pakistan. J Biosoc Sci 2013;46:698−701.

[10] Bittles A. Consanguinity and its relevance to clinical genetics. Clin Genet 2001;60:89−98.

[11] Bittles A, Hamamy H. Endogamy and consanguineous marriage in Arab populations. In: Teebi AS, editor. Genetic disorders among Arab populations. 2nd ed. Heidelberg: Springer; 2010.

[12] Khoury SA, Massad D. Consanguineous marriage in Jordan. Am J Med Genet 1992;43:769−75.

[13] Khlat M. Consanguineous marriages in Beirut: time trends, spatial distribution. Soc Biol 1988;35:324−30.

[14] Vardi-Saliternik R, Friedlander Y, Cohen T. Consanguinity in a population sample of Israeli Muslim Arabs, Christian Arabs and Druze. Ann Hum Biol 2002;29:422−31.

[15] Bittles AH, Mason WM, Greene J, Rao NA. Reproductive behavior and health in consanguineous marriages. Science 1991;252:789−94.

[16] Sandridge AL, Takeddin J, Al-Kaabi E, Frances Y. Consanguinity in Qatar: knowledge, attitude and practice in a population born between 1946 and 1991. J Biosoc Sci 2010;42:59−82.

[17] Assaf S, Khawaja M. Consanguinity trends and correlates in the Palestinian Territories. J Biosoc Sci 2009;41:107−24.

[18] Shaw A. Negotiating risk: British Pakistani experiences of genetics. New York, NY: Berghahn Books; 2010.

[19] Bhopal RS, Petherick ES, Wright J, Small N. Potential social, economic and general health benefits of consanguineous marriage: results from the Born in Bradford cohort study. Eur J Public Health 2014;24:862—9.

[20] Shaw A. Kinship, cultural preference and immigration: consanguineous marriage among British Pakistanis. J R Anthropol Inst 2002;7:315—34.

[21] Cavalli-Sforza LL, Maroni A, Zei G. Consanguinity, inbreeding and genetic drift in Italy. Monographs in population biology 39. Princeton, NJ: Princeton University Press; 2004.

[22] Hamamy H, Jamhawi L, Al-Darawsheh J, Ajlouni K. Consanguineous marriages in Jordan: why is the rate changing with time? Clin Genet 2005;67:511—16.

[23] Barbour B, Salameh P. Consanguinity in Lebanon: prevalence, distribution and determinants. J Biosoc Sci 2009;41:505—17.

[24] Sirdah M, Bilto YY, el Jabour S, Najjar K. Screening secondary school students in the Gaza strip for Beta-Thalassaemia trait. Clin Lab Haematol 1998;20:279—83.

[25] Na'amnih W, Romano-Zelekha O, Kabaha A, et al. Continuous decrease of consanguineous marriages among arabs in israel. Am J Hum Biol 2015;27:94—8.

[26] Audinarayana N, Krishnamoorthy S. Contribution of social and cultural factors to the decline in consanguinity in south India. Soc Biol 2000;47:189—200.

[27] Bener A, Alali KA. Consanguineous marriage in a newly developed country: the Qatari population. J Biosoc Sci 2006;38:239—46.

[28] Jurdi R, Saxena PC. The prevalence and correlates of consanguineous marriages in Yemen: similarities and contrasts with other Arab countries. J Biosoc Sci 2003;35:1—13.

[29] Givens B, Hirschman C. Modernization and consanguineous marriage in Iran. J Marriage Fam 1994;56(4):820—34. Available from: <http://www.jstor.org/stable/353595>.

[30] Akrami SM, Montazeri V, Shomali SR, Heshmat R, Larijani B. Is there a significant trend in prevalence of consanguineous marriage in Tehran? A review of three generations. J Genet Couns 2009;18:82—6.

[31] Islam MM. The practice of consanguineous marriage in Oman: prevalence, trends and determinants. J Biosoc Sci 2012;44:571—94.

[32] Imaizumi Y. A recent survey of consanguineous marriages in Japan. Clin Genet 1986;30:230—3.

[33] Bras H, Van PF, Mandemakers K. Relatives as spouses: preferences and opportunities for kin marriage in a Western society. Am J Hum Biol 2009;21:793—804.

[34] Grjibovski AM, Magnus P, Stoltenberg C. Decrease in consanguinity among parents of children born in Norway to women of Pakistani origin: a registry-based study. Scand J Public Health 2009;37:232—8.

[35] Awwad R, Veach PM, Bartels DM, LeRoy BS. Culture and acculturation influences on Palestinian perceptions of prenatal genetic counseling 19. J Genet Couns 2008;17:101—16.

[36] Ciceklioglu M, Ergin I, Demireloz M, Ceber E, Nazli A. Sociodemographic aspects of consanguineous marriage in an urban slum of a metropolitan area in Izmir, Turkey. Ann Hum Biol 2013;40:139—45.

[37] Khemir S, El AM, Sanhaji H, et al. Phenylketonuria is still a major cause of mental retardation in Tunisia despite the possibility of treatment. Clin Neurol Neurosurg 2011;113:727—30.

[38] Warsy AS, Al-Jaser MH, Albdass A, Al-Daihan S, Alanazi M. Is consanguinity prevalence decreasing in Saudis? A study in two generations. Afr Health Sci 2014;14:314—21.

[39] Joseph N, Pavan KK, Ganapathi K, et al. Health awareness and consequences of consanguineous marriages: a community-based study. J Prim Care Community Health 2015;6:121—7.

[40] Heidari F, Dastgiri S, Akbari R, Khamnian Z, Hanlarzadeh E, Baradaran M, et al. Prevalence and risk factors of consanguineous marriage. Eur J Gen Med 2014;11(4):248—55.

[41] Jalal Abbasi-Shavazi M, McDonald P, Hosseini-Chavoshi M. Modernization or cultural maintenance: the practice of consanguineous marriage in Iran. J Biosoc Sci 2008;40:911—33.

[42] Abbasi-Shavazi M, Torabi F. Inter-generational differences of consanguineous marriage in Iran. Iran Sociol J 2007;7(4):116—43.

[43] Al-Farsi OA, Al-Farsi YM, Gupta I, et al. A study on knowledge, attitude, and practice towards premarital carrier screening among adults attending primary healthcare centers in a region in Oman. BMC Public Health 2014;14:380.

[44] Bittles AH, Black ML. The impact of consanguinity on neonatal and infant health. Early Hum Dev 2010;86:737−41.

[45] Bennett R, Motulsky A, Bittles A, Hudgins L, Uhrich S, Doyle D, et al. Genetic counseling and screening of consanguineous couples and their offspring: recommendations of the National Society of Genetic Counselors. J Genet Couns 2002;11:97−119.

[46] Sheridan E, Wright J, Small N, et al. Risk factors for congenital anomaly in a multiethnic birth cohort: an analysis of the Born in Bradford study. Lancet 2013;382:1350−9.

[47] Bhatta DN, Haque A. Health problems, complex life, and consanguinity among ethnic minority Muslim women in Nepal. Ethn Health 2014;14:1−17.

[48] Tadmouri G, Nair P, Obeid T, Hamamy H. Community health implications of consanguinity. In: Kumar D, editor. Arab populations. OUP USA Oxford Monographs on Medical Genetics Genomics and Health in the Developing World; 2012. p. 625−42.

[49] Do QT, Iyer S, Joshi S. The economics of consanguineous marriages. Available from: <www.cid. harvard.edu/neudc07/docs/neudc07_s1_p13_do.pdf>; 2008.

[50] Bittles AH. A community genetics perspective on consanguineous marriage. Community Genet 2008;11:324−30.

[51] Mobarak AM, Kuhn R, Peters C. Consanguinity and other marriage market effects of a wealth shock in Bangladesh. Demography 2013;50:1845−71.

[52] Hussain R. Community perceptions of reasons for preference for consanguineous marriages in Pakistan. J Biosoc Sci 1999;31:449−61.

[53] Salway S, Ali P, Ratcliffe G, Bibi S. Understandings related to consanguineous marriage and genetic risk: findings from a community level consultation in Sheffield and Rotherham. NIHR CLAHRC for South Yorkshire; 2012.

[54] Kanaan ZM, Mahfouz R, Tamim H. The prevalence of consanguineous marriages in an under-served area in Lebanon and its association with congenital anomalies. Genet Test 2008;12:367−72.

[55] Basel-Vanagaite L, Taub E, Halpern GJ, et al. Genetic screening for autosomal recessive nonsyndromic mental retardation in an isolated population in Israel. Eur J Hum Genet 2007;15:250−3.

[56] World Health Organization. Preventing intimate partner and sexual violence against women: taking action and generating evidence. Geneva: World Health Organization; 2010.

[57] Shaikh MA, Kayani A, Shaikh IA. Domestic violence and consanguineous marriages—perspective from Rawalpindi, Pakistan. East Mediterr Health J 2014;19(Suppl. 3):S204−7.

[58] el-Zanaty F, Way A. Egypt Demographic and Health Survey 2005. Cairo, Egypt: Ministry of Health and Population, National Population Council, el-Zanaty and Associates, and ORC Macro. Egypt: Ministry of Health and Population, National Population Council, el-Zanaty and Associates, and ORC Macro; 2006.

[59] Clark CJ, Hill A, Jabbar K, Silverman JG. Violence during pregnancy in Jordan: its prevalence and associated risk and protective factors. Violence Against Women 2009;15:720−35.

[60] Okour AM, Badarneh R. Spousal violence against pregnant women from a Bedouin community in Jordan. J Womens Health (Larchmt) 2011;20:1853−9.

[61] Fikree FF, Jafarey SN, Korejo R, Afshan A, Durocher JM. Intimate partner violence before and during pregnancy: experiences of postpartum women in Karachi, Pakistan. J Pak Med Assoc 2006;56:252−7.

[62] Maziak W, Asfar T. Physical abuse in low-income women in Aleppo, Syria. Health Care Women Int 2003;24:313−26.

[63] Khawaja M, Tewtel-Salem M. Agreement between husband and wife reports of domestic violence: evidence from poor refugee communities in Lebanon. Int J Epidemiol 2004;33:526−33.

[64] Shahidi S, Zaal B, Mazaheri MA. Forgiveness in relation to attachment style and consanguine marriage in Iranian married individuals. Psychol Rep 2012;110:489−500.

[65] Hamamy H, Bittles AH. Genetic clinics in arab communities: meeting individual, family and community needs. Public Health Genomics 2009;12:30−40.

[66] Saadat M. Association between consanguinity and survival of marriages. Egypt J Med Hum Genet. Available from: <www.ejmhg.eg.net>, Article in Press; 2014.

[67] Hussien FH. Endogamy in Egyptian Nubia. J Biosoc Sci 1971;3:251−7.

[68] Abbas WA, Azar NG, Haddad LG, Umlauf MG. Preconception health status of Iraqi women after trade embargo. Public Health Nurs 2008;25:295−303.

[69] Tierney J. The struggle for Iraq: traditions; Iraqi family ties complicate American efforts for change. New York Times. Available from: <www.nytimes.com>; September 9, 2003.

[70] Mansour N, Chatty D, El-Kak F, Yassin N. They aren't all first cousins: Bedouin marriage and health policies in Lebanon. Ethn Health 2014;19:529−47.

[71] Na'amnih W, Romano-Zelekha O, Kabaha A, et al. Prevalence of consanguineous marriages and associated factors among Israeli Bedouins. J Community Genet 2014;5:395−8.

[72] Kerkeni E, Monastiri K, Saket B, et al. Association among education level, occupation status, and consanguinity in Tunisia and Croatia. Croat Med J 2006;47:656−61.

[73] Koc I. Prevalence and sociodemographic correlates of consanguineous marriages in Turkey. J Biosoc Sci 2008;40:137−48.

[74] Imaizumi Y. A recent survey of consanguineous marriages in Japan: religion and socioeconomic class effects. Ann Hum Biol 1986;13:317−30.

[75] Sirdah MM. Consanguinity profile in the Gaza Strip of Palestine: large-scale community-based study. Eur J Med Genet 2014;57:90−4.

[76] Posch A, Springer S, Langer M, et al. Prenatal genetic counseling and consanguinity. Prenat Diagn 2012;32:1133−8.

[77] Lisa A, Astolfi P, Zei G, Tentoni S. Consanguinity and late fertility: spatial analysis reveals positive association patterns. Ann Hum Genet 2015;79:37−45.

[78] Hussain R, Bittles AH. Consanguineous marriage and differentials in age at marriage, contraceptive use and fertility in Pakistan. J Biosoc Sci 1999;31:121−38.

[79] Helgason A, Palsson S, Gudbjartsson DF, Kristjansson T, Stefansson K. An association between the kinship and fertility of human couples. Science 2008;319:813−16.

[80] Hampshire KR, Smith MT. Consanguineous marriage among the Fulani. Hum Biol 2001;73:597−603.

[81] Al-Thakeb F. The Arab family and modernity: evidence from Kuwait. Curr Anthropol 2015;26: 575−80.

[82] Akrami SM, Osati Z. Is consanguineous marriage religiously encouraged? Islamic and Iranian considerations. J Biosoc Sci 2007;39:313−16.

[83] Zaoui S, Biemont C. Frequency of consanguineous unions in the Tlemcen area (West Algeria). Sante 2002;12:289−95.

[84] Hafez M, El-Tahan H, Awadalla M, et al. Consanguineous matings in the Egyptian population. J Med Genet 1983;20:58−60.

[85] Othman H, Saadat M. Prevalence of consanguineous marriages in Syria. J Biosoc Sci 2009;41: 685−92.

CHAPTER 19

The International Law and Regulation of Medical Genetics and Genomics

Atina Krajewska
School of Law, Cardiff University, Cardiff, Wales, UK

Contents

INTRODUCTION

In 2011, the Polish Ministry of Science set up a Committee on Genetic Testing and Biobanking whose aim was to put forward legislative proposals concerning the use of genetic data and biological samples in health care and research—an area remaining fundamentally beyond state regulation. The Committee decided to use OECD Guidelines as a primary point of reference for their work. Polish law was to be harmonized with Guidelines for Quality Assurance in Molecular Genetic Testing (2007) [1] and for Human Biobanks and Genetic Research Databases (HBGRDs) (2009) [2]. By the end of 2012, the Committee concluded their work by submitting two white papers on genetic testing and on biobanks for research, which are still awaiting government's approval in order to be referred to Parliament for further deliberations. This move toward the regulation of genetic testing and biobanks for research should be viewed positively as it will hopefully bring certainty and transparency for patients, health care professionals, and researchers in terms of their rights, obligations, and procedural standards. Nevertheless, an intriguing question arises with regard to Committee's choice of the normative basis and the point of reference for their proposals. Namely, what was the rationale behind the decision to choose the OECD Guidelines over other existing international documents, of which the most obvious choice would be the Council of Europe's Additional Protocol to the Convention on

Genomics and Society
DOI: http://dx.doi.org/10.1016/B978-0-12-420195-8.00019-7

Human Rights and Biomedicine on Genetic Testing for Health Purposes (2008),[1] any of the UNESCO Declarations on Human Genome and Human Rights (1997), Human Genetic Data (2003), and Bioethics (2005),[2] or finally the WMA Statement on Genetics and Medicine (2005 and 2009).[3]

The reasons might, of course, be manifold. The most plausible—albeit formalistic—answer would be that Poland is under no obligation to follow the provisions of the Additional Protocol to the Oviedo Convention, as it has neither ratified, not signed it. In fact, the Protocol, despite being open for signature since 2008, has only been signed by seven member states of the Council of Europe, and ratified by three, and therefore has not yet met the requirements for its entry into force stipulated in its Art. 24.[4] Similarly, the Polish government was under no obligation to implement any of the three UNESCO declarations as they are not legally binding political documents. Of course, it could also be argued that the Polish Ministry of Science and Higher Education, which initiated the legislative process, had undertaken a detailed evaluation of different international instruments and chose the OECD Guidelines because of their content and scope, values reflected in the document, as well as legitimacy and transparency of the processes that led to its adoption. The OECD Guidelines might indeed meet most of the above criteria. They stipulate quality standards necessary for safe and professional provision of genetic testing services and the establishment and managing of biobanks for research. Nevertheless, anecdotal evidence suggests that the political process leading to the adoption of this particular instrument as the basis for the new legislation was far less conspicuous. First, it stemmed from the quest to avoid ethical and religious controversies, which could jeopardize the whole legislative initiative. As the guidelines refer to quality assurance processes, they do not touch upon the issue of the legal status of the embryo/fetus in preimplantation and prenatal genetic testing, which remains highly controversial in Poland and has prevented the introduction of assisted reproduction technologies regulation. Second, it was underpinned by a questionable assumption that the Polish

---

[1] Additional Protocol to the Convention for the protection of Human Rights and dignity of the human being with regard to the application of biology and medicine: Convention on Human Rights and Biomedicine 4/4/1997 concerning Genetic Testing for Health Purposes (2008) (ETS 203).

[2] Universal Declaration on the Human Genome and Human Rights (unanimously and by acclamation on November 11, 1997), International Declaration on Human Genetic Data (October 16, 2003), Universal Declaration on Bioethics and Human Rights (October 19, 2005); available at: http://www.unesco.org/new/en/social-and-human-sciences/themes/bioethics/about-bioethics/.

[3] WMA, Statement on Genetics and Medicine, Santiago, Chile, October 2005 and amended by the WMA General Assembly, New Delhi, India, October 2009, replaced the Declaration on the Human Genome Project, adopted by the 44th World Medical Assembly Marbella, Spain, September 1992; available at: http://www.wma.net/en/30publications/10policies/g11/index.html.

[4] According to Art. 18 of the Vienna Convention on the Law of the Treaties 1969, these seven signatories, who have not ratified the Protocol, will merely be under the obligation to refrain from acts which would defeat its object and purpose. See Ref. [3].

government is required to implement the OECD Guidelines.[5] Domestic legal and sociopolitical conditions underlying the implementation of international law into the Polish and other national legal systems constitute a fascinating topic for further academic studies.

However, this chapter proposes to focus on another significant and pressing issue illustrated by the Polish example that occurs in the area of international biomedical law. Namely, it aims to investigate the *global* phenomena that allow for the practice of "cherry picking" of international instruments by national governments; international, regional, or national professional organizations; public and private bodies offering health services or conducting research. This chapter seeks to provide systematic analysis of the recent processes of fragmentation,[6] proliferation, and juridification[7] of international biomedical law. These phenomena, along with constitutionalization,[8] globalization [12], pluralization,[9] and privatization,[10] have dominated legal debates in the twenty-first century. Interestingly, international biomedical law remains at the periphery of these discussions. This might be due to the fact that medical law has traditionally been theorized and practiced as a part of domestic private law or due to the

---

[5] Guidelines are a form of OECD recommendations. Recommendations are not legally binding, but practice accords them great moral force as representing the political will of Member countries and there is an expectation that Member countries will do their utmost to fully implement a Recommendation. See http://www.oecd.org/legal/legal-instruments.htm.

[6] The phenomenon of fragmentation of international law is related with the increasing specialization and related autonomization of parts of society, called by sociologists "functional differentiation." Such differentiation has been accompanied by the emergence of specialized and (relatively) autonomous rules, legal institutions, and spheres of legal practice that have no clear relationship to each other. International law is fragmented along functionally defined issue areas such as human rights law, trade law, environmental law, humanitarian law, criminal law, and law concerning scientific and technological cooperation. See International Law Commission, Study Group on Fragmentation of International Law: Difficulties Arising from the Diversification and Expansion of International Law, 58th Session, Geneva May–August 2006 (A/CN.4/L.702), para. 483; available at: http://untreaty.un.org/ilc/guide/1_9.htm. Also see Ref. [4].

[7] "Juridification is a contested concept. It can be understood as 'the proliferation of law' or as 'the tendency towards an increase in formal (or positive, written) law'. It can also mean 'the expansion of judicial power' and some quite generally link juridification to the spread of rule guided action or the expectation of lawful conduct, in any setting, private or public." Juridification is related to two other concepts, judicialization and legalization, concepts that are sometimes used interchangibly or at least overlap with the concept of juridification. See Ref. [5].

[8] The concept of "constitutionalization" describes the emergence of constitutional law within a given legal order. It implies that a constitution or constitutional law can come into being in a process extended through time. It may be, in short, a constitution-in-the-making. See Refs. [6–11].

[9] Legal pluralism is the existence of multiple legal systems within one (human) population and/or geographic area. Plural legal systems are particularly prevalent in former colonies, where the law of a former colonial authority may exist alongside more traditional legal systems. See Refs. [13–15].

[10] The old understanding of international law as something created solely by and for sovereigns is defunct. Today the production and enforcement of international law increasingly depends on private actors and networks, not traditional political authorities. See Refs. [16,17].

fact that it has entered the international arena relatively recently, therefore escaping the attention of international lawyers. This is indeed unfortunate, because it provides a great case study for analyzing the "the Great Legal Complexity of the World" [18], and it should be of relevance not only to international theorists, but also biomedical lawyers, life scientists, and health care professionals who in one way or another will be affected by legal phenomena at the global level.

For the purpose of the analysis, international biomedical law and governance will be understood as the (emerging) system of international norms regulating the application of new technologies and scientific advances in medicine.[11] Particular attention will be given to rules governing the genetic and genomic technologies in medical research and practice, including genetic testing, biobanking, data sharing, or reproductive medicine. This chapter first summarizes the recent regulatory developments in the area of biomedicine and identifies the most pressing problems that have emerged at the international level. It is argued that the problems stem from two seemingly opposite, yet closely interwoven processes, that is, the institutional and normative proliferation *coupled with* scarcity of international legally binding ("hard law") norms and reflexive (second-level) rules guiding normative conflicts as well as absence of an adjudicative body—in other words, a lack of a constitutional framework containing the basic principles governing international biomedical law. Second, it proposes an explanation to these problems rooted in sociological theories of functional differentiation. This approach enables an analysis that goes beyond normative claims about fragmentation and institutional or normative proliferation and it allows us to observe these processes as an integral part of the rising autonomy of the global health system. It also provides analytical tools for coherent conceptualization of the *prima facie* chaotic expansion of legal, ethical, and professional norms created by public, private, and semiprivate actors. Finally, it attempts to draw some conclusions from these approaches and provide an alternative interpretation of the phenomena discussed in the first part of this chapter.

MAIN CHALLENGES IN THE FIELD OF INTERNATIONAL BIOMEDICAL LAW

Lack of Harmonization

In their paper published in Human Genetics in 2014, Bartha Knoppers et al. raised serious concerns about current data sharing models and policies which are not working [20]. They have identified four main contributing factors: (i) the inadequacy of

---

[11] Defined in this manner, international biomedical law forms part of global health law system defined as "a field of practice, research and education focussed on health and the social, economic, political and cultural forces that shape it across the world...[and a] discipline (...) concerned with health-related issues that transcend national boundaries and the differential impacts of globalisation." See Ref. [19].

regulatory systems, which were not designed (nor updated) to deal with the consequences of globalization and foster widespread cross-study collaboration and transborder open sharing of data; (ii) the lack of harmonization between the policies and procedures of research initiatives, funders, biobank practices, as well as financial, organizational, technical, and governance structures and resources; (iii) the legal and ethical hurdles, such as informed consent, privacy, and confidentiality requirements which may render data sharing difficult or even impossible; (iv) the cultural and behavioral considerations which may negatively affect data sharing attitudes among researchers. These problems are closely related to the two main developments that have recently taken place in biomedical research, that is, first, the increasing connectivity and mobility of data, researchers, and participants; and second, fundamental changes in the nature of biomedical research [21]. The character of biomedical research is shifting from classic, physically risky, small-scale, single-site, or one country-specific disease studies on "human subjects" toward large-scale multipurpose studies involving in international, large-scale, collaborative, longitudinal, or remote analyses of samples and data to better understand complex disease etiology [22]. The universal influence of new technologies has undermined conventional territorial boundaries and emerged as global and common objects of regulation. The ethics codes adopted in the mid-to-late twentieth century by the World Medical Association, UNESCO, or the WHO had been designed to deal with the former type of medical research and do not seem sufficient to address all the complexities of transnational collaboration and data processing. Consequently, the pressure has been rising to set new biomedical standards predominantly at supranational and international levels. [23] In particular, the need for harmonization and interoperability of privacy and data protection laws across the globe and for reforms and/or implementation of antidiscrimination provisions has been expressed on numerous occasions.

Similarly, Jane Kaye have argued that our current governance system of biobanks for research is unable to provide all of the oversight and accountability mechanisms that are required for this new way of doing research that is based upon flows of data across international borders. She rightly pointed out that the "current governance framework for research is nationally based, with a complex system of laws, policies and practice that can be unique to a jurisdiction. It is also evident that many of the nationally based governance bodies in this field do not have the legal powers or expertise to adjudicate on the complex issues, such as privacy and disclosure risks..." [24]. Studies of various levels of biobanks governance at the international, European, and national levels show that several regulatory models have been adopted to govern biobanks, but none of them supersedes the others [25]. International consortia, such as International Cancer Genome Consortium (ICGC) launched in 2008, are virtual in nature and although "management" offices are located in one country, the consortium itself is not bound by the legislation of any one country. Thus, decision-making powers usually rest with an executive committee, comprising representatives of funding agencies in countries where projects are based. Decisions are negotiated between the members, who use a patchwork of (commonly nonbinding) international and national instruments. However, not having a single national regulatory structure in place can potentially leave gaps in oversight. Member studies may use varying scientific methodologies, regulatory and ethical review, consent, anonymization, re-contacting, or general benefit-sharing procedures. If no top-down compliance mechanism exists, the international consortium is fundamentally based on trust, that its studies are being conducted in accordance with local regulations, national and international law, and ethics policies, and have appropriate institutional oversight [26]. These problems are by no means new. The reasons for the adoption of the UNESCO Declaration on Bioethics and Human Rights (2005) to address ever more complex social, political, and moral questions stemming from global forces of the techno-scientific economy which give rise to increasing uncertainty [27]. These concerns referring to the regulation of new biotechnologies resonate with a much broader criticism of global health law (an area encompassing biomedical law), which has been said to be developing in a fragmented, uncoordinated, amorphous, hence inconsistent, inefficient, and incomplete manner [28,29]. Therefore, it has long been argued that "due to proliferation of biobanks..., the question of international coordination and organisation arises, both on a scientific and ethical level.... Harmonisation has therefore become a priority, although it poses significant challenges" [30]. This claim for international harmonization and infrastructure was further supported and developed by Ruth Chadwick and Heather Strange, who argue for harmonization understood as an ongoing process accommodating a the interplay of multiplicity of voices, especially those less often heard [31]. The biggest challenge in achieving harmonization is the extraordinary expansion and proliferation of: (i) actors (international organizations (IOs), nongovernmental organizations (NGOs), informal professional and/or research networks (Public Population Project in Genomics and Society (P3G)), private—public partnerships and consortia (e.g., GlaxoSmithKline and Wellcome Trust, ICGC, Tissue Bank Consortium in Asia, Biobanking and Biomolecular Resources Research Infrastructure), and private entities (e.g., pharmaceutical companies); (ii) normative practices (treaties, declarations, IOs' regulations, informal rules and agreements or private contracts); and (iii) levels of norm production (international, regional, national, transnational) that occurred in recent years in the area of health, medicine, and science.

Institutional and normative proliferation

Since the adoption of the WHO Constitution in 1948, health governance at the international (now global) level has been usually associated with the WHO as the undisputed leader in the field. As stated in its Constitution, the WHO is destined to "act as the directing and co-ordinating authority on international health work."[12] It is often claimed that it would act as an umbrella health agency convening legal and nonlegal activities of different organizations providing thereby a more effective collective management [32]. The Constitution was intended to set out formal foundations for the operation of the WHO and guarantee coherency of the global health system. However, the internal coherency turned out to be illusory, which was later exacerbated by globalization of medical research and health care services and remarkable (all-encompassing) digitization of science and medicine. The emergence of the *global* health paradigm has been marked by structural changes in stakeholders involved in the field. Market-driven global economic policies, notably those associated with extended intellectual property rights and structural adjustment programs implemented by the World Bank and the International Monetary Fund, have been pointed as reasons for increased commercialization in health care systems and trade openness which could hinder health reforms in particular in low-income countries [33]. The reconceptualization of health politics and funding has undermined some of the traditional structures designed to address cross-border health concerns. Instead of reforming and strengthening existing organizations with explicit health mandates, greater energy has gone into creating new actors and expanding the mandates of others [34].

Consequently, the new international health framework is no longer dominated by a few intergovernmental organizations, but consists of numerous global health actors; some with finance-policy-operational functions,[13] and others with professional-standards-setting functions. Nonstate actors are equally active at regional levels.[14] This complex network offers an institutional alternative which offsets the weight of states and intergovernmental organizations. At the same time, rapid advances in science, coupled with strong promises of local medical and economic gains, reveal competing ambitions between nonstate actors. Such competition is well reflected in the regulatory activities of several NGOs undertaken at the end of the twentieth—and the beginning of the twenty-first century. For instance, the Council for International

---

[12] Art. 2 (a) of the Constitution of the World Health Organization as adopted by the International Health Conference, New York (June 19–22, 1946) (Official Records of the World Health Organization, no. 2, p. 100), entered into force on April 7, 1948.

[13] The World Economic Forum has sponsored deliberations about HIV/AIDS, tobacco control, and vaccines to obesity. The UN Security Council has been involved in issues related to HIV/AIDS. Finally, the private sector, for- and nonprofit organizations, developed as a force in international health as relatively new, yet important players. See Ref. [35].

[14] For example, regional nonstate regulators in Europe include for instance: European Forum for Good Clinical Practice, European Medical Research Councils, European Society of Human Reproduction and Embryology, European Committee for Standardization.

Organizations of Medical Sciences (CIOMS)[15] and the World Medical Association (WMA),[16] apart from adopting general rules concerning medical research involving human subjects, have both engaged with the developments in the science of genetics issuing respectively, the Declaration on Inuyama on Human Genome Mapping, Genetic Screening and Gene Therapy (1990), and the Statement on Genetics and Medicine (2005 and 2009).[17] At the same time, the Ethical, Legal, and Social Issues Committee of the Human Genome Organization prepared the Statement on the principled conduct of genetics research (1996) which had also played a role in shaping the debate and research practices about genetics [36]. Examples of governmental organizations that have taken interest in issues of health and biomedicine are many. While the UN Commission for Human Rights has adopted resolutions pertaining to human rights and bioethics with implications for public health and biomedicine,[18] UNESCO was drafting the three major Declarations on Human Genome and Human Rights 1997, Human Genetic Data 2003, and Bioethics and Human Rights 2005.[19] At the same time, in the years of the Human Genome Project, the WHO started preparing separate, competing reports on genomics, health, and intellectual property rights.[20] It was not until 2003 that the WHO and UNESCO established a special Inter-Agency

[15] Council for International Organizations of Medical Sciences (CIOMS). (2002). *International Ethical Guidelines for Biomedical Research Involving Human Subjects* (enumerating a set of ethical guidelines for biomedical research dealing with topics like confidentiality, informed consent, and the duty to provide health services).

[16] World Medical Association. (1964). *Ethical Principles for Medical Research Involving Human Subjects*, Declaration of Helsinki (as amended by 59th WMA General Assembly, Seoul, October 2008) (providing guidelines for medical professionals around the world who conduct experimental research with human subjects), http://www.wma.net/e/.

[17] World Medical Association. (2009). *Statement on Genetics and Medicine*, Santiago, Chile, October 2005 and amended by the WMA General Assembly, New Delhi, India, replaced the *Declaration on the Human Genome Project*, adopted by the 44th World Medical Assembly Marbella, Spain, September 1992: http://www.wma.net/en/30publications/10policies/g11/index.html.

[18] UN Commission for Human Rights, Resolution on Human Rights and Bioethics 2003/69, 2001/71, 1999/63, 1997/71, 1993/91: http://www.unhchr.ch/huridocda/huridoca.nsf/(Symbol)/E.CN.4. RES.2001.71.En?Opendocument.

[19] United Nations Educational, Scientific, and Cultural Organisation (UNESCO), Universal Declaration on the Human Genome and Human Rights (unanimously and by acclamation on November 11, 1997), International Declaration on Human Genetic Data (October 16, 2003), Universal Declaration on Bioethics and Human Rights (October 19, 2005): http://www.unesco.org/new/en/social-and-human-sciences/themes/bioethics/about-bioethics/.

[20] For example, WHO (1996). *Control of hereditary disorders: Report of WHO Scientific Meeting*, World Health Assembly *Statement on "Cloning in Human Reproduction"* 1997, WHO, Statement of the WHO Expert Consultation on New Developments in Human Genetics, 2000, WHO/HGN/WG/00.3; available at: http://whqlibdoc.who.int/hq/2000/WHO_HGN_WG_00.3.pdf. Report of the Advisory Committee on Health Research: *Genomics and World Health* 2002, Review of Ethical Principles in Medical Genetics 2003; available at: http://www.who.int/topics/genetics/en/.

Committee on Bioethics[21] to address potential tensions and to facilitate dialogue and compliance between the documents issued by both organizations. Other conflicts, this time between the WTO and the WHO, have been revealed in the context of disputes over access to medicines emerging in relation to the Doha Declaration[22] on the Trade-related Aspects of Intellectual Property Rights (TRIPS) Agreement[23] and Public Health.[24] These disputes serve as an example of tensions between different rationalities, intellectual traditions—and corporate logic—of different organizations within the ever expanding UN system, characterized by a multiplication of committees and sub-committees and new administrative structures such as programs and funds.[25] These complex structures are then supplemented by regional organizations, such as the Council of Europe and the European Union which thus far have probably been the most active and influential regulators of biomedicine and biotechnology worldwide. Finally, as mentioned earlier national legislators, governments, and courts continue to play a vital law in the process of law production. This overall complex normative structure governing medicine and science (globally) can only be described as *transnational* biomedical law.

This proliferation of multilateral institutions with overlapping ambitions and legal authority has resulted in the serious criticism that the multitude of IOs reflects an increasingly fragmented and incongruent global health agenda [41]. The appearance of standard-setting organizations raises closely related questions with regard to the scope of regulatory authority, as well as the legitimacy of grounds and enforcement [42—44]. The health sector has resisted attempts to measure its efficiency and effectiveness, avoiding close monitoring and accountability. As rightly summarized by Alan Taylor: "dramatic advances in the field of biomedical science have recently triggered numerous, uncoordinated regional and global initiatives, which, while undertaken without meaningful consultation, coordination or planning, obscure rather than rationalised the global legal framework" [28]. This development can be seen as an impediment rather than an incentive for international scientific cooperation, making it extremely difficult for doctors, researchers, and private companies to determine which

---

[21] Member organizations are listed at: www.who.int/ethics/about/unintercomm/en/.

[22] Doha Declaration on the TRIPS Agreement and Public Health, adopted in Doha, Qatar on November 14, 2001, WT/MIN(01)/DEC/2; available at: http://www.wto.org/english/thewto_e/minist_e/min01_e/mindecl_trips_e.htm.

[23] The TRIPS Agreement is Annex 1C of the Marrakesh Agreement Establishing the World Trade Organization, signed in Marrakesh, Morocco, April 15, 1994; available at: http://www.wto.org/english/tratop_e/trips_e/t_agm0_e.htm.

[24] See Ref. [37]. For a detailed discussion on the medicines decision: see Refs. [38,39].

[25] The latest example of this institutional incoherence is the fight against HIV/AIDS. The Joint UN Programme on HIV and AIDS, UNAIDS, which was recently established, has been given responsibilities which should normally fall within the framework of the WHO mandate. Subsequently, the Global Fund, dedicated to the same goals, has been recently created outside the UN. See Ref. [40]

governance regime is applicable in any particular case, and which prevails when conflicts occur. It may also become confusing for national legislators, who, when trying to regulate particular problems, look for clear and coherent international guidelines. Many differences in the legal requirements at the national level remain and militate against networking. Consequently, the proliferation and specialization of laws leads to a multiplication of standards and terminology [45–47]. Quite apart from the varying interests of different societies, there are often "different starting points for the very idea of regulation."[26] However, this is only a part of the overall picture of the international governance of biomedicine. Parallel to the multitude of actors, whose status can only inadequately be reassigned to well-known public/private distinctions, the biomedical world is witnessing the emergence of a large, decentralized body of norms lacking a common legal point of reference.

Scarcity of international law, second-level rules, and adjudication mechanisms

What characterizes biomedical law at the international level is a general lack of legally binding norms—the so-called hard law—alongside an overwhelming proliferation of soft law instruments. For the mainstream international lawyer the term "hard law" refers to legally binding obligations that are both formal and enforceable. Reliance on coercion—understood as diplomatic measures, reprisals, or dispute settlement—is crucial [50]. In addition to requiring commitment to a background set of legal norms—including engagement in established legal processes and discourse—legalization provides actors with a means to instantiate normative values.[27] Despite the WHO's astounding normative powers, modern international health law is remarkably thin. There currently exist three legally binding international health instruments: the International Classification of Disease 1948, International Health Regulations (IHRs) [52], and the WHO Framework Convention on Tobacco Control 2003 (FCTC),[28] and two of them predate the WHO.[29] These instruments are often characterized by "structural weaknesses—e.g., vague standards, ineffective monitoring, and weak enforcement—and a "statist" approach that insufficiently harnesses the creativity and resources of nonstate actors and civil society [53]." Their impact of is severely hindered by the fact that they

---

[26] M.D. Kirby. (1999). *Human Freedom and Human Genome: The Ten Rules of Valencia*, cited by Ref. [48]. Also see Ref. [49].

[27] Treaties are by definition always hard law, because they are always binding, although an agreement involving states may still be binding in the absence of a treaty, so the distinction between soft law and hard law is not simply synonymous with the distinction between treaties and nontreaties. See Ref. [51].

[28] WHO Framework Convention on Tobacco Control (WHO FCTC) adopted by the World Health Assembly (May 21, 2003), entered into force on February 27, 2005: http://whqlibdoc.who.int/publications/2003/9241591013.pdf.

[29] Depending on the scope of the definition of health it is possible to include also the UN Convention of the People with Disabilities 2006.

provide no financial or technical support to do so, member states may reject or submit reservations to their provisions, and most importantly they lack enforcement mechanisms for addressing compliance failure. For instance, one analysis of the IHRs effectiveness over their 56-year history concluded that, due to poor national surveillance systems and protection measures, they had been relatively ineffective in achieving their main goals.[30] The FCTC remains the first and only legally binding global *health* treaty and it was criticized with regard to its formation and content [54]. Together with other WHO instruments it still neglects rights-based terminology [55] and it covers an extremely limited subject matter. The only other legally binding instruments that might have a bearing on the field of biomedicine are the TRIPS Agreement and less directly the Convention on Biological Diversity[31] accompanied by the Cartagena Protocol on Biosafety.[32] The only legally binding norms directly concerning biomedicine have been adopted by the Council of Europe[33] and the European Union.[34] However, as both

[30] D. Fidler, L.O. Gostin, see Refs. [46,47,59].

[31] The Convention on Biological Diversity (CBD) adopted May 22, 1992 at the Conference for the Adoption of the Convention on Biological Diversity, Nairobi, Kenya, entered into force on December 29, 1993.

[32] The Cartagena Protocol on Biosafety adopted at First Extraordinary Meeting of the Conference of the Parties, Cartagena, Colombia February 22–23, 1999 and Montreal, Canada, January 24–28, 2000, entered into force September 11, 2003.

[33] The Convention for the protection of Human Rights and dignity of the human being with regard to the application of biology and medicine: Convention on Human Rights and Biomedicine 4/4/1997 (ETS 164). Accompanied by Additional Protocols concerning: the Prohibition of Cloning of Human Beings, 1998 (ETS 168); Transplantation of Organs and Tissues of Human Origin, 2002 (ETS 186); Biomedical Research, 2005 (ETS 195); Genetic Testing for Health Purposes, 2008 (ETS 203). As mentioned earlier, some of these protocols have not yet come into force, which means they could not be invoked in courts.

[34] Primary law now includes the Charter of Fundamental Rights of the EU and directives binding as to the effect, for example: Directive 2004/23/EC of the European Parliament and of the Council of March 31, 2004 on setting standards of quality and safety for the donation, procurement, testing, processing, preservation, storage, and distribution of human tissues and cells; Directive 2001/20/EC of the European Parliament and of the Council of April 4, 2001 on the approximation of the laws, regulations and administrative provisions of the Member States relating to the implementation of good clinical practice in the conduct of clinical trials on medicinal products for human use; Directive 98/79/EC of the European Parliament and of the Council of October 27, 1998 on *in vitro* diagnostic medical devices; Directive 98/44/EC of the European Parliament and of the Council of July 6, 1998 on the legal protection of biotechnological inventions; Directive 95/46/EC of the European Parliament and of the Council of October 24, 1995 on the protection of individuals with regard to the processing of personal data and on the free movement of such data. These provisions are further developed by of the Court of Justice of the EU, which has issued controversial decisions in the area of genetics and genomics, for example, Netherlands v European Parliament & Council (C-377/98) [2001] E.C.R. I-7079, Brüstle v Greenpeace e. V. (C-34/10) [2012] 1 C.M.L.R. 41, Google Spain SL and Google Inc. v AEPD and Mario Costeja González (C-131/12) [2014].

organizations operate at a regional, rather than global, level they can be ignored by countries and individuals outside Europe.

Interestingly, it is not the WHO instruments, but the International Covenants on Civil and Political Rights and Social Economic and Cultural Rights[35] that so far seems to have had the biggest impact on the development of the health care and biomedical law system. The right to health, stipulated in Art. 12 ICCSEC, has been often proclaimed to be a well-established part of international law and has been reproduced in different "shapes and forms" in numerous documents, some of which certainly carry a constitutional status [56]. The right to health has constituted the basis for most global health actions at the global and national level (e.g., in disputes with WTO about TRIPS Agreements, or in access to medicines and health care services litigation in South Africa, India, Brazil, Columbia [57]). In addition to Art. 12 of the ICESCR, Art. 15 (1) of the same act guarantees the right "of everyone … to enjoy the benefits of scientific progress and its applications" and the right "of everyone … to benefit from the protection of the moral and material interests resulting from any scientific … production of which he is the author." As a recent UN report notes, the "benefits" of science encompass not only scientific results and outcomes but also the scientific process, its methodologies and tools [58]. According to Knoppers et al. "the right 'to benefit from', i.e., to enjoy the benefits of scientific progress and its applications, clearly implies a right that all can exercise to have access to and share in both the development and fruits of science across the translation continuum, from basic research through practical, material application (e.g., diagnostics and therapeutics)." Although the article advocates an approach based on rights enshrined in international human rights law, it acknowledges that the content of these rights is neither self-evident nor promoted in the world of genomics. Indeed, they remain extremely underdeveloped as a whole and have been cited in only a handful of court decisions around the world, and each time only in passing reference.[36] Similarly, despite many recent positive developments, the justiciability of the right to health (i.e., the ability to rely on it successfully in courts) is still being disputed by many governments and academic scholars.[37] Furthermore, ambiguities persist about the nature and scope of the right to health. For instance, the UN Special Rapporteur on the Right to Health noticed that duties of pharmaceutical companies with regard to access to medicines were not articulated as peremptory duties.[38]

[35] United Nations, International Covenants on Civil and Political Rights and International Covenant on Social Economic and Cultural Rights, adopted by General Assembly resolution 2200A (XXI) of December 16, 1966, entered into force January 3, 1976.

[36] See Refs. [59,60].

[37] For summary of these debates see Ref. [61]. See also Refs. [57,62].

[38] UN, Report of the Special Rapporteur on the right of everyone to the enjoyment of the highest attainable standard of physical and mental health, A/63/263; available at: http://www.who.int/medicines/areas/human_rights/A63_263.pdf.

These considerations suggest that the field of biomedical law at the global level is predominantly "soft," in that it is dominated by declarations, communications, recommendations, resolutions, codes of practice, guidelines, notices, and positions.[39] The realm of "soft law" is much more elusive as it begins once legal arrangements are weakened along one or more of the dimensions of obligation, precision, and delegation. The most common distinction between hard and soft law norms is that soft law lacks the possibility of legal sanctions [63]. Soft law rules do not have in common a uniform standard of intensity as far as their legal scope is concerned, but they do share a desire to influence the practice of states, IOs, and individuals. They contain an element of law-making intention and progressive development, but without containing international enforceable (and justiciable) rights and obligations.[40] In the field of biomedical law, lack of legally binding international instruments enables interested parties such as tourists-scientists, patients, and industries to "pick and choose" from a patchwork of regulation containing a wide spectrum of possible options. Such freedom may contribute to the more efficient diffusion of controversial technologies and new regulatory regimes without conceding time for their cultural elaboration [66]. Health risks and ethical problems are complex and the reaction to them may be culturally based. The result may be both a lack of constructive diversity and a lack of cultural and democratic legitimacy. It may also lead to larger inequalities and a kind of moral imperialism. Paradoxically, it might be easier to impose norms and values through ethical and professional codes of practice rather than through binding international treaties, which have to be negotiated in accordance with formal procedures and which treat all parties as sovereign equals.

Furthermore, what has been acutely missing, are the secondary legal norms prescribing how the identification, setting, amendment, and regulation of competences for the issuing and delegating of primary norms are to occur. There are still no visible and clear rules defining authority, competences, and mechanisms for the global health law-making. It is uncertain how violations should be handled, and how third parties should be included. Secondary rules enclosed in the WHO Constitution concerning its interpretation and dispute resolution have never been enforced. The interpretative role has been partially taken over by the UN Committee on Economic, Social, and Cultural Rights that issued the General Comment No. 14: The Right to the Highest Attainable Standard of Health,[41] and it is now the Office of the High Commissioner for Human Rights, that adopts Reports on various aspects of the right to health or

---

[39] This proliferation is well illustrated, for instance, on the EU Science in Society Portal which enumerates only the legal instruments related to research involving medical intervention; available at: http://ec.europa.eu/research/science-society/index.cfm?fuseaction = public.topic&id = 1428.

[40] See Refs. [64,65].

[41] No General Comment has been adopted on the right to enjoy the benefits of scientific progress. However, General Comment 17, the right of everyone to benefit from the protection of the moral and material interests resulting from any scientific, literary, or artistic production of which he or she is the author, Article 15, paragraph 1 (c) of the ICCPR (i.e., intellectual property rights) has been adopted in 2005. See: UN Doc. E/C.12/GC/17, 12 January.

the right to enjoy the benefits of scientific progress, which aim to advance clarity regarding the scope and content of these rights.[42] However, as the Human Rights Council mechanisms are predominantly soft and spread across a whole range of areas, a question arises to what extent the Council can and will assume mediating authority to the crowded global health landscape. In other words, what hinders biomedical research, the application of scientific advances in health care, and/or the protection of patients and research (data) subjects is "the incommensurability of authority claims— in particular of the discrete claims to final authority over the interpretation and extent of jurisdiction of the various political units" [67]. They legal and regulatory systems governing biomedicine where there is a convincing rule of recognition has not been formulated, yet. The institutions for "adjudication" are often nonjudicial or absent, and the processes of change are not easily articulated in terms of rules [68]. The multiplicity of political, legal, and other forms of public decision-making, developed to address the complexities of specialized knowledge, tend to lack some of the qualities of democratic participation, transparency, stability, accountability, and effectiveness. This is why so many lawyers feel extremely uneasy when confronted with the notion of soft law where it indicates a lack of institutionalized adoption procedures. And this is why there is a structural and technical demand for harmonization and compatibility [23]. However, as such harmonization has still not been achieved despite numerous and persistent efforts, the question arises if perhaps there are other explanations for the above-mentioned phenomena occurring in the field of transnational biomedical law. The analysis of medicine, health care and public health as functional regimes based on systems theory traces the recent changes back to the intrinsic characteristics of the system itself.

Health and biomedicine and a functional system

According to Luhmann, modern societies are seen as emerging via generalized functional and communicative systems which traverse society—irrespective of territorial and institutional boundaries—such as economics, politics, law, sciences, religion, and art. None of the communicative systems are privileged or seen as the center of society [69,70]. They are simultaneously characterized by both normative autonomy and complex relations and their interdependencies. At the global level, the combination of the globalization of markets, international treaties, and new technologies (e.g., telecommunication and genetics) may lead to more comprehensive global dynamics in several fields than previously existed. Science and knowledge and the discourses related to these are part of what might be called "transnational dynamics" [71].

---

[42] See: http://www.nesri.org/resources/general-comment-no-14-the-right-to-the-highest-attainable-standard-of-health#sthash.deGP2egv.dpuf and http://www.ohchr.org/EN/Issues/Health/Pages/AnnualReports.aspx.

Medicine and health care[43] clearly constitutes a separate functional system despite the fact that it is often coupled with other systems including science, economy, and law. The criterion for differentiation is its code ill/healthy or health and illness, which is used in the communication between the doctor and the patient, with illness being the value side of the distinction. Only illness is instructive for the doctor, as he can only act when the latter is identified. Health gives nothing to work with, as it only describes what one feels in the absence of illness. Consequently, there are many illnesses and only one health. This is why the notion of health poses so many difficulties and it is both, problematic and empty. From medicine's point of view, healthy people are those who are not yet ill, or not ill anymore, or those who are ill, but asymptomatic. Thus, the main aim and function of medicine is to free from disease, the move from positive toward the negative value [74]. Once this aim (i.e., the reflexive value of health) has been achieved, there is nothing more, upon which to reflect. All the communication between the doctor and the patient is about present illness and, hence, unnecessary when the illness disappears and the patient is cured. No second-order observation is necessary; the absence of illness speaks for itself. For Luhmann (and later also for J. Bauch) this explains why, in contrast to other function systems, medicine and health care has developed neither a specific generalized symbolic means of communication, nor a complex theory of reflection based purely on its function.[44] However, this is not to say that there is no communication at all. It just observes that the system has so far not developed a reflection theory, although this is changing with the expansion of the notion (and positive value) of illness, which focuses on the different classifications of disease and discovery (distinction) of ever new conditions.

Luhmann's system-theoretical analysis of health care has been criticized for its narrow (almost biomedical) understanding of health as the absence of health, by J. Bauch and J. Pelikan [77]. Bauch criticized Luhmann for not taking into account newer health related developments and therefore proposed a much wider binary code; that of hindering versus promoting health. The reformulation of the code in this way enables the inclusion of public health into the health system. Pelikan, on the other hand, emphasizes that clinical medicine and public health seem to be two extremely differing approaches in dealing with health policy. For Pelikan clinical medicine is oriented to treat actual, manifest, and severe ill health of single individuals, whereas public health is oriented at avoiding future, possible ill health of abstract populations. For medical interventions partly standardized technical solutions are possible, which lead to marketable individual goods, products, and services

---

[43] To define a specific function system dealing with health in modern society, Luhmann interchangeably used different terms: "treatment of disease," "treatment of ill persons" or "medicine," "health care," not carefully distinguishing between the three terms. See Refs. [72,73].

[44] See Refs. [75,76].

for big populations, forming the basis for continuous growth, in contrast public health depends heavily upon social interventions in social conditions, processes, and behaviors which are much less stable and predictable. Therefore, according to Pelikan, public health has not developed into a function system and can be seen as a social movement which still has not fully adapted to the conditions of late modernity. This professed deficiencies of Luhmann's account of medicine and health care lie at the background of the fact that his analysis in this respect has not been taken up much in sociological or medical sociological literature. However, his analysis might have been dismissed too easily.

Despite Luhman's—perhaps too—rigorous definition of the medical code as ill/ healthy, he acknowledged that the shift from communicable diseases toward life style illnesses has considerably broadened the scope of the system (spreading over the whole lifespan of a person). Similarly to other functional systems, the complexity of medicine and health care increases with new classifications, new images, and new understandings of disease. He also mentions the effect of genetic knowledge, as a result of which what has thus far been seen as healthy will come under scrutiny as potentially ill in future, depending on yet another distinction along the lines of genetically acceptable/ genetically questionable. Advances in life sciences such as genetics[45] have blurred the relevant distinction between prevention and treatment. Therefore, for Luhmann genetics brought about lack of clarity and confusion about the criteria of differentiation and consequently, even more uncertainty about anthropological and social consequences. This has been taken up by Fuchs who described the development of medicine and health care as expansionist and hence typical to functional systems. The expansion stems from the fact that illness is seen not simply as disturbance of the body (and more recently also of "consciousness," i.e., the mind), which constitutes a necessary environment for all communication systems in society, but as an immediate threat to every "virtual full inclusion" [78], a threat that the system tries to stabilize, eliminate, or prevent through anticipatory means of health care. As a result, over the years the health care system has proliferated so massively that it has not only identified illnesses of the body, but also "created" new diseases, to which it can react, which in turn led to even more proliferation of diagnostic, therapeutic, and prophylactic procedures, medicines, and treatments [79].

Like other global regimes freed from the framework and territorial boundaries of the nation state global health care exhibits tendency of expansion. As mentioned above, as a result of scientific advancement new illnesses are being identified, new

---

[45] For instance, genetic testing for breast cancer (BRCA1&2) can be seen as part of prevention as it reveals susceptibility to the illness, but also a part of cancer treatment, where diagnosis of genetic preconditions can influence the course of treatment. Furthermore, there are strong links between genetics and epidemiology through genetic screening and genetic testing.

conditions medicalized, and new public health areas brought at the forefront of policy-making. This creates a pressing need for "cure," that is, more health care, more pharmaceutical products, treatments, and more medical research which in turn generates new categories of disease. Similarly, in science, research generates uncertainties which can only be solved by more research, in turn producing even more uncertainties. However, at the global level there are no state structures to balance, limit, and bring stability to these developments. Consequently, both regimes—medicine and science—are undergoing processes of self-reproduction and maximization of their particular rationalities which become excessive. This specific growth compulsion can lead to a catastrophe. The experience of immediate crisis brings to the crucial realization that self-restraint is vital. This is when the "constitutional moment" occurs.

Constitutional moment of transnational biomedical law

Constitutions provide a basis for a stabilization of the systems in question through legal means, as well as for the establishment of reflexive mechanisms capable of ensuring that they exercise self-restraint to a degree which leads to a reduction in negative externalities, asymmetries, and crowding-out effects *vis-à-vis* other systems [80]. All the documents concerning genetics and biomedicine, in particular the Oviedo Convention 1997 with its additional protocols and the UN and UNESCO Declarations (1997, 2003, 2005), complemented by a growing body of national laws[46] represent an attempt to introduce rules restraining the centrifugal tendencies of the biomedical subarea of the health care. This attempt includes the search for appropriate scientific and medical equivalents of traditionally state-related human rights. This is how concepts such as prohibition of genetic discrimination, the right to genetic privacy, the right to reproductive autonomy, the right to biological origins, and the right to die in dignity have been coined. These rights are not clearly and unequivocally directed against the totalizing tendencies of the political power of the state. Rather they are called upon to set boundaries and to protect the

---

[46] For example, The Swiss Federal Law on the Genetic Testing of Humans (Bundesgesetz über genetische Untersuchungen beim Menschen (GUMG), 8.10.2004, AS 2007 635, entered into force 1.04.2007, at: http://www.admin.ch/ch/d/sr/c810_12.html); the US Genetic Information Non-discrimination Act (GINA 2008), at: http://www.govtrack.us/congress/billtext.xpd?bill = h110-493&show-changes = 0&page-command = print; the German *human genetic examination act* (Gesetz über genetische Untersuchungen bei Menschen (Gendiagnostikgesetz - GenDG), 31.07.2009 BGBl. I S. 2529, 3672; in force 1.02.2010, at: http://www.buzer.de/gesetz/8967/index.htm); the Portuguese Law no. 12/2005 on personal genetic information and information regarding health (Privileged Project. "Privacy in law, ethics and genetic data," *Genetic Databases and Biobanks by Country*, at: http://www.privileged.group.shef.ac.uk/projstages/stage-2-genetic-databases-and-biobanks/genetic-databases-and-biobanks-by-country/portugal).

individual's mental and bodily integrity against the expansive tendencies of social institutions (e.g., knowledge, medicine, technology). Therefore, in order to make their protection effective, attempts are made to readjust them to the rationality and normativity of the subsystem. Of course, it is equally possible that this process of redefinition of human rights is primarily orientated at the reduction of intrusions of other actors and competing domains [81], thus strengthening and giving preponderance to the transnational biomedical law.[47] It might also be seen to perform another important function, namely, the formation of a unifying "myth of origin" for transnational biomedical law.

Although functional systems are not represented by a group of individuals or a collective actor necessary in the traditional constitutional theory for the creation of a constitution, it has been argued that "transnational regimes still create their own individual myths of origin, by developing and constantly referring to fictional explanations of the beginnings."[48] A myth in the context of this paper should be understood as a "motif" or a "legitimizing narrative" which would provide frames of understanding and justification for regime's operations. Institutions or associations such as states, nations, the church, or a company do not "possess" a memory but "produce" it by means of symbols, thereby creating a type of identity for themselves [84]. This narrative provides justification for the fully spontaneous development and expansion of the system, but also offers a basic sense of its coherency. Such coherency, as we know, is only illusionary, and therefore, a "myth." A myth of unity that is nevertheless necessary for law. There are a few candidates here, such as the right to health and the general well-being (not only of the individual, but of the global population), the duty of care, or the principle of consent. The last two notions are more convincing as they illustrate two dimensions of health care: the public (public health) and the individual (medicine).

The principle of "care" has been concealed over centuries in the notions of "care of the self" (*epimeleia eautou*) [85], non-malevolence, benevolence, and more recently also in the ethics of care [86,87]. Although extending far beyond health and medicine into the field of family, criminal, tort, and employment law, it seems to be the primary underlying justification for all the communication appearing in the health care regime at the national and global level. It has been subject of multiple readings in theorizations of relations between individual citizens and institutional actors. It has been

---

[47] At the same time it is worth mentioning Teubner's approach to human rights expressed elsewhere: "How can society ever 'do justice' to real people if people are not its parts but stand outside communication, if society cannot communicate with them but at most about them, indeed not even reach them but merely either irritate or destroy them? In the light of grossly inhuman social practices the justice of human rights is a burning issue, but one which has no prospect of resolution." This has to be said in all rigour [82].

[48] The idea is close to Ref. [83].

heavily criticized over the years.[49] Nevertheless, it paved its way to various professional codes and the law[50] and if we look more closely, we can find it in the reasoning for different paradigmatic shifts in the patient—doctor relationship (be it paternalism or autonomy). The ancient maxim of *primum non nocere*—carrying quasi mystical authority—forms the center of the Hippocratic Oath and has been since called—"medicine's most cherished principle" [92,93], "a dictum which has governed the practice of medicine since the time of Hippocrates" [94], a major tenet that all doctors should learn in medical school and that they all try to observe in clinical practice [95,96], and so on. Many centuries later "care" served as ideological justification (and "obscuration") for the establishment of modern medicine, standardization, and medicalization of modern society. It could be said to have strengthened the biomedical standpoint by assuming individuals to be vulnerable in addressing their own health concerns and situating practitioners in the position of power by asserting their knowledge and expertise.

The idea of looking after others has been translated into concepts such as duty of care or the right to equitable health care. The reason why it might not be visible as a separate sector-specific constitutional narrative (or why it does not surpasses existing laws) is because health care has become historically strongly intertwined with state structures and has already been integrated into binding law (e.g., criminal or tort law) and other principles, such as equality or justice, which became more visible. According to critical ethics of care discourse, welfare states offered citizens care in the form of "cradle-to-grave economic security," while neoliberal citizens are expected to resort to practices of self-care.[51] Therefore, for some authors, the law remains fixated on individualized legal personhood and individualized models of rights, while care—based on "respect, responsibility, and relationality"—remains a myth. This makes society much more vulnerable to "fantasies of care and security," and dreams elicited by biotechnologies such as synthetic biology where "perpetual plenitude sidelines the need to manage contingency with unending [self] control."[52] Thus, throughout the centuries the motif of benevolence and non-malevolence has supported the emergence and expansion of health care professions, the health care services, hence the health care system. It continues to support the self-foundation of transnational

---

[49] An interesting analysis of the "ethics of care" based on Foucauldian and Agamenian readings of biopolitics is to be found in Refs. [88—91].

[50] General Medical Council (2006). *Good Medical Practice*, (Muster-)Berufsordnung für die in Deutschland tätigen Ärztinnen und Ärzte - MBO-Ä 1997 - in der Fassung der Beschlüsse des 114. Deutschen Ärztetages 2011; Code de déontologie médicale sur le site de l'Ordre national des médecins 8.08.2004, figurant dans le Code de la Santé Publique sous les numéros R.4127-1 à R.4127-112), Décret 2012-694 du 07.05.2012; Kodeks Etyki Lekarskiej 2.01.2004 (Biuletyn NRL z 2004 r. Nr 1(81)).

[51] R. Mackenzie, see Refs. [69,70]; footnote 27.

[52] J. Herring, see Refs. [92,93]; footnote 1.

biomedical law directly spelled out in international professional codes of practice[53] and indirectly incorporated in the new developing principle of global social justice, which is implicitly based on values of responsibility and care for the less wealthy parts of the world [97]. The principle of care did not disappear after the Second World War, but has been supplemented by and redefined through principles of human dignity, personal integrity, and autonomy. Within these concepts shaping the idea of a "unifying myth of origin" a special place is reserved for the principle of informed consent.

Grounded in the philosophical concept of autonomy and human dignity, it protects bodily inviolability and personal integrity. It is now a key marker for ethical practice and is seen as necessary (and by some as sufficient) ethical justification for action that affects others, including medical treatment, research on human subjects, and uses of human tissues. The practice of informed consent in the clinical arena evolved primarily through the medical profession's responses to various decisions by the courts (establishing civil and criminal liability for unauthorized medical interventions), but in some legal systems it has now, together with the right to bodily inviolability, gained the status of a fundamental right.[54] Due to the link to principle of autonomy, dignity, and the human rights discourse, that practice began to be perceived as a paradigm of medical ethics and it now penetrates human rights instruments concerning biomedicine and bioethics and relevant documents of medical practice, including the Council of Europe Convention on Human Rights and Biomedicine (i.e., the Oviedo Convention), UNESCO Declarations, the Ethical Principles for Medical Research Involving Human Subjects were established by the Nuremberg Code in 1947, and further developed in the subsequent WMA Declaration of Helsinki. The same practice carries much less weight in public health law, which to a large degree is concerned with the legal powers and duties of the state and the constraints of autonomy, privacy, liberty, property, or other interests for common good. However, the fact that the limits of state power in cases such as immunization or mental health law are now being carefully studied indicates that the principle of consent (expressing values of dignity and autonomy) has also begun to penetrate public health law. Consent is now perceived as an individual entitlement that has belonged to mankind since times immemorial. It has entered the realm of myth and by doing so it contributed to the progressing constitutionalization of transnational biomedical law.

---

[53] World Medical Association. *Ethical Principles for Medical Research Involving Human Subjects*, Declaration of Helsinki, June 1964 (as amended by 59th WMA General Assembly, Seoul, October 2008) (providing guidelines for medical professionals around the world who conduct experimental research with human subjects); available at: http://www.wma.net/e/.

[54] For example, Art. 3 EU Charter of Fundamental Rights, Art. 5 Convention on Biomedicine and Human Rights, Art. 3 ECHR.

Constitutional norms develop incrementally remaining embedded in the ensemble of legal norms, forming the constitution in long concealed evolutionary processes. In a nation state as well as on a global scale, they also exist latently and are peculiarly invisible. We might find that they are hidden in particular treaty provisions, codes of practice, or even research agreements and consent forms. Hence, the main task of medical law scholars and policy-makers could be not to insist on further hierarchical codification, but to start "revealing" those constitutional norms hidden and concealed in transnational medical practice. Consequently, constitutional status could be ascribed to rules such as, the respect of human life in all stages of its development, the duty to obtain informed consent before medical interventions, the obligation to seek approval of an ethics committee before medical research, the prohibition of human reproductive cloning or any modification in the genome of the descendants, or the currently developing principles of data sharing. It has been, therefore, argued that the effects of these recent developments in life sciences on legal institutions, such as personhood, rights, citizenship, and legitimacy, have been so profound that they have redrafted established boundaries between science and law, and state and society. They have redefined constitutional frameworks as they have radically restructured state—society relations. They are latent in the sense that to date they have not been incorporated in any national or global constitutional texts. Nevertheless, it is arguable that these redefined rights and principles can acquire constitutional status, especially that they will be followed as binding by the different global players and will form the basis of their actions and transnational agreements. Such an approach enables a theorization of transnational biomedical law as a societal autonomous regime, a transnational regulatory regime, a functional order, or a global fragment.

CONCLUSIONS

This chapter aimed at analyzing recent developments concerning the legal regulation of genetics and genomics at the international level that lead to the common practice of "cherry picking" of international norms by decision- and policy-makers (governments, professional organization, and individual researchers). The analysis identified three main phenomena that might and do indeed hinder the application of genetic and genomic science in medical research and health care, namely, institutional and normative proliferation accompanied by the absence of hard law instruments and mechanisms resolving norm collisions. However, instead of embarking on a quest for unifying principles that could be applied globally, the chapter sought to provide an in-depth nonlegal explanation of fragmentation, based on systems theory developed by Niklas Luhmann and his followers, that the constitutional moment and subsequent self-constitutionalization occurred in the area of biomedicine, positioned at the intersection of science and medicine. As a result, recent developments in the area of genetics and genomics could be viewed as a constitutional moment in the process of growing autonomy of transnational biomedical law. This might seem somewhat surprising, because the technological developments are so recent (whereas "constitution" is traditionally conceptualized as an old event of mature orders), but most importantly because *prima facie*, global or transnational biomedical law (in the sense of binding law) is difficult to identify, especially in the field of genetics and genomics. Here too an important discovery has been made. A closer look into this issue reveals that constitutional norms are not captured in one document or a treaty, but that they are hidden in different international and transnational acts. This finding could have potentially significant consequences. It could shift the pressures to legalize certain issues at the international level in a top-down manner and redirect the efforts toward attempts to reveal self-founding and self-limitative rules that already exist within the system. This gradual and systematic "unconcealment" of the transnational biomedical *law* might be one of the most important tasks not only for future lawyers, judges, or policy-makers, but also for researchers and health care professionals.

REFERENCES

[1] OECD Guidelines for Quality Assurance in Molecular Genetic Testing, <http://www.oecd.org/sti/biotech/38839788.pdf>; 2007.
[2] OECD. Human Biobanks and Genetic Research Databases (HBGRDs), <http://www.oecd.org/science/biotech/44054609.pdf>; 2009.
[3] United Nations, Vienna Convention on the Law of Treaties, 23 May 1969, United Nations, Treaty Series, vol. 1155, 331. Available from: <http://www.refworld.org/docid/3ae6b3a10.html>.
[4] Koskenniemi M. The fate of public international law: between technique and politics. Mod Law Rev 2007;70:1–30.
[5] Chr Blichner L, Molander A. Mapping juridification. Eur Law J 2008;14(1):36–54.
[6] Loughlin M, Walker N. The paradox of constitutionalism: constituent power and constitutional form. OUP; 2007.
[7] Klabbers J, Peters A, Ulfstein G. The constitutionalization of international law. OUP; 2009.
[8] Krisch N. Beyond constitutionalism. The pluralist structure of postnational law. OUP; 2010.
[9] Dobner P, Loughlin M. Twilight of constitutionalism. OUP; 2011.
[10] Thornhill C. A Sociology of constitutions: constitutional and state legitimacy in historical sociological perspective. CUP; 2011.
[11] Teubner G. Constitutional fragments. Societal constitutionalism and globalisation. OUP; 2012.
[12] MacGillivary A. A brief history of globalisation. London: Robinson; 2006. Provides a useful overview.
[13] Twining W. Globalisation and legal theory. London: Butterworths; 2000.
[14] de Sousa Santos B. Towards a new legal common sense. London: Butterworths; 2002.
[15] Michaels R. Global legal pluralism. Annu Rev Law Soc Sci 2009;5:243–62.
[16] Slaughter A-M. A new world order. Princeton, NJ: Princeton University Press; 2004.
[17] Tully S. Corporations and international lawmaking. Martin Nijhoff; 2007.
[18] Delmas-Marty M. Ordering pluralism: a conceptual framework for understanding the transnational legal world. Hart Publishing; 2009. p. 13.
[19] Rowson M, Willott C, Hughes R, Maini A, Martin S, Miranda J, et al. Conceptualising global health: theoretical issues and their relevance for teaching. Global Health 2012;8(36).

[20] Knoppers BM, Harris JR, Budin-Ljøsne I, Dove ES. A human rights approach to an international code of conduct for genomic and clinical data sharing. Hum Genet 2014;133(7):895—903.

[21] Dove ES, Knoppers BM, Zawati MS. Towards an ethics safe harbor for global biomedical research. J Law Biosci 2014;1(1):3—51.

[22] Budin-Ljøsne I, Isaeva J, Maria KB, Marie TA, Shen HY, McCarthy MI, Harris JR. Data sharing in large research consortia: experiences and recommendations from ENGAGE. Eur J Hum Genet 2014;22:317—21.

[23] Sand I-J. Polycontextuality as an alternative to constitutionalism. In: Joerges C, Sand I-J, Teubner G, editors. Transnational governance ad constitutionalism. Hart Publishing; 2004. p. 55.

[24] Kaye J. From single biobanks to international networks: developing e-governance. Hum Genet 2011;130(3):377—82.

[25] Rial-Sebbag E, Cambon-Thomsen A. The emergence of biobanks in the legal landscape: towards a new model of governance, J Law Soc 2012;39(1):113—30.

[26] Wallace S, Knoppers B. Harmonised consent in international research consortia: an impossible dream? Genomics, Soc Policy 2011;7:35—46.

[27] Kim Y. A common framework for the ethics of the 21st century. Paris: UNESCO; 1999, III.

[28] Taylor A. Governing globalization of public health. J Law, Med Ethics 2004;500—8. p. 504.

[29] Ruger JP, Yach D. The global role of the world health organization. Global Health Governance 2008/2009;2:2.

[30] Cambon-Thomsen A, Sallée C, Rial-Sebbag E, Knoppers BM. Populational genetic databases: is a specific ethical and legal framework necessary? GenEdit 2005;1(3):6.

[31] Chadwick R, Strange H. Harmonisation and standardisation in ethics and governance: conceptual and practical challenges. In: Widdows H, Mullen C, editors. The governance of genetic information: who decides? CUP; 2009. p. 213.

[32] Buse K, Walt G. Global public—private health partnerships: part I—a new development in health? Bull World Health Organ 2000;78:549—61.

[33] Labonté R, Mohindra K, Schrecker T. The growing impact of globalisation for health and public practice. Annu Rev Public Health 2011;32:263.

[34] Sridhar D. Seven challenges in international development assistance for health and ways forward. J Law Med Ethics 2010;38(3):459—69. p. 462.

[35] Ruger JP, Yach D. The global role of the world health organization. Global Health Governance 2008/2009;2(2):1—11.

[36] Human Genome Organisation. Ethical, legal, and social issues committee; statement on the principled conduct of genetics research: <http://www1.umn.edu/humanrts/instree/geneticsresearch.html>; 1996.

[37] Fischer-Lescano A, Teubner G. Regime collisions: the vain search for legal unity in the fragmentation of global law. Mich J Int Law 2004;25:999—1045.

[38] Abbott FM. The WTO medicines decision: world pharmaceutical trade and the protection of public health. Am J Int Law 2005;99:317—58.

[39] Gervais DJ. Trips, Doha and traditional knowledge. J World Int Prop 2003;6:403—19.

[40] de Sanarclens P. The United Nation as a social and economic regulator. In: de Sanarclens P, Kazancigli A, editors. Regulating globalisation, critical approaches to global governance. United Nations University; 2007. p. 8—35.

[41] Gostin LO, Taylor AL. Global health law: a definition and grand challenges. Public Health Ethics 2008;1(1):53—63.

[42] Brunsson N, Jacobsson B. A world of standards. OUP; 2000.

[43] Schepel H. Constituting private governance regimes: standards bodies in American law. In: Joerges C, Sand I-J, Teubner G, editors. Transnational governance and constitutionalism. Oxford: Hart Publishing; 2004. p. 161—88.

[44] Zumnbasen P. Transnational law. In: Smits J, editor. Encyclopedia of comparative law. Edward Elgar; 2006.

[45] Knoppers BM, Abdul-Rahman MH, Bédard K. Genomic databases and international collaboration. King's Law J 2007;18:291—311.

[46] Cambon-Thomsen A, Sallée C, Rial-Sebbag E, Knoppers BM. Populational genetic databases: is a specific ethical and legal framework necessary? GenEdit 2005;3(1):1–13.

[47] Gibbons SMC, Kaye J. Governing genetic databases: collection, storage and use 18 King's Law J 2007;201–8.

[48] Brownsword R. Rights, regulation, and the technological revolution. Oxford: University Press; 2008. p. 295.

[49] Kirby MD. The human genome project – promise and problems. J Contemp Health Law Policy 1994;11(1).

[50] Spiliopoulou Åkermark S. Soft law and international financial institutions – issues of hard and soft law from a lawyer's perspective. In: Mörth U, editor. Soft law in governance and regulation – an interdisciplinary analysis. Edward Elgar; 2004. p. 61–101, 77–78.

[51] Boyle AE. Some reflections on the relationship between treaties and soft law. Int Comp Law Q 1999;48:901–13: 901.

[52] Fidler D, Gostin LO. The new international health regulations: an historic development for international law and public health. J Med Ethics 2006;34:85–94.

[53] Gostin LO. Public health law: power, duty, restraint. Berkley, Los Angeles, London: University of California Press; 2008. p. 240.

[54] Mason Meier B, Shelley D. The fourth pillar of the framework convention on tobacco control: harm reduction and the international human right to health. Public Health Rep 2006;121(5):494–500.

[55] Kuppuswamy C. The international legal governance of the human genome. Routledge; 2009.

[56] Wollf J. The human right to health. New York, London: W.W. Northon and Company; 2012.

[57] Yamin AE, Gloppen S. Litigating health rights: can court bring more justice to health. Harvard University Press; 2011.

[58] Shaheed F. Report of the special rapporteur in the field of cultural rights: the right to enjoy the benefits of scientific progress and its applications. United Nations: Human Rights Council, Twentieth Session. 2012.

[59] Knoppers BM, Harris JR, Budin-Ljosne I, Dove ES. A human rights approach to an international code of conduct for genomic and clinical data sharing. Hum Genet 2014;133(7):895–903.

[60] Donders Y. The right to enjoy the benefits of scientific progress: in search of state obligations in relation to health. Med Health Care Philos 2011;43:371–81.

[61] Lagdford M. Social rights jurisprudence: emerging trends in international and comparative law. Cambridge: CUP; 2008.

[62] Gauri V, Brinks D. Courting social justice: judicial enforcement of social and economic rights in the developing world. Cambridge: CUP; 2008.

[63] Mörth U. Soft law in governance and regulation—an interdisciplinary analysis. Edward Elgar; 2004.

[64] Sztucki J. Reflections on international soft law. In: Ramberg L, Bring O, Mahmoudi S, editors. Festskrift till Lars Hjerner. Norstedts; 1990. p. 549–75, 573.

[65] Chinkin Ch. Normative development in the international legal system. In: Shelton D, editor. Commitment and compliance: the role of non-binding norms in the international legal system. OUP; 2000. p. 21–42.

[66] Stoltzfus Jost T. Comparative and international health law. Health Matrix 2004;141:14.

[67] Walker N. The idea of constitutional pluralism. Mod Law Rev 2002;65:317–59, 338.

[68] Kingsbury B. The concept of 'law' in global administrative law. Eur J Int Law 2009;20(1):23–57, 29.

[69] Luhmann N. The differentiation of society. Columbia University Press; 1982.

[70] Luhmann N. Gesellschaft der Gesellschaft. Suhrkamp; 1997.

[71] Beck U. What is globalism. Cambridge: Polity Press; 2000.

[72] Luhmann N. Anspruchsinflation im Krankheitssystem. Eine Stellungnahme aus gesellschaftstheoretischer Sicht. In: Herder-Dorneich P, Schuller A editors. Die Anspruchsspirale: Schicksal oder Systemdefekt? 3. Koelner Kolloquium. Stuttgart: Kohlhammer; 1983.

[73] Luhmann N. Medizin und Gesellschaftstheorie. MMG 1983;8:168–75.

[74] Luhmann N. Der medizinische Code. In: Luhmann N, editor. Soziologische Aufklärung 5. Konstruktivistische Perspektiven. Opladen: Westdeutscher Verlag; 1990. p. 1176–88.

[75] J. Bauch *Gesundheit als sozialer Code. Von der Vergesellschaftung des Gesundheitswesens zur Medikalisierung der Gesellschaft.* Weinheim (Munchen): Juvents; 1990.

[76] Bauch J. Selbst- und Fremdbeschreibung des Gesundheitswesens. In: de Berg H, Schmidt J, editors. Rezeption und Reflexion. Zur Resonanz der Systemtheorie Niklas Luhmanns außerhalb der Soziologie. Frankfurt/Main: Suhrkamp; 2000.

[77] Pelikan JM. Understanding differentiation of health in late modernity—by use of sociological system theory. In: McQueen DV, Kickbusch IS, editors. Health and modernity: the role of theory in health promotion. New York (NY): Springer; 2007. p. 74−102.

[78] Fuchs P. Das Gesundheitssystem ist niemals verschnupft. In: Bauch J, editor. Gesundheit als System. Systemtheoretische Betrachtungen des Gesundheitswesens. Konstanz: Hartung-Gorre-Verlag); 2006. p. 21−38, 24.

[79] Scambler G. Habermas, critical theory and health. London: Routledge; 2001.

[80] Kjaer P. The concept of the political in the concept of transnational constitutionalism. A sociological perspective. In: Joerges Ch, Ralli T, Editors. After globalization − new patterns of conflict and their sociological and legal reconstruction. Oslo: Arena Report Series; 2011. p. 285−321, 314.

[81] Dilling, et al. Responsible business. Self-governance and law in transnational economic transactions. Oxford: Hart Publishing; 2008. p. 8.

[82] Teubner G. The anonymous matrix: human rights violations by 'private' transnational actors. Mod Law Rev 2006;69:327−46.

[83] Dunoff JL. New approach to regime interaction. In: Young MA, editor. Regime interaction in international law. Cambridge: CUP; 2011. p. 136−74.

[84] Assmann A cited in: Hansen-Magnusson H., Wiener A. Studying contemporary constitutionalism: memory, myth and horizon. J Common Mark Stud 2006;48(1):30.

[85] Foucault M. History of sexuality: introduction, the uses of pleasure, and care of the self [R. Hurley, Trans.]. New York (NY): Vintage Books; 1988, 1978. p. 90.

[86] Slote MA. The ethics of care and empathy. London: Routledge; 2007.

[87] Held V. Ethics of care. Oxford: OUP; 2006. p. 64.

[88] Mackenzie R. Care as cornucopia: a critical ethics of care and fantasies of security in the neoliberal affective economy In: Priaulx N, editor. Ethics, law, and society: vol. 5. Ashgate, 2013.

[89] Gillon R. "Primum non nocere" and the principle of non-maleficence. Br Med J 1985;291: 130−1.

[90] Brewin TB. How much ethics is needed to make a good doctor? Lancet 1993;341:161−3.

[91] Gifford RW. Primum non nocere. J Am Med Assoc 1977;238:589−90.

[92] Goodyear-Smith F. Political correction or primum non nocere? Z Med J 1993;106:416.

[93] Miller WT. Primum non nocere. Semin Roentgenol 1993;28:291−2.

[94] Brouillette JN. Primum non nocere. J Fla Med Assoc 1991;78:527−8.

[95] Eisdorfer C, Kessler DA, Spector AN. Caring for the elderly. Baltimore (MD): Johns Hopkins University Press; 1987.

[96] Hurwitz B, Richardson R. Swearing to care: the resurgence in medical oaths. Br Med J 1997;315:1671−4.

[97] Daniels N. Just health. CUP; 2008. p. 145.

GLOSSARY: "GENOMICS AND SOCIETY"*

ACRONYMS AND ABBREVIATIONS

CIOMS	Council for International Organizations of Medical Sciences
DCE	Discrete choice experiment
DTC	'Direct-to-consumer' genetic or genomic testing
FCTC	Framework Convention on Tobacco Control
GCOS-24	Genetic Counseling Outcome Scale
HC	Hereditary cancer
HD	Huntington Disease
HGP	Human Genome Project
HUGO	Human Genome Organisation
HVP	Human Variome Project
ICCPR	International Covenant on Civil and Political Rights
ICER	Incremental Cost-Effectiveness Ratio
ICESCR	International Covenant on Economic, Social and Cultural Rights
IF	Incidental (unanticipated) findings (from a test)
IO	International/Intergovernmental Organisation
MIC	Minimal important change
MID	Minimal important difference
NGO	Non-Governmental Organisation
NHS	National Health Service
NICE	National Institute of Health and Care Excellence (formally National Institute of Health and Clinical Excellence)
OECD	Organisation of Economic Cooperation and Development
PRO	Patient Reported Outcomes
PROM	Patient-reported outcome measures
QALY	Quality-Adjusted Life Years
TRIPS	Trade-Related Aspects of Intellectual Property Rights
UN	United Nations
UNESCO	United Nations Educational, Scientific, and Cultural Organisation
VUS	Variant of unknown clinical significance
WHO	World Health Organisation
WMA	World Medical Association
WTO	World Trade Organisation
WTP	Willingness to pay

---

* Compiled and Edited by Dhavendra Kumar.
This section of the book includes definitions, phrases, mini-statements, acronyms and abbreviations collated from a number of sources including authors who have contributed chapters in the current edition. The publisher and editor do not claim any ownership and deny breach of any copyrights issues arising from inclusion in Glossary.

TERMS AND PHRASES

Allele   An alternative form of a gene at the same chromosomal locus.

Allelic heterogeneity   Different alleles for one gene.

Annotation   The descriptive text that accompanies a sequence in a database method.

Anticipation   A phenomenon in which the age of onset of a disorder is reduced and/or severity of the phenotype is increased in successive generations.

Apoptosis   Programmed cell death.

Autosome   Any chromosome other than a sex chromosome (X or Y) and the mitochondrial chromosome.

Autozygosity   In an inbred person, homozygosity for alleles identical by descent; autozygosity mapping—a form of genetic mapping for autosomal recessive disorders in which affected individuals are expected to have two identical disease alleles by descent.

Bioinformatics   An applied computational system that includes development and utilization of facilities to store, analyze, and interpret biological data.

Biotechnology   The industrial application of biological processes; for example, recombinant DNA technology and genetic engineering.

Blastocyst   The mammalian embryo at the stage at which it is implanted into the wall of the uterus.

BRCA1/2   BRCA1 and BRCA2 are two genes that when mutated cause hereditary breast/ovarian cancer.

Candidate gene   Any gene which by virtue of a known property (function, expression pattern, chromosomal location, structural motif, etc.), is considered as a possible locus for a given disease.

Carrier   A person who carries an allele for a recessive disease (see heterozygote) without the disease phenotype, but can pass it on to the next generation; also used to denote a female carrying the mutation on one of the two X-chromosomes for an X-linked recessive disorder.

Carrier testing   Carried out to determine whether an individual carries one copy of an altered gene for a particular recessive disease.

"Central Dogma"   A term proposed by Francis Crick in 1957—DNA is transcribed into RNA, which is translated into protein.

cDNA (complementary DNA)   A piece of DNA copied in vitro from mRNA by a reverse transcription enzyme.

Centromere   The constricted region near the center of a chromosome that has critical role in cell division.

Chromosome   Subcellular structures that contain and convey the genetic material of an organism.

Chromosome painting   Fluorescent labeling of whole chromosomes by a FISH procedure in which labeled probes consist of a complex mixture of different DNA sequences from a single chromosome.

Clinical proteomics   Translational application of new protein-based technologies in clinical medicine.

Clinical sensitivity   The proportion of persons with a disease phenotype who test positive.

Clinical specificity   The proportion of persons without a disease phenotype who test negative.

Clinical genetics services   Specialized NHS service, which offer diagnosis of genetic conditions, genetic risk information including information about specific genetic conditions and their inheritance and risks to unaffected and unborn family members, genetic testing, and supportive counseling to help the family make decisions and cope better with the genetic condition in their family. The service is offered to all members of a family in which a genetic condition may be present, not just those who have the condition.

Coding DNA (sequence)   The portion of a gene that is transcribed into mRNA.

Codon   A three-base sequence of DNA or RNA that specifies a single amino acid.

Comparative genomics   The comparison of genome structure and function across different species in order to further an understanding of biological mechanisms and evolutionary processes.

Comparative genome hybridization (CGH)   Replaced use of competitive fluorescence *in situ* hybridization (FISH) to detect chromosomal regions that are duplicated or deleted by employing genomic arrays of such regions (*arrayCGH)* is now increasingly applied in clinical genetics.

Complex diseases   Diseases characterized by risk to relatives of an affected individual that is greater than the incidence of the disorder in the population.

Complex trait   One which is not strictly Mendelian (dominant, recessive, or sex-linked) and may involve the interaction of two or more genes to produce a phenotype, or may involve gene—environment interactions.

Congenital   Any trait, condition, or disorder that exists from birth.

Consanguinity   Marriage between two individuals having common ancestral parents, commonly between first cousins; an approved practice in some communities who share social, cultural, and religious beliefs. In genetic terms, two such individuals could be heterozygous by descent for an allele expressed as coefficient of relationship, and any off spring could be therefore homozygous by descent for the same allele expressed as coefficient of inbreeding.

Conserved sequence   A base sequence in a DNA molecule (or an amino acid sequence in a protein) that has remained essentially unchanged throughout evolution.

Copy-number   The number of different copies of a particular DNA sequence in a genome.

Copy number variation (CNV)   Variation in set of copy numbers of particular DNA sequence.

Cytoplasm   The internal matrix of a cell. The cytoplasm is the area between the outer periphery of a cell (the cell membrane) And the nucleus (in a eukaryotic cell).

Denaturation   Dissociation of complementary strands to give single stranded DNA and/or RNA.

Delphi survey   A research method used to address issues that are not supported by a strong evidence base and usually involve at least two rounds that collect views from a panel of experts. After each round an anonymous summary of the experts' views or judgments from the previous round is circulated to panel members who are encouraged to revise their previous answers in light of the responses of the whole panel. The theory is that the group will converge towards the "correct" (or consensus) answer.

Demographic transition   The change in the society from extreme poverty to a stronger economy, oft en associated by a transition in the pattern of diseases from malnutrition and infection to the intractable conditions of middle and old age; for example obesity, diabetes, and cancer.

Determinism (genetic)   Philosophical doctrine that human action is not free but determined by genetic factors.

Diploid   a genome (the total DNA content contained in each cell) that consists of two homologous copies of each chromosome.

Disease   A fluid concept influenced by societal and cultural attitudes that change with time and in response to new scientific and medical discoveries. The human genome sequence will dramatically alter how we define, prevent, and treat disease. Similar collections of symptoms and signs (phenotype) may have very different underlying genetic constitutions (genotype). As genetic capabilities increase, additional tools might become available to subdivide disease designations that are clinically identical (phenocopy).

Disease etiology   Any factor or series of related events directly or indirectly causing a disease. For example, the genomics revolution has improved our understanding of disease determinants and provided a deeper understanding of molecular mechanisms and biological processes.

Disease expression   When a disease genotype is manifested in the phenotype.

Disease management   A continuous, coordinated healthcare process that seeks to manage and improve the health status of a patient over the entire course of a disease. The term may also apply to a patient population. Disease management services include disease prevention efforts as well as patient management.

Diversity, genomic   The number of base differences between two genomes divided by the genome size.

DNA (deoxyribonucleic acid)   The chemical that comprises the genetic material of all cellular organisms.

DNA cloning   Replication of DNA sequences ligated into a suitable vector in an appropriate host organism (see cloning vector).

DNA fingerprinting   Use of hypervariable mini-satellite probe (usually those developed by Sir Alec Jeffreys) on a Southern blot to produce an individual-specific series of bands for identification of individuals or relationships. The technique is now extended to employ genomic polymorphisms (copynumber variation and single nucleotide polymorphisms) using quantitative PCR and next generation sequencing methods.

DNA library   A collection of cell clones containing different recombinant DNA clones.

DNA sequencing   Technologies through which the order of base pairs in a DNA molecule can be determined; the Sanger method is commonly used, but now replaced by the next generation sequencing (NGS) techniques.

Domain   A discrete portion of a protein with its own function. The combination of domains in a single protein determines its overall function.

Dominant   An allele (or the trait encoded by that allele) which produces its characteristic phenotype when present in the heterozygous form.

Dominant negative mutation   A mutation in heterozygous form that results in a mutant gene product inhibiting the function of the wild type gene product; see *Gain of Function* mutation.

Dosage effect   The number of copies of a gene; variation in the number of copies can result in aberrant gene expression or can be associated with disease phenotype.

Embryonic stem cells (ES cells)   A cell line derived from undifferentiated, pluripotent cells from the embryo.

Environmental factors   May include chemical, dietary factors, infectious agents, physical and social factors.

Enzyme   A protein which acts as a biological catalyst that controls the rate of a biochemical reaction within a cell.

Epigenetic   A term describing nonmutational phenomena, such as methylation and histone modification that modify the expression of a gene; epigenomics is also used in the same context, but refers to several genomic regions having similar role or function.

Euchromatin   The fraction of the nuclear genome that contains transcriptionally active DNA and which, unlike heterochromatin, adopts a relatively extended conformation.

Eukaryote   An organism whose cells show internal compartmentalization in the form of membrane-bounded organelles, such as a nucleus (includes multicellular organisms—animals, plants, fungi, and algae).

Exon   The sections of a gene that code for its functional product. Eukaryotic genes may contain many *exons* interspersed with non-coding *introns*. An *exon* is represented in the mature mRNA product—the portions of an mRNA molecule that is left after all *introns* are spliced out, which serves as a template for protein synthesis.

Expression sequences tag (EST)   Partial or full complement DNA sequences which can serve as markers for regions of the genome which encode expressed products.

Family history   An essential tool in clinical genetics. Interpreting the family history can be complicated by many factors, including small families, incomplete or erroneous family histories, consanguinity, variable penetrance, and the current lack of real understanding of the multiple genes involved in polygenic (complex) diseases.

Fluorescence *in situ* hybridization (FISH)   A form of chromosome *in situ* hybridization in which nucleic acid probe is labeled by incorporation of a *flurophore*, a chemical group that fluoresces when exposed to UV irradiation.

Founder effect   Changes in allelic frequencies that occur when a small group is separated from a large population and establishes in a new location.

Frame-shift mutation   The addition (duplication) or loss (deletion) of a number of DNA bases that is not a multiple of three, thus causing a shift in the reading frame of the gene. This shift leads to a change in the reading frame of all parts of a gene that are downstream from the mutation leading to a premature stop codon, and thus to a truncated protein product.

Functional genomics   The development and implementation of technologies to characterize the mechanisms through which genes and their products function and interact with each other and with the environment.

Gain-of-function mutation   A mutation that produces a protein that takes on a new or enhanced function.

Gene   The fundamental unit of heredity; in molecular terms, a gene comprises a length of DNA that encodes a functional product, which may be a polypeptide (a whole or constituent part of a protein or an enzyme) or a ribonucleic acid. It includes regions that precede and follow the coding region as well as introns and exons. The exact boundaries of a gene are often ill-defined since many promoter and enhancer regions dispersed over many kilobases may influence transcription.

Gene-based therapy   Refers to all treatment regimens that employ or target genetic material. This includes (i) *transfection* (introducing cells whose genetic make-up is modified) (ii) *antisense* therapy, and (iii) *naked DNA* vaccination (*see Gene therapy*).

Gene expression   The process through which a gene is activated at a particular time and place so that its functional product is produced such as, transcription into mRNA followed by translation into protein.

Gene expression profile   The pattern of changes in the expression of a specific set of genes that is relevant to a disease or treatment. The detection of this pattern depends upon the use of specific gene expression measurement technique.

Gene family   A group of closely related genes that make similar protein products.

Gene knockouts   A commonly used technique to demonstrate the phenotypic effects and/or variation related to a particular gene in a model organism, for example in mouse (*see Knock-out*); absence of many genes may have no apparent effect upon **phenotypes** (though stress situations may reveal specific susceptibilities). Other single knockouts may have a catastrophic effect upon the organism, or be lethal so that the organism cannot develop at all.

Gene regulatory network   A functional map of the relationships between a number of different genes and gene products (proteins), regulatory molecules, etc. that define the regulatory response of a cell with respect to a particular physiological function.

Gene therapy   A therapeutic medical procedure that involves either replacing/manipulating or supplementing non-functional genes with healthy genes. Gene therapy can be targeted to somatic (body) or germ (egg and sperm) cells. In **somatic gene therapy** the recipient's genome is changed, but the change is not passed along to the next generation. In **germ-line gene therapy**, the parent's egg or sperm cells are changed with the goal of passing on the changes to their offspring.

Genetics   Refers to the study of heredity, gene and genetic material. In contrast to **genomics**, the genetics is traditionally related to lower-throughput, smaller scale emphasis on single genes, rather than on studying structure, organization and function of the whole genome.

Genetic architecture   Refers to the full range of genetic effects on a trait. Genetic architecture is a moving target that changes according to gene and genotype frequencies, distributions of environmental factors, and such biological properties as age and sex.

Genetic code   The relationship between the order of nucleotide bases in the coding region of a gene and the order of amino acids in the polypeptide product. It is universal, triplet, non-overlapping code such that each set of three bases (termed a codon) specifies which of the 20 amino acids is present in the polypeptide chain product of a particular position.

Genetic counselling Genetic counseling has been defined as integrating interpretation of family and medical histories to assess the chance of disease occurrence or recurrence, education about inheritance, testing, management, prevention, resources and research and counselling to promote informed choices and adaptation to the risk or condition.

Genomic sequencing Sequencing of the entire genome (all of the DNA sequence) or exome (DNA sequences which are transcribed and remain part of mature RNA after introns are spliced out; mutations in the exome probably account for about 85% of those with a large effect).

Genomic testing This is a more general term that the term "genomic sequencing". Genomic testing includes genomic sequencing and other whole genome tests such as array comparative genomic hybridization (arrayCGH) which is a method for the detection of chromosomal aberrations in the form of copy number imbalances. ArrayCGH does not involve DNA sequencing. Instead, the method is based on co-hybridization of sample and control genomic DNAs to detect copy number variants.

Genomic (genetic) susceptibility testing DNA testing that provides probabilistic genetic risk information for common chronic diseases, rather than for strongly genetic conditions that "run in families" in a recognizably Mendelian fashion.

Genetic screening Testing a population group to identify a subset of individuals at high risk for having or transmitting a specific genetic disorder.

Genetic susceptibility Predisposition to a particular disease due to the presence of a specific allele or combination of alleles in an individual's genome.

Genetic test An analysis performed on human DNA, RNA, genes and/or chromosomes to detect heritable or acquired genotypes. A genetic test also is the analysis of human proteins and certain **metabolites**, which are predominantly used to detect heritable or acquired **genotypes, mutations or phenotypes**.

Genetic testing Strictly refers to testing for a specific chromosomal abnormality or a DNA (nuclear or mitochondrial) mutation already known to exist in a family member. This includes diagnostic testing (postnatal or prenatal), presymptomatic or predictive genetic testing or for establishing the carrier status. The individual concerned should have been offered full information on all aspects of the genetic test through the process of '*nonjudgemental and nondirective*' genetic counseling. Most laboratories require a formal fully informed signed consent before carrying out the test. Genetic testing commonly involves DNA/RNA-based tests for single gene variants, complex genotypes, acquired mutations and measures of gene expression. Epidemiologic studies are needed to establish clinical validity of each method to establish sensitivity, specificity, and predictive value.

Genome The complete set of chromosomal and extra-chromosomal DNA/RNA of an organism, a cell, an organelle or a virus.

Genome annotation The process through which landmarks in a genomic sequence are characterized using computational and other means; for example, genes are identified, predictions made as to the function of their products, their regulatory regions defined and intergenic regions characterized (*see Annotation*).

Genome project The research and technology development effort aimed at mapping and sequencing the entire genome of human beings and other organisms.

Genomics The study of the structure and function of whole genome of an organism. The term is commonly used to refer large-scale, high-throughput molecular analyses of multiple genes, gene products, or regions of genetic material (DNA and RNA). The term also includes the comparative aspect of genomes of various species, their evolution, and how they relate to each other (*see comparative genomics*).

Genotype The genetic constitution of an organism; commonly used in reference to a specific disease or trait.

Genomic profiling   Complete genomic sequence of an individual including the expression profile; this may be targeted to specific requirements, for example most common complex diseases (diabetes, hypertension and coronary heart disease).

Germ-line cells   A cell with a haploid chromosome content (also referred to as a gamete); in animals, sperm, or egg and in plants, pollen, or ovum.

Germline mosaic (germinal mosaic, gonadal mosaic, gonosomal mosaic)   An individual who has a subset of germline cells carrying a mutation which is not found in other germline cells.

Germline mutation   A gene change in the body's reproductive cells (egg or sperm) that becomes incorporated into the DNA of every cell in the body of offspring; germline mutations are passed on from parents to offspring, also called *hereditary mutation*.

Haploid   Describing a cell (typically a gamete) which has only a single copy of each chromosome (i.e., 23 in man).

Haplotype   A series of closely linked loci on a particular chromosome which tend to be inherited together as a block.

Heterozygote   Refers to a particular allele of a gene at a defined chromosome locus. A heterozygote has a different allelic form of the gene at each of the two homologous chromosomes.

Heterozygosity   The presence of different alleles of a gene in one individual or in a population—a measure of genetic diversity.

Homology   Similarity between two sequences due to their evolution from a common ancestor, often referred to as *homologs*.

Homozygote   Refers to same allelic form of a gene on each of the two homologous chromosomes.

Human Genome Project   A programme to determine the sequence of the entire 3 billion bases of the human genome.

Health literacy   The ability to obtain, read, understand, and use health information to make appropriate informed health and healthcare decisions and follow treatment instructions.

Health locus of control   The extent to which individuals believe they can control their own health. The "locus" (place) can be internal (the person believes they themselves can control their own health) or external (the person believes their health is controlled by external factors over which they have no influence, for example, powerful others (carers, healthcare professionals), chance, God or fate).

Health status   A holistic concept representing health, and includes functional, physical and mental wellbeing.

Heuristics   Experience-based problem solving and learning methods that finds solutions that although may not be optimal, are good enough for a given purpose.

Interventions   Formalized efforts by healthcare providers to promote health, relieve disease symptoms or promote behavior to either improve mental and physical health, discourage behavior that puts health at risk, or reframe the beliefs and attitudes of those with health conditions or risks. Can be surgical, pharmaceutical, educational, or psychological.

Identity by descent (IBD)   Alleles in an individual or in two people that are identical because they have been inherited from the same common ancestor, as opposed to *identity by state (IBS)*, which is coincidental possession of similar alleles in unrelated individuals (*see Consanguinity*).

Immunogenomics   Refers to the study of organization, function and evolution of vertebrate defense genes, particularly those encoded by the Major histocompatibility complex (MHC) and the Leukocyte receptor complex (LRC). Both complexes form integral parts of the immune system.

Intron   A noncoding sequence within eukaryotic genes that separates the exons (coding regions). Introns are spliced out of the messenger RNA molecule created from a gene after transcription, prior to protein translation (protein synthesis).

Isoforms/isozymes   Alternative forms of protein/enzyme.

In situ **hybridization**   Hybridization of a labelled nucleic acid to a target nucleic acid which is typically immobilized on a microscopic slide, such as DNA of denatured metaphase chromosomes (as in *fluorescent in situ hybridization* (FISH)) or the RNA in a section of tissue (as in *tissue in situ hybridization* (TISH)).

In vitro   (Latin) literally "in glass," meaning outside of the organism in the laboratory, usually is a tissue culture.

In vivo   (Latin) literally "in life," meaning within a living organism.

Locus   The specific site on a chromosome at which a particular gene or other DNA landmark is located.

Loss-of-function mutation   A mutation that decreases the production or function (or both) of the gene product.

Meiosis   Reductive cell division occurring exclusively in testis and ovary and resulting in the production of haploid cells, including sperm cells and egg cells.

Mendelian genetics   Classical genetics, focuses on **monogenic** genes with high **penetrance**. The mendelian genetics is a true **paradigm** and is used in discussing the mode of inheritance (*see Monogenic disease*).

Microarrays- diagnostics   A rapidly developing tool increasingly used in pharmaceutical and genomics research and has the potential for applications in high-throughput diagnostic devices. Microarrays can be made of DNA sequences with known gene mutations, polymorphisms, as well as selected protein molecules, used in proteomics.

Mitochondria   Cellular organelles present in eukaryotic organisms which enable aerobic respiration and generate the energy to drive cellular processes. Each mitochondria contains a small amount of circular DNA encoding a small number of genes (approximately 50).

Mitosis   Cell division in somatic cells.

Molecular genetic testing   Molecular genetic testing for use in patient diagnosis, management, and genetic counseling; this is increasingly used in pre-symptomatic (predictive) genetic testing of "at-risk" family members using a previously known disease-causing mutation in the family.

Molecular genetic screening   Screening a section of the population known to be at a higher risk to be heterozygous for one of the mutations in the gene for a common autosomal recessive disease, for example, screening for cystic fibrosis in the North-European populations and beta-thalassemia in the Mediterranean and Middle-East population groups.

Multifactorial disease   Any disease or disorder caused by interaction of multiple genetic (polygenic) and environmental factors.

Mutation   A heritable alteration in the DNA sequence.

Natural selection   The process whereby some of the inherited genetic variation within a population will affect the ability of individuals to survive to reproduce (*fitness*).

Newborn screening   Performed in newborns in state public health programs to detect certain genetic diseases for which early diagnosis and treatment are available.

Noncoding sequence   A region of DNA that is not translated into protein. Some noncoding sequences are regulatory portions of genes, others may serve structural purposes (telomeres, centromeres), while others may not have any function.

Nonsense mutation   Substitution of a single DNA base that leads in a stop codon, thus leading to the truncation of a protein.

Nucleotide   A subunit of the DNA or RNA molecule. A nucleotide is a base molecule (adenine, cytosine, guanine, and thymine in the case of DNA), linked to a sugar molecule (deoxyribose or ribose) and phosphate groups.

OMIM   Acronym for McKusick's **O**n-line **M**endelian **I**nheritance in **M**an, a regularly updated electronic catalog of inherited human disorders and phenotypic traits accessible on NCBI network. Each entry is designated by a number (*MIM number*).

Lynch syndrome   A hereditary cancer predisposing syndrome that causes high risks of colorectal and endometrial cancer and increased risks of a range of other cancers.

Mendelian   Inherited in a pattern clearly consistent with Mendel's laws: the law of segregation and the law of independent assortment.

Mortality   A measure of the number of deaths in a given time or the proportion of deaths in a given population (e.g., of patients). Often used "numbers per 1000" as a healthcare outcome measure.

Morbidity   The incidence of illness (number of people ill) in a given population (e.g., of patients). Often used as a healthcare outcome measure.

Mutation   A permanent change of the DNA sequence.

Non-directive   A person-centered approach to healthcare that supports the individual to come to their own decision, rather than telling them what action to take.

Penetrance   The likelihood that a person carrying a particular mutant gene will have an altered phenotype (*see Phenotype*).

Pharmacogenomics   A broad term (*see also* Pharmacogenetics), now increasingly applied to the identification of genes along with all regulatory sequences that could lead to new drug discovery, drug development, and assessing individual's variation in the efficacy or toxicity to a drug.

Phenotype   The clinical and/or any other manifestation or expression, such as a biochemical immunological alteration, of a specific gene or genes, environmental factors, or both.

Polymerase chain reaction (PCR)   A molecular biology technique developed in the mid-1980s through which specific DNA segments may be amplified selectively.

Polymorphism   The stable existence of two or more variant allelic forms of a gene within a particular population, or among different populations.

Predictive testing   Determines the probability that a healthy individual with or without a family history of a certain disease might develop that disease.

Predisposition, genetic   Increased susceptibility to a particular disease due to the presence of one or more gene mutations, and/or a combination of alleles (haplotype), not necessarily abnormal, that is associated with an increased risk for the disease, and/or a family history that indicates an increased risk for the disease.

Predisposition test   A test for a genetic predisposition (incompletely **penetrant** conditions). Not all people with a positive test result will manifest the disease during their lifetimes.

Preimplantation genetic diagnosis (PIGD)   Used following *in vitro* fertilization to diagnose a genetic disease or condition in a preimplantation embryo.

Prenatal diagnosis   Used to diagnose a genetic disease or condition in a developing fetus.

Pre-symptomatic test   Predictive testing of individuals with a family history. Historically, the term has been used when testing for diseases or conditions such as Huntington's disease where the likelihood of developing the condition (known as **penetrance**) is very high in people with a positive test result.

Protein   A protein is the biological effector molecule encoded by sequences of a gene. A protein molecule consists of one or more polypeptide chains of aminoacid subunits. The functional action of a protein depends on its three-dimensional structure, which is determined by its aminoacid composition.

Proteome   All of the proteins present in a cell or organism.

Proteomics   The development and application of techniques to investigate the protein products of the genome and how they interact to determine biological functions.

Perceived personal control   The amount of control a person believes themselves to have. Has been defined as comprising cognitive control ("knowledge is power"), decisional control (able to make decisions) and behavioral control (able to take action to relieve a stressor).

Pre-symptomatic (predictive) genetic testing   DNA testing used to detect mutations associated with genetic conditions that appear after birth, often in adult life; used to determine whether a person will develop the condition before any signs or symptoms appear.

Prenatal testing   Testing for a genetic conditions in a fetus or embryo before it is born.

Psychological distress   Emotions such as anxiety, depression, worry that are interfering with activities of daily living.

Psychometric validation   Methods for defining the measurement properties of questionnaires (sometimes referred to as scales, surveys, or instruments). See Table 1.

Qualitative research   Exploratory research used to gain more understanding of peoples' reasons, opinions, and motivations. It is useful for developing ideas or hypotheses for potential quantitative research, or for uncovering trends in peoples' attitudes and beliefs about a problem. Qualitative data is usually collected using unstructured or semi-structured focus groups (group discussions), group or individual interviews, and participant observation. Sample sizes in qualitative research are typically small, and findings are not usually generalizable beyond the sample.

Quality of life   The general well-being of individuals and societies. Health-related quality of life describes quality of life in the health domain.

Randomized controlled trial (RCT)   A research study, usually in healthcare, that allocates people at random (by chance alone) to receive one of several clinical interventions. One of these interventions is a standard of comparison or control, often usual care or sometimes a placebo ("sugar pill"). RCTs are considered the gold standard in health intervention research.

Recurrence risk   The probability of an event recurring.

Recessive   An allele that has no phenotypic effect in the heterozygous state; a defined phenotype could be expected in the homozygous or compound heterozygous state.

Recombinant DNA technology   The use of molecular biology techniques such as restriction enzymes, ligation, and cloning to transfer genes among organisms (see genetic engineering).

Ribonucleic acid (RNA)   A single stranded nucleic acid molecule comprising a linear chain made up from four nucleotide subunits (A, C, G and U). There are three types of RNA: messenger, transfer, and ribosomal.

Risk communication   An important aspect of genetic counselling which involves pedigree analysis, interpretation of the inheritance pattern, genetic risk assessment, and explanation to the family member (or the family).

Risk perception   The subjective judgment a person makes about the characteristics and severity of a specific risk.

Screening   Carrying out of a test or tests, examination(s) or procedure(s) in order to expose undetected abnormalities, unrecognized (incipient) diseases, or defects: examples are early diagnosis of cancer using mass X-ray mammography for breast cancer and cervical smears for cancer of the cervix.

Screening   Population-based methods used to identify apparently healthy people who may be at increased risk of a disease or condition.

Service evaluation   In healthcare, an exercise designed to answer the question "what standard does this service achieve?"

Sensitivity (of a screening test)   Extent (usually expressed as a percentage) to which a method gives results that are free from false negatives; the fewer the false negatives, the greater the sensitivity. Quantitatively, sensitivity is the proportion of truly diseased persons in the screened population who are identified as diseased by the screening test.

Sex chromosome   The pair of chromosomes that determines the sex (gender) of an organism. In man one X and one Y chromosomes constitute a male compared to two X chromosomes in a female.

Sex selection   Preferential selection of the unborn child on the basis of the gender for social and cultural purposes. However, this may be acceptable for medical reasons, for example, to prevent the birth of a male assessed to be at risk for an x-linked recessive disease. For further information visit: http://www.bioethics.gov/topics/sex_index.html.

Single-nucleotide polymorphism (SNP)   A common variant in the genome sequence; the human genome contains about 10 million SNPs.

Somatic   All of the cells in the body which are not gametes (germ-line).

Splicing   A process, prior to transcription by mRNA, by which *introns* are removed and the *exons* adjoined.

Stem cell   A cell which has the potential to differentiate into a variety of different cell types depending on the environmental stimuli it receives.

Stop codon   A codon that leads to the termination of a protein rather than to the addition of a amnioacid. The three stop codons are TGA, TAA, and TAG.

Systems biology   A bioinformatics tool; refers to simultaneous measurement of thousands of molecular components (such as transcripts, proteins, and metabolites) and integrate these disparate data sets with clinical end points, in a biologically relevant manner; this model can be applied in understanding the etiology and pathogenesis of disease.

Transcription   The process through which a gene is expressed to generate a complementary RNA molecule on a DNA template using RNA polymerase.

Tumor suppressor gene   A gene which serves to protect cells from entering a cancerous state; according to Knudson's "two-hit" hypothesis, both alleles of a particular tumor suppressor gene must acquire a mutation before the cell will enter a transformed cancerous state.

X-chromosome inactivation   Random inactivation of one of the two X chromosomes in mammals by a specialized form of genetic imprinting (*see Lyonization*).

INDEX

Printed in the United States
By Bookmasters